Applications of Fibonacci Numbers

Applications of Fibonacci Numbers

Volume 9

Proceedings of The Tenth International Research Conference on Fibonacci Numbers and Their Applications

edited by

Frederic T. Howard
Wake Forest University,
Winston-Salem, North Carolina, U.S.A.

Springer-Science+Business Media, B.V.

A C.I.P. Catalogue record for this book is available from the Library of Congress.

ISBN 978-90-481-6545-2 ISBN 978-0-306-48517-6 (eBook)
DOI 10.1007/978-0-306-48517-6

Cover figure by John C. Turner

Printed on acid-free paper

Originally published by Kluwer Academic Publishers 2004
Softcover reprint of the hardcover 1st edition 2004

TABLE OF CONTENTS

A REPORT ON
THE TENTH INTERNATIONAL CONFERENCE
ON
FIBONACCI NUMBERS AND THEIR APPLICATIONS

The Tenth International Conference on Fibonacci Numbers and Their Applications held at Northern Arizona University in Flagstaff, Arizona from June 24-28, 2002 found over 70 enthusiastic Fibonacci number lovers from Australia, Canada, England, Germany, Italy, Japan, Mexico, New Zealand, Poland, Romania, Scotland and the USA gathered together to hear over 50 excellent presentations. The gathering was attended by both old and new Fibonacci friends, but it was sadly noted that several regulars were unable to be with us this year. They were both warmly remembered and greatly missed. A special thanks goes to organizer Cal Long and all the folks at Northern Arizona University for their hospitality and generosity for hosting this outstanding conference.

Monday through Wednesday morning found us savoring a variety of talks on things theoretical, operational and applicable of a Fibonacci and related nature, with members sharing ideas while renewing old friendships and forming new ones.

Later on Wednesday the group was doubly treated. After the morning talks, we were entertained by mathemagician Art Benjamin's most impressive presentation; displaying his skills and cleverness by mentally performing challenging mathematical manipulations and zapping out magic squares as if (yes) by magic. After graciously sharing some of the secrets of his wizardry with us, he dazzled one and all by mentally and accurately multiplying two five place numbers to terminate his mesmerizing performance.

That afternoon we were bussed to our second wonder of the day: The Grand Canyon. Here we were able to spend several hours gazing at nature's wondrous spectacle. Oh to be a condor for an hour! In the evening a steak dinner was catered for us as we exchanged social and mathematical dialogue to the background of exquisite scenic wonder at the edge of the canyon. On the way back to campus we were able to witness a magnificent display of stars but an arm length away in the clear Arizona night sky.

On Thursday and Friday it was back to many more interesting, informative presentations and during the breaks we were treated to Peter Anderson's marvelous computer display of the many photographs he took of association members and their families enjoying the Canyon.

The closing banquet on Friday night terminated with a special tribute to Calvin T. Long for his distinguished career of 50 years as a teacher, mentor, and researcher, as well as valued friend, contributor and supporter of The Fibonacci Association. He was both praised and roasted by President Fred T. Howard and former editor Jerry Bergum. After much laughter and tears, Cal received a standing ovation from his proud and grateful group of his friends and colleagues.

After over an hour of cordial good-byes everyone eventually drifted away, vowing that, Lord willing, we'll all meet again in Braunschweig, Germany in 2004.

Charles K. Cook

AUTHORS, COAUTHORS AND OTHER CONFERENCE PARTICIPANTS

Professor Arnold Adelberg (pp. 1-8)
Mathematics Department
Grinnell College
Grinnell, IA 50112
USA

Octavian Agratini (pp. 9-18)
Faculty of Mathematics and Informatics
"Babes-Bolyai" University
str. Kogalniceanu, Nr. 1
3400 Cluj-Napoca
ROMANIA

Professor Peter G. Anderson
School of Computer Science and Technology
Rochester Institute of Technology
102 Lomb Memorial Drive
Rochester, NY 14623-5608
USA

Professor Arthur T. Benjamin (pp. 19-24; 25-28)
Department of Mathematics
Harvey Mudd College
Claremont, CA 91771
USA

Professor Kenneth S. Berenhaut
Department of Mathematics
Wake Forest University
Winston-Salem, NC 27109
USA

Gerald Bergum
1029 Circle Drive
Brookings, SD 57006
USA

Marjorie Bicknell-Johnson (pp. 29-38; 39-42)
665 Fairlane Avenue
Santa Clara, CA 95051
USA

Paul Bien
204 West Jefferson Avenue
Fairfield, IA 52556
USA

Mr. Nathan Blecke (pp. 43-62)
4201 Menton
Flint, MI 48507
USA

Jens-P. Bode (pp. 63-67)
Diskrete Mathematik
Technische Universität Braunschweig
D-38023 Braunschweig
GERMANY

Tom C. Brown
Department of Mathematics and Statistics
Simon Fraser University
Burnaby BC, V5A 1S6
CANADA

Paul Bruckman
3025 Shattuck Avenue, Apt. C.
Berkeley, CA 94705
USA

Barbara Mae Brunner
P.O. Box 762
Sedona, AZ 86339
USA

Dr. Colin M. Campbell (pp. 69-85)
The Mathematical Institute
University of St. Andrews
St. Andrews KY 16 9SS
Fife, SCOTLAND

P.P. Campbell (pp. 69-85)
The Mathematical Institute
University of St. Andrews
St. Andrews KY 16 9SS
Fife, SCOTLAND

Walter Carlip
University of California, Santa Barbara
408 Harvard Street
Vestal, NY 13850
USA

Jeff Chamberlein
Department of Mathematics and Computer Science
Ithaca College
Ithaca, NY 14850
USA

Professor Hei-Chi Chan
Mathematical Sciences Department
University of Illinois at Springfield
P.O. Box 19243
Springfield, IL 62794
USA

Dr. Charles Cook (pp. 87-99)
Mathematics Department
University of South Carolina at Sumter
Sumter, SC 29150-2498
USA

Professor Curtis Cooper (pp. 267-287)
Department of Mathematics and Computer Science
Central Missouri State University
Warrensburg, MO 64093-5045
USA

Professor Karl Dilcher
Department of Mathematics, Statistics and Computer Science
Dalhousie University
Halifax NS, B3H 3J5
CANADA

H. Doostie (pp. 69-85)
Mathematics Department
University for Teacher Education
49 Mofateh Avenue
Tehran, 15614
IRAN

Eric S. Egge
Department of Mathematics
Gettysburg College
Gettysburg, PA 17325
USA

Michele Elia
Dipt. di Elettronica
Politecnico di Torino
C.so Duca degli Abruzzi 24
I-10129 Torino
ITALY

Larry Ericksen
P.O. Box 172
Millville, NJ 08332-0172
USA

Professor Emeritus Dan Fielder (pp. 101-108)
School of Electrical and Computer Engineering
Georgia Institute of Technology
Atlanta, GA 30332-0250
USA

Dr. Kristen Flemming (pp. 43-62)
Department of Mathematics
Central Michigan University
Mt. Pleasant, MI 48859
USA

C. T. Flynn
8021 Cobblestone
Des Moines, IA 50322
USA

George Grossman (pp. 43-62)
Department of Mathematics
Central Michigan University
Mt. Pleasant, MI 48859
USA

Professor Helen Grundman
Department of Mathematics
Bryn Mawr College
Bryn Mawr, PA 19010-2899
USA

Ben Gum
Grinnell College
1905 Sunset Street
Grinnell, IA 50112
USA

Professor Dr. Heiko Harborth (pp. 63-67)
Diskrete Mathematik
Technische Universität Braunschweig
D-38023 Braunschweig
GERMANY

Russell Jay Hendel
Towson University
Box 47581
Baltimore, MD 21244
USA

Nate Higgins
Department of Mathematics and Computer Science
Ithaca College
Ithaca, NY 14850
USA

Professor Robert Hochberg (pp. 109-119)
Department of Computer Science
East Carolina University
Greenville, NC 27858
USA

Professor A. F. Horadam (pp. 213-224)
Department of Mathematics, Statistics and Computer Science
The University of New England
Armidale, N.S.W. 2351
AUSTRALIA

Professor Fredric T. Howard (pp. 121-135; 297-308)
Department of Mathematics
Wake Forest University
Winston-Salem, NC 27109
USA

Professor Richard Hudson (pp. 177-185)
Department of Mathematics
University of South Carolina
Columbia, SC 29208
USA

Professor Glenn Hurlbert (pp. 109-119)
Department of Mathematics and Statistics
Arizona State University
Tempe, AZ 85287-1804
USA

Professor Naotaka Imada
Department of Mathematics
Kanazawa Medical University
Uchinada, Ishikawa 920-0293
JAPAN

Professor John H. Jaroma
School of Mathematics and Computer Science
Lake Superior State University
Sault Sainte Marie, MI 49783
USA

Professor Clark Kimberling (pp. 137-144)
University of Evansville
1800 Lincoln Avenue
Evansville, IN 47722
USA

Dr. Ron Knott
1 Upperton Road
Guildford, Surry, GU2 7NB
UNITED KINGDOM

Ed Korntved
Department of Mathematics and Computer Science
Northwest Nazarene University
623 Holly Street
Nampa, Idaho 83686
USA

Professor Harris Kwong
Department of Mathematics and Computer Science
SUNY at Fredonia
Fredonia, NY 14063
USA

Mr. Jack Lee (pp. 145-157)
280 86th Street, Apt. D2
Brooklyn, NY 11209
USA

Professor Aihua Li (pp. 159-166)
Mathematics Department
Loyola University - Campus Box 35
New Orleans, LA 70118
USA

Calvin Long
Northern Arizona University
2120 N. Timberline Drive
Flagstaff, AZ 86004
USA

Jean Loyola
University of Philippines
1621 Almanor Street
Oxnard, CA 93030
USA

Professor Florian Luca (pp. 167-175)
Instituto de Matematicas de la UNAM
Campus Morelia
Apartado Postal 61-3 (Xangari) CP 58 089
Morelia, Michoacan
MEXICO

Quincy Magby
Arizona Western College
PO Box 929
Yuma, AZ 85365
USA

Toufik Mansour
LaBRI (UMR 5800)
Université Bordeaux 1
33405 Talence Cedex
FRANCE

Jonathan Mason (pp. 177-185)
Department of Mathematics
University of South Carolina
Columbia, SC 29208
USA

Professor Wayne McDaniel
University of Missouri - St Louis
1958 Duello Road
O'Fallon, MO 63366
USA

Shigeru Nakamura
Tokyo University of Mercantile Marine
2-1-6, Etchujima, Koto-Ku
Tokyo, 135-8533
JAPAN

Professor R. L. Ollerton (pp. 187-199)
University of Western Sydney
Nepean, 2747
AUSTRALIA

Jennifer J. Quinn (pp. 19-24)
Department of Mathematics
Occidental College
Los Angeles, CA 90041
USA

Professor Neville Robbins (pp. 201-211)
Department of Mathematics
San Francisco State University
San Francisco, CA 94132
USA

E. F. Robertson (pp. 69-85)
The Mathematical Institute
University of St. Andrews
St. Andrews KY 16 9SS
Fife, SCOTLAND

Kenneth R. Roden
Letourneau University
Box 7001
Longview, TX 75607
USA

Professor Andrzej Rotkiewicz
Polish Academy of Sciences
ul. Sniadeckich 8, skr. poczt. 137
00-950 Warszawa
POLAND

Jeremy A. Rouse (pp. 19-24; 25-28)
Department of Mathematics
Harvey Mudd College
Claremont, CA 91711
USA

Professor A. G. Shannon (pp. 187-199; 213-224)
Warrane College
The University of New South Wales
P.O. Box 123, Kensington, NSW 1465
AUSTRALIA

Iekata Shiokawa
Department of Mathematics
Keio University
3-14-1 Hiyoshi, Kohoku-ku
Yokohama, 223-8522
JAPAN

Peter Shiue
Department of Mathematical Sciences
University of Nevada, Las Vegas
Las Vegas, NV 89154-4020
USA

Professor Lawrence Somer (pp. 225-234)
Department of Mathematics
The Catholic University of America
Washington, DC 20064
USA

Professor Pantelimon Stanica (pp. 167-175)
Mathematics Department
Auburn University at Montgomery
Montgomery, AL 36124-4023
USA

Professor Paul Stockmeyer
Department of Computer Science
P.O. Box 8795
College of William and Mary
Williamsburg, VA 23187
USA

Professor K. B. Stolarsky
Mathematics Department
University of Illinois
Urbana, IL 61801
USA

Professor M. V. Subbarao (pp. 201-211)
Department of Mathematical Sciences
University of Alberta
Edmonton, CANADA

Dr. John C. Turner (pp. 235-245; 247-261)
Department of Mathematics and Statistics
University of Waikato
Private Bag 3105
Hamilton
NEW ZEALAND

Professor Sindhu Unnithan (pp. 159-166)
Mathematics Department
Xavier University of Louisiana
New Orleans, LA 70125
USA

Professor Gary Walsh
Mathematics and Statistics
University of Ottawa
Ottawa, ON K1N 6N5
CANADA

Professor William A. Webb (pp. 263-266)
Department of Pure and Applied Mathematics
Washington State University
Pullman, WA 99164-3113
USA

Michael Wiemann (pp. 267-287)
Department of Mathematics and Computer Science
Central Missouri State University
Warrensburg, MO 64093-5045
USA

Dr. Yongzhi (Peter) Yang (pp. 289-295)
Department of Mathematics
Mail #201 OSS
2115 Summit Avenue
St. Paul, MN 55105-1079
USA

Professor Osman Yürekli
Department of Mathematics and Computer Science
Ithaca College
Ithaca, NY 14850
USA

Chizhong Zhou (pp. 297-308)
Yueyang University
Yueyang City
Hunan 414000
P. R. CHINA

FOREWORD

This book contains 28 research articles from among the 49 papers and abstracts presented at the Tenth International Conference on Fibonacci Numbers and Their Applications. These articles have been selected after a careful review by expert referees, and they range over many areas of mathematics. The Fibonacci numbers and recurrence relations are their unifying bond.

We note that the article "Fibonacci, Vern and Dan", which follows the Introduction to this volume, is not a research paper. It is a personal reminiscence by Marjorie Bicknell-Johnson, a longtime member of the Fibonacci Association. The editor believes it will be of interest to all readers.

It is anticipated that this book, like the eight predecessors, will be useful to research workers and students at all levels who are interested in the Fibonacci numbers and their applications.

March 16, 2003 The Editor

Fredric T. Howard
Mathematics Department
Wake Forest University
Box 7388 Reynolda Station
Winston-Salem, NC 27109

THE ORGANIZING COMMITTEES

LOCAL COMMITTEE

Calvin Long, Chairman

Terry Crites

Steven Wilson

Jeff Rushal

INTERNATIONAL COMMITTEE

A. F. Horadam (Australia), Co-Chair

A.N. Philippou (Cyprus), Co-Chair

A. Adelberg (U.S.A.)

C. Cooper (U.S.A.)

H. Harborth (Germany)

Y. Horibe (Japan)

M. Bicknell-Johnson (U.S.A.)

P. Kiss (Hungary)

J. Lahr (Luxembourg)

G.M. Phillips (Scotland)

J. Turner (New Zealand)

Conference Participants

LIST OF CONTRIBUTORS TO THE CONFERENCE

*ADELBERG, ARNOLD, "Universal Bernoulli Polynomials and p-adic Congruences."
*AGRATINI, OCTAVIAN, "A Generalization of Durrmeyer-Type Polynomials."
BENJAMIN, ART, "Mathemagics."
*BENJAMIN, ARTHUR, (coauthor Jeremy A. Rouse), "Recounting Binomial Fibonacci Identities."
*BENJAMIN, ARTHUR, (coauthors Jennifer Quinn, Jeremy A. Rouse), "Fibinomial Identities."
BERENHAUT, KEN, "Recurrences with Restricted Coefficients."
*BICKNELL-JOHNSON, MARJORIE, "On Purple Parrots, Fibonacci Numbers and Color Theory."
*BICKNELL-JOHNSON, MARJORIE, "The Fibonacci Diatomic Array Applied to Fibonacci Representations."
BIEN, PAUL, "A Numbered Icosahedron from India: Hidden Approximations."
*BLECKE, NATHAN, (coauthor George Grossman), "Finding Fibonacci in Fractals."
BROWN, TOM C., (coauthors Peter Shiue, Alan R. Freedman), "Progressions of Squares."
BRUNNER, BARBARA, "The Composition of Number."
*CAMPBELL, C.M., (coauthors P.P. Campbell, H. Doostie, E.F. Robertson), "On the Fibonacci Length of Powers of Dihedral Groups."
CHAN, HEI-CHI, "On Random Fibonacci-type Sequences."
*COOK, CHARLES, "Some Sums Related to Sums of Oresme Numbers."
*COOPER, CURTIS, (coauthor Michael Wiemann) "Divisibility of an F-L Type Convolution."
DILCHER, KARL, (coauthor K.B. Stolarsky), "Resultants of Chebyshev and Related Polynomials.
EGGE, ERIC, "Restricted Permutations, Fibonacci Numbers, and k-Generalized Fibonacci Numbers."
ELIA, MICHELE, "A Class of Triangles with Minimal Area Related to Fibonacci Numbers."
*FIELDER, DANIEL C., "Some Thoughts on Rook Polynomials on Square Chessboards."
FIELDER, DANIEL C., "Lexicographic Counting of Combinations."
FLYNN, C.T., "Precise Approximations."
*HARBORTH, HEIKO, (coauthor Jens-P. Bode), "Positive Integers $(a^2 + b^2)/(ab + 1)$ are Squares."
HARMAN, JAYDEN D., "Application of the Fibonacci Sequence to High Efficiency Design Geometry."

*The asterisk indicates that the paper is included in this book.

HENDEL, RUSSELL JAY, "Consequences of Explicit Factorizations of Sums of Fibonacci Numbers."
HORADAM, A.F., "Chebyshev and Pell Connections."
*HOWARD, FRED, "A General Lacunary Recurrence Formula."
*HOWARD, FRED, (coauthor Chizhong Zhou), "F-L Representation of Division of Polynomials over a Ring."
*HURLBERT, GLENN, (coauthor Rob Hochberg), "Pythagorean Quadruples."
IMADA, KAOTAKA, "Relationships Between Convolutions of Signed Ballot Numbers and Catalan Numbers."
JAROMA, JOHN H., "On the Rank of Apparition of Primes of the form $2^k p \pm 1$."
*KIMBERLING, CLARK, "Ordering Words and Sets of Numbers: The Fibonacci Case."
*LEE, JACK, "Some Basic Properties of a Fibonacci Line-Sequence."
*LI, AIHUA, (coauthor Sindhu Unnithan), "A Sequence Constructed From Fibonacci Numbers."
LUCA, FLORIAN, "Palindromic Fibonacci Numbers."
*MASON, JONATHAN F., (coauthor Richard Hudson), "A Generalization of Euler's Formula and its Connection to Fibonacci Numbers."
MC INTOSH, RICHARD, "On the Search for Fibonacci-Wieferich Primes."
NAKAMURA, SHIGERU, "The Book Proof that the Golden Number is Irrational."
NOGUEIRA, JOAQUIM, "The Least Period of the Ratio Sequence."
NORTHSHIELD, SAM, "On Certain Sequences Arising from Apollonian Circle Packings."
*OLLERTON, R.L., (coauthor Tony Shannon), "Extended Generalized Binomial Coefficients."
PETERBURGSKY, IRINA, "Fibonacci Numbers in Undergraduate Mathematics."
*ROBBINS, NEVILLE, (coauthor M.V. Subbarao), "Some Parity Results Regarding t-core Partitions."
ROTKIEWICZ, ANDRZEJ, "Fibonacci and Frobenius Pseudoprimes."
*SHANNON, TONY, (coauthor A.F. Horadam), "Generalized Pell Numbers and Polynomials."
SHANNON, TONY, (coauthors V.K. Atanassova, T. Atanassov), "On Fibonacci "Bang" or a New Extension of the Fibonacci Sequence."
SHIOKAWA, IEKATA, "Irrationality and Transcendence Results on Reciprocal Sums of Fibonacci Numbers."
SOFO, ANTHONY, "From the Lambert Series to Abel Polynomials."
*SOMER, LARRY, "A Further Note on Lucasian Numbers."
*STANICA, PANTELIMON, (coauthors Florian Luca), "Cullen Numbers in Second Order Recurrent Sequences."
STOCKMEYER, PAUL, "The Stealth Fractal."
*TURNER, JOHN C., "Some Applications of Triangle Transformations in Fibonacci Geometry."
*TURNER, JOHN C., "Some Constructions and Theorems in Goldpoint Geometry."
WALSH, GARY, "Squares in Lucas-Lehmer Sequences and Classical Quartic Diophantine Equations."
*WEBB, WILLIAM A., "Cryptography and Lucas Sequence Discrete Logarithms."
*YANG, YONGZHI, "Using Pascal Matrix Decomposition to Develop Row - Generating Functions of Geometric - Progression Arrays."
YÜREKLI, OSMAN, (coauthors Jeff Chamberlein, Nate Higgins), "M-Bonacci Numbers and their Finite Sums."

INTRODUCTION

The Fibonacci numbers

$$1, 1, 2, 3, 5, 8, 13, 21, 34, 55, 89, \ldots,$$

were first mentioned in 1202 in the *Liber Abaci*, a book written by Leonardo of Pisa to introduce the Hindu-Arabic numeral system to western Europe. Leonardo, perhaps the greatest mathematician of the Middle Ages, wrote under the name of Fibonacci - - a contraction of "filius Bonacci" (son of Bonacci). In *Liber Abaci* the numbers appeared in the famous rabbit problem, but they were not called "Fibonacci numbers" until the nineteenth century, when the French mathematician Edouard Lucas used that term. Lucas studied the Fibonacci numbers extensively, and the simple generalization

$$2, 1, 3, 4, 7, 11, 18, 29, 47, 76, 123, \ldots,$$

bears his name.

The rich and interesting history of the Fibonacci numbers in the eighteenth and nineteenth centuries can be found in L. E. Dickson's *History of the Theory of Numbers, Volume 1.* During the last half of the twentieth century, interest in the numbers and their applications increased dramatically, and in 1963, Verner E. Hoggatt, Jr., and his associates founded The Fibonacci Association. They began publishing *The Fibonacci Quarterly*, and they organized the first Fibonacci conference. Some interesting personal recollections about these activities and about Professor Hoggatt can be found in the article "Fibonacci, Vern and Dan", which appears in this volume immediately after this introduction.

In 1984, the First International Conference on Fibonacci Numbers and Their Applications was held in Patras, Greece, and the proceedings from that conference were published. Since then, conferences have been held every other year in the following locations: (1986) San Jose, California, (1988) Pisa, Italy, (1990) Winston-Salem, North Carolina, (1992) St. Andrews, Scotland, (1994) Pullman, Washington, (1996) Graz, Austria, (1998) Rochester, New York, (2000) Luxembourg City, Luxembourg, (2002) Flagstaff, Arizona. With the exception of the 2000 conference, proceedings for all of the above have been published. The eleventh conference is scheduled for 2004 in Braunschweig, Germany.

It is impossible to overemphasize the importance and relevance of the Fibonacci numbers in the mathematical and physical sciences, as well as other areas of study. It is believed that the contents of this book, like its predecessors, will prove useful to anyone interested in this important branch of mathematics.

The editor would like to acknowledge The Fibonacci Association and Northern Arizona University for their financial and other assistance in making the conference a success. He would also like to thank Calvin Long, Art Benjamin and Marjorie Bicknell-Johnson for their help, advice and contributions to the conference. Finally, the editor thanks the technical typist, Patricia Solsaa, for her excellent work.

The Editor

FIBONACCI, VERN AND DAN

Marjorie Bicknell-Johnson
665 Fairlane Avenue, Santa Clara, CA 95051

"Buon giorno. Per favore, dove è la statua di Fibonacci?"
"Fibonacci? ... Si, Fibonacci. ... Un momento ... Non lo so, mi dispiace."
"Grazie."

The location of Fibonacci's statue must be the best kept secret of Pisa. A medical student, a policeman, the clerk at the information office, everyone asked someone else, with much pointing and gesturing. Our phrasebook Italian wouldn't have withstood the test if someone had given us directions, but everyone understood our question. Fibonacci, or Leonardo di Pisa, should have 'most favored son status', since he was the foremost mathematician of Europe in 1202. Leonardo's *Liber Abbaci* introduced Hindu-Arabic numerals with methods of calculation to Europe, a major scientific breakthrough for a world using Roman numerals and the abacus. Leonardo was most proud of his other works, solving problems posed by court mathematicians for Emperor Frederick II, and books on number theory and geometry. Now Fibonacci is mostly remembered for a sequence bearing his name, $1, 1, 2, 3, 5, 8, 13, \ldots$,, arising from a problem in the *Liber Abbaci*, counting hypothetical rabbits. Vern wanted a picture of Fibonacci's statue.

We were there, so we climbed the Leaning Tower, braving eight smooth marble terraces with no railing on the low side, a heart-pounding, dizzy walk with a panoramic view, above a 296-step staircase. The white marble *campanile* (bell tower) of the cathedral in Pisa is 52 feet in diameter at the base and 170 feet high, 183 feet if it stood up straight. The tower began its tilt as soon as it was first erected in 1174; architects tried to compensate, strengthening the 3rd and 5th stories and tapering the walls. Fibonacci might have climbed it; he certainly watched its construction, which was not finished until 1350. There was no wait in 1978, and only one tourist kiosk for sodas, souvenirs, and tourist information. My husband Frank took the obligatory tourist shot of me holding up the tower before we started back for our *Europe on $5 a Day* accommodations in Florence.

While making our empty-handed return to the train station, the tourist map showed 'Via Fibonacci' nearby. Outside of 1 Via Fibonacci, an old woman was working in her garden. "Buon Giorno. Per favore, ..."; she ran inside and locked her door.

Across the street we noticed a high wall, with a gate leading to a green area. There, in the corner of Giardino Scotto within the Fortezza Campo Santo, but rapidly being walled in, was Leonardo Fibonacci's nineteenth century statue, depicting him in traveler's robes. Leonardo called himself Leonardo Bigollo (the traveler), or Leonardus Pisanus in Latin; he was nicknamed 'Fibonacci' (son of Bonaccio) by eighteenth century Italian mathematicians. One snapshot shows me with the statue and some framework, but there was no way to get a clean shot, until my husband Frank climbed fifteen feet above the ground on the rickety scaffolding to get a close-up of the face. We had snapped our prize, a 1978 photograph that graced Vern's office, my office, the cover of *California Mathematics* (April, 1984), the conference program for the 1988 Conference on Fibonacci Numbers in Pisa, and several websites by 2002.

While I was a student in mathematics at San Jose State College, my thesis advisor, Professor Verner E. Hoggatt, Jr., started the *Fibonacci Quarterly*, an academic journal devoted to properties of the integers. Such an exciting idea - I was in love with Fibonacci. In October, 1962, I spent 20 hours proofreading the article, "The First 571 Fibonacci Numbers;" F_{571} has 120 digits, a monster number at the end of the range for computers then. The article appeared with two 50-digit errors! The printer's apprentice at *Recreational Mathematics* confessed that he had dropped a tray of individual lead characters (yes, that was how type was set in those days). The same magazine announced the soon-to-appear *Fibonacci Quarterly*, with an annual subscription rate of $4.00.

Vern, nicknamed 'Professor Fibonacci', loved research problems and supervised more thesis students than the entire rest of the department, or, $39 = 3F_7$ masters' students, as he would have put it; he especially liked to encourage young women to go into mathematics. As an undergraduate, I found all possible determinant values using the integers 1 through 9; $9! = 362,880$ possible arrays. No, I didn't have to write them all down; I found an entirely different way to solve the problem. When I wrote up my solution for publication, Vern said, "There's only one thing wrong with this paper. It sounds like it belongs in *Ladies' Home Journal*." I rewrote "An Investigation of Nine-Digit Determinants" for *Mathematics Magazine*, and a second paper as well, "64 Ways to Write 64 Using Four 4's" for *Recreational Mathematics*. The nine-digit determinant problem, as originally posed, listed some "impossible" values. I found two of them. I wrote to the author, who was terminally ill; when he sent me a box of his complete notes, I had a thesis topic.

I wrote four dozen articles with Vern as co-author, mostly on properties of Pascal's triangle, convolution arrays, and representations; of course, all related to the Fibonacci sequence in some way. When Vern passed away after editing the Fibonacci Quarterly for 18 years, it was one of the saddest days of my life. I could not mention Fibonacci numbers for months without choking up. Vern was depressed and felt that he was losing his creativity; he took his own life at age 59. I felt responsible. He had a sabbatical leave to write a book on Fibonacci numbers, but I could not take an unpaid semester off from the high school. He often worked late at night, and while I lived only two miles away, he sent me handwritten 8 or 10 page mathematical letters at least twice a week. After he died, I kept dreaming that I received a letter from Vern, with the usual cryptic PS scrawled on the envelope, but the envelope was empty. Afterwards, I would awake sobbing. Vern never learned to type. I typed more than 5000 pages for him on my manual typewriter, including a joint paper he wrote with Paul Erdös, the most prolific and renowned mathematician of the 1970's.

Vern shared his creative thoughts with me, my mathematical father, so to speak. Mathematics is unraveling an intricate and delicious puzzle, but is not good coffee time conversation. Often when I told another woman that I taught high school mathematics, I heard, "I never was very good at math", or "I always hated math". I even heard, "Mathematicians are a little crazy ..."; not necessary or sufficient, but it helps. Once my own son asked, "How much cake is left?" My reply, "Four-sevenths remains," led to 'Oh, Mother! Why can't you say half of it like a normal mother?"

In August, 1980, incoming *Fibonacci Quarterly* editor Gerald E. Bergum and I cleaned out Vern's desk and office, a curious mixture of manuscripts, letters, notes, a decade of Christmas cards and paid utility bills, but no indication of which papers were accepted for publication, or of replies to correspondence. We boxed up the whole mess and sent it to South Dakota. I wrote to some of Vern's other co-authors, including Daniel C. Fielder, Georgia Institute of Technology, who began a long correspondence with me, sometimes sending me papers in progress for critique.

The Fibonacci Association held its Third International Conference on Fibonacci Numbers and Applications in Pisa in 1988. Frank and I arrived two days early; I was in charge of registration. Dan Fielder, then professor emeritus at age 70, also came a day early; he said, "I've wanted to meet you for years." What energy; he could walk circles around most younger men, and he spoke so quickly to get out his overflowing thoughts that he would sometimes stutter. Frank and I took him for a long walk to visit Fibonacci's statue, no longer hiding behind scaffolding. The Leaning Tower tilted even more; we three climbed it, clinging to the inner wall on the underside. In 1988, the wait to climb the tower was more than an hour, and there were six kiosks catering to milling tourists. But Vern, who never liked to travel, missed all of this; I arranged for his widow to attend the Pisa conference.

The Italians rolled out the red carpet for the Fibonacci Association, hosting a reception in the mayor's chambers, in a 14th century castle with gold leaf on the moldings, museum pieces on the walls, and marble statuary along the red carpeted halls. A special commemorative one-day postmark was struck for our postcards, and two postal officials sat in the lobby of our convention. Fibonacci's statue was dusted off, and a plaque in a public building named Leonardo Fibonacci, Pisa's favorite son. Even the newspaper had a glowing report. I took a group from the conference to visit Fibonacci's statue, and Frank organized an excursion to Florence for spouses, visiting all of his favorite places. I looked ten pounds heavier around the middle with the pile of registration fees in my money belt. I paid the conference expenses in cash, including the largest restaurant bill I have ever seen: 3,100,000 Lira (about $2400) for the conference banquet.

For Dan, 'retirement' from Georgia Tech after teaching forty years meant part-time teaching and full-time research. His last published journal paper (number $110 = 2F_{10}$), was written with me, "The Smallest Integer Having 331 Fibonacci Representations," *Fibonacci Quarterly*, November 2001. In 1996 in Graz, Austria, Dan handed me a 66-page computer print-out on counting the number of Fibonacci representations of integers, saying, "Let's crack this thing." We worked intently for six years, putting out seven joint papers, and I wrote another six papers on related topics. Dan sent me at least 10 reams of paper that ran through his computers, a thirteen PC stable in his basement. I wore out two dozen pencils. Dan's letters to me were packages, usually an inch thick.

Dan contributed our joint paper, "A Fast Algorithm for a Special Expansion," to the International Computer Science Convention on Soft Computing in Genoa, Italy, June, 1999. In May, 1999, Dan's doctor told him to stay home; he mailed the lecture transparencies and handouts to me. At the same time, I had an unexpected hysterectomy; I didn't want Dan to worry. So twelve days after my surgery, I flew to Genoa, expecting to present our paper. Dan arrived at the last minute and gave the lecture after all. We spent a few pleasant days with Dan, who got a less expensive room, in a hotel six miles from the meetings, and carried his laptop computer back and forth to the worst part of the city late at night. I worried about a mugging or a heart attack; Dan said, "I'm a young man of 81." I also remember a train-trip from Graz, Dan's air ticket hanging precariously out of a shirt pocket, while he talked passionately about research problems. Dan loved to go to meetings anywhere, and he always brought his guardian angel.

Frank and I stopped in Pisa in 1999; Piazza Duomo was paved with tourists. The Leaning Tower was strapped and supported with all sorts of cables, and no one was climbing it. Getting older, just like the rest of us, carrying on with engineering help.

Dan came to every Fibonacci meeting, including Flagstaff, June, 2002. He would say, "I'm no mathematician, but ...," always thinking of himself as an electrical engineer. Mathematically, Dan never met a series he didn't like; he passed away in October, 2002, at age 84. Ironically, that same week the *Fibonacci Quarterly* accepted my latest research paper (number $89 = F_{11}$), and twenty years earlier, Vern's last journal paper (number $144 = F_{12}$), co-written posthumously with me. Forty years ago, the forthcoming *Fibonacci Quarterly* was announced; 800 years ago, Fibonacci had just finished his *Liber Abbaci*.

Dan's problem lives on, unsolved: Find the smallest integer having n Fibonacci representations. The sequence of solutions begins $1, 3, 8, 16, 24, 37, 58, \ldots, A_n, \ldots$, where A_n is the smallest integer having n Fibonacci representations. For example, $A_4 = 16$ has four representations, $13+3, 13+2+1, 8+5+3$, or $8+5+2+1$, the number of ways to write each number as the sum of distinct Fibonacci numbers. We can calculate how many representations that any integer N has, or we can create numbers N having exactly n Fibonacci representations. We do not have a formula for the smallest integer having n Fibonacci representations except for special values of n. Dan computed the first 330 consecutive solutions, but as N gets larger, the computer eventually runs out of memory and some values for A_n will be missing. And N gets very large indeed; $A_{331} = 327,367$ while $A_{330} = 134,406$. Perhaps, in another 800 years, ...

REFERENCES

[1] Fibonacci, L.P. The Book of Squares. Orlando, FL: Academic Press, 1987.

[2] Gies, J., and Gies, F. Leonard of Pisa and the New Mathematics of the Middle Ages. New York: Thomas Y. Crowell Company, 1969.

[3] Hoggatt, V.E., Jr. "Number Theory: The Fibonacci Sequence." Encyclopedia Britannica, Yearbook of Science and the Future, 1977, pp. 178-191. Chicago: Britannica, 1977.

[4] Horadam, A.F. "Eight Hundred Years Young." *The Australian Mathematics Teacher*, Vol. *31* (1975): pp. 123-134.

[5] Livio, M. The Golden Ratio: The Story of Phi, the World's Most Astonishing Number. New York: Broadway Books, 2002.

[6] Smith, D.E. History of Mathematics. New York: Dover Publications, 1958.

AMS Classification Numbers: 01A70, 11B39

Statue Leonardo Fibonacci and Marjorie Bicknell-Johnson.

(taken in 1978.)

UNIVERSAL BERNOULLI POLYNOMIALS AND P-ADIC CONGRUENCES

Arnold Adelberg

1. INTRODUCTION

The *classical Bernoulli numbers* B_n are defined by

$$\frac{t}{e^t - 1} = \sum_{n=0}^{\infty} B_n \frac{t^n}{n!}.$$

A fundamental property of these numbers is the Clausen-von Staudt result that the denominator of B_n is the product of distinct primes.

The *classical* l^{th} *order Bernoulli numbers* $B_n^{(l)}$ are defined by

$$\left(\frac{t}{e^t - 1}\right)^l = \sum_{n=0}^{\infty} B_n^{(l)} \frac{t^n}{n!},$$

and the *classical* l^{th} *order Bernoulli polynomials* $B_n^{(l)}(x)$ are defined by

This paper is in final form and no version of it will be submitted for publication elsewhere.

F.T. Howard (ed.), Applications of Fibonacci Numbers, Volume 9: Proceedings of the Tenth International Research Conference on Fibonacci Numbers and their Applications, 1–8.

$$\left(\frac{t}{e^t-1}\right)^l e^{tx} = \sum_{n=0}^{\infty} B_n^{(l)}(x)\frac{t^n}{n!},$$

which is the associated Appel sequence.

The special case $l = 1$ gives the ordinary Bernoulli numbers and polynomials, and $B_n^{(l)}(0) = B_n^{(l)}$. It is easy to show that if $B_n^{(l)}(x) = \sum_{i=0}^n a_i x^{n-i}$ then $a_i = \binom{n}{i}B_i^{(l)}$, which is the characteristic Appel property.

L. Carlitz proved in [5, Theorem 14] and gave a more direct proof in [6, Theorem A] that if the base p representation of a positive integer l has r non-zero digits, then the highest power of the prime p in the denominator of $B_n^{(l)}$ is at most r, i.e. $p^r B_n^{(l)} \in \mathbf{Z}_p$. The proofs involve the complicated theory of Hurwitz series. The above Appel formula then shows that the same result holds for all coefficients of $B_n^{(l)}(x)$.

F. Clarke defined universal Bernoulli numbers $\hat{B}_n$ in [7] which depend on parameters $c_1, c_2, \ldots$. They are important for studying universal formal groups. We generalized these numbers in [3] to universal l^{th} order Bernoulli numbers $\hat{B}_n^{(l)}$. Our principal tool for the study of these numbers is the use of certain explicit representations coming from Lagrange inversion and related to the theory of partitions, which we first noted in [1].

In this paper we define *universal l^{th} order Bernoulli polynomials* $\hat{B}_n^{(l)}(x)$ which generalize the classical polynomials for $c_i = (-1)^i$, using the Lagrange inversion terms. We do not have a generating function for these polynomials, nor are the coefficients simple functions of lth order universal Bernoulli numbers, but it is still true that $\hat{B}_n^{(l)}(0) = \hat{B}_n^{(l)}$. Other specializations that may be of interest in the context of this conference are to the Fibonacci ($c_i = F_{i+1}$) or Lucas ($c_i = L_{i+1}$) formal groups.

The basis for our definition of the universal Bernoulli polynomials is the classical inversion formula [1]

$$\frac{B_n^{(l)}(x)}{n!} = [t^n](1+t)^{x-1}\left(\frac{\log(1+t)}{t}\right)^{l-n-1}.$$

We have taken a concrete, combinatorial generalization of this formula as the definition (2), but an equivalent umbral formulation is also given by (6).

It should be noted that our universal l^{th} order Bernoulli polynomials are not an Appel sequence, e.g., they are not monic. In [10] N. Ray defined universal first order Bernoulli polynomials as an Appel sequence. This definition has some good functorial properties, but we do not believe that his polynomials are as useful as ours. In particular, our polynomials are isobaric in the variables $c_1, c_2, \ldots$, while his are not, and ours satisfy Kummer congruences for p-adic integer values and his do not (Theorem 4).

We prove that the Carlitz bound noted above holds for the coefficients of $\hat{B}_n^{(l)}(x)$ (Theorem 3) and show that the bound is attainable, and in fact, $\nu\left(\hat{B}_n^{(l)}\right) = -r$ for suitable n. We believe that our proofs are conceptually simpler than Carlitz's, and show his result is really a special case of our Theorem 1. This is the first publication of our simple proof.

Since our analysis of the p-adic pole structure of $B_n^{(l)}(x)$ was based entirely on the Lagrange inversion terms, the statements and proofs carry over essentially without change to the universal polynomials (Theorems 1, 2). In particular, the down-sloping portion of the Newton polygon of $\hat{B}_n^{(l)}(x)$ is identical with that of $B_n^{(l)}(x)$, and indeed with all specializations where the c_i are p-adic units.

Finally, we consider the universal (ordinary) Bernoulli polynomials, where $l = 1$. We show that if $a \in \mathbf{Z}_p$ and $p - 1 \nmid n$ then $\hat{B}_n(a)/n \in \mathbf{Z}_p[c_1, c_2, \ldots, c_n]$, and extend the Kummer congruences for $n \not\equiv 0, 1 (\text{mod } p-1)$ to the universal case (Theorem 4) for values of $\hat{B}_n(x)/n$. This generalizes our work on universal Kummer congruences in [3, 4].

2. UNIVERSAL ARBITRARY ORDER BERNOULLI POLYNOMIALS

Consider the formula from [1, §3]

$$B_n^{(l)}(x) = n! \sum_{i=0}^{n} (-1)^{n-i} \binom{x-1}{i} \sum_{w(u)=n-i} \frac{\binom{s}{d}\binom{d}{u}}{\Lambda^u} \tag{1}$$

where $s = l - n - 1$, the summation is over all non-negative integer sequences $(u) = (u_1, u_2, \ldots, u_n)$, with $w(u) = \sum_i iu_i$ and $d = d(u) = \sum_i u_i$,

$$\binom{d}{u} = \binom{d}{u_1\ u_2 \ldots u_n}, \quad \Lambda^u = 2^{u_1} 3^{u_2} \ldots (n+1)^{u_n}, \text{ and } \binom{s}{d} \text{ and } \binom{x-1}{i}$$

are binomial coefficients.

Note that (u) is a partition of $w(u)$ into $d(u)$ parts, where u_i = number of occurrences of i in the partition. With the same notations, let $\tau_u(s) = \binom{s}{d}\binom{d}{u}/\Lambda^u$.

Definition: Let $c_1, c_2, \ldots$ be indeterminates over $\mathbf{Q}$, let $c_0 = 1$, and let $c^u = c_1^{u_1} c_2^{u_2} \ldots c_n^{u_n}$. Then

$$\hat{B}_n^{(l)}(x) = n! \sum_{i=0}^{n} (-1)^i c_i \binom{x-1}{i} \sum_{w(u)=n-i} \tau_u(l-n-1)c^u. \tag{2}$$

From (1), the specialization $c_i = (-1)^i$ gives the classical $B_n^{(l)}(x)$.

Some low degree examples are $\hat{B}_0^{(l)}(x) = 1, \hat{B}_1^{(l)}(x) = -c_1(x - l/2)$, and $\hat{B}_2^{(l)}(x) = c_2x^2 - (3c_2 + (l-3)c_1^2)x + (2lc_2/3 + l(l-3)c_1^2/4)$.

Note that all coefficients of $\hat{B}_n^{(l)}(x)$ are isobaric polynomials of weight n in $\mathbf{Q}[l][c_1, c_2, \ldots, c_n]$, where $wt(c_i) = i$, and the highest coefficient is $(-1)^n c_n$. For most of our applications, l will be a positive integer, a p-adic integer, or a variable.

The critical cases are when l is an integer in the range $0 \leq l \leq n+1$. For $l = n+1$, obviously

$$\hat{B}_n^{(n+1)}(x) = (-1)^n c_n (x-1)_n = (-1)^n c_n (x-1)(x-2)\ldots(x-n). \tag{3}$$

In the classical case $B_n^{(0)}(x) = x^n$, but there is no corresponding simple formula for $\hat{B}_n^{(0)}(x)$.

It is true that $\hat{B}_n^{(0)}(0) = 0$ if $n > 0$, and we will show (Corollary 2 to Theorem 3) that all

coefficients of $\hat{B}_n^{(0)}(x)$ are in $\mathbf{Z}[c_1, c_2, \ldots, c_n]$.

The classical $B_n^{(l)}(x)$ is skew-symmetric about $x = l/2$, but there is no symmetry in the universal case, and no obvious root if $n > 1$ and n is odd, unlike the classical case.

From [3, Corollary 2.3], we get the key formulas

$$\hat{B}_n^{(l)}(0) = \hat{B}_n^{(l)} \tag{4}$$

and

$$\hat{B}_n^{(l+1)}(1) = \frac{l-n}{l} \hat{B}_n^{(l)}. \tag{5}$$

There is an umbral way of representing $\hat{B}_n^{(l)}(x)$, namely let $F(t) = \sum_{i=0}^{\infty} c_i t^{i+1}/(i+1)$ be the logarithm of the universal formal group law, so $(t/F^{-1}(t))^l$ generates the universal l^{th} order Bernoulli numbers [3]. Then

$$\frac{\hat{B}_n^{(l)}(x)}{n!} = [t^n](1 - Ct)^{x-1} \left(\frac{F(t)}{t} \right)^{l-n-1}, \tag{6}$$

where $C^i = c_i$.

We do not take (6) as the definition, because of the presence of the index n in the "umbral generating function." We suspect that an umbral form of Lagrange inversion is involved. Our definition gives us the concrete terms that we need for analysis.

If J is the ideal generated by all $c_i - c_1^i$ in $\mathbf{Q}[l][c_1, \ldots, c_n]$, the umbral representation leads naturally to the easily verified recursive congruence

$$\hat{B}_n^{(l)}(x+1) - \hat{B}_n^{(l)}(x) \equiv -nc_1\hat{B}_{n-1}^{(l-1)}(x) \pmod{J}.$$

However, since the specialization $c_i = c_1^i$ clearly implies $\hat{B}_n^{(l)}(x) = (-c_1)^n B_n^{(l)}(x)$, this is equivalent to the classical recursion

$$B_n^{(l)}(x+1) - B_n^{(l)}(x) = nB_{n-1}^{(l-1)}(x).$$

3. P-ADIC CONSIDERATIONS

Let p be prime. Extend the standard p-adic valuation $\nu = \nu_p$ of $\mathbf{Q}_p$ to $\mathbf{Q}_p[c_1, c_2, \cdots]$ by $\nu(\sum a_I c^I) = \min\{\nu(a_I)\}$, so in particular a polynomial $f(c)$ is integral if and only if all coefficients of f are integral. If $b > 0$ and $\nu(f(c)) = -b$, we say $f(c)$ has a *pole of order b*.

Since the analysis of the pole structure of $B_n^{(l)}(x)$ carried out in [1] entirely involved single terms $\tau_u(s)$ where (u) is concentrated in place $p-1$, i.e., such that $u_i = 0$ if $i \neq p-1$, all the results carry over without change to $\hat{B}_n^{(l)}(x)$. For ease of reference, we summarize these results using the versions stated in [2].

Let $n = \sum_{i=0}^m a_i p^i$ be the base p expansion of n. Then $S(n) = \sum_i a_i$ is the digit sum. If $S(n) \geq p-1$, then $N(n)$ is the smallest $t > 0$ such that $p-1|t$ and $p\nmid\binom{n}{t}$. Thus $N(n)$ is the smallest t such that $p\nmid\binom{n}{t}$ and $S(t) = p-1$.

Theorem 1: Let $l \in \mathbf{Z}_p$. If $S(n) < p-1$, then the coefficients of $\hat{B}_n^{(l)}(x)$ have no poles. If $S(n) \geq p-1$, the successively higher order poles of the coefficients, from degree n down, are determined as follows: the first pole is simple (order one) and occurs in degree $n - N_1$, where N_1 is minimal satisfying $N_1 = N(n - l_1)$ for some bottom segment l_1 of n (possibly $l_1 = 0$) such that $p\nmid\binom{l-n-1}{N_1/(p-1)}$. Similarly, the next higher order pole is double, and occurs in degree $n - N_1 - N_2$ where N_2 is minimal satisfying $N_2 = N(n - l_1 - N_1 - l_2)$ for some bottom segment l_2 of $n - l_1 - N_1$ such that $p\nmid\binom{l-n-1}{N_2/(p-1)}$, etc.

If $f(x) = \sum_{i=0}^{n} a_i x^{n-i} \in \mathbf{Q}_p[c_1, c_2, \ldots][x]$, consider the *spots* $(i, \nu(a_i))$ where $a_i \neq 0$ as lattice points in the (x, y)-plane. The *Newton polygon* of $f(x)$ is the lower boundary of the convex null of the set of spots [2].

The preceding theorem characterizes the downward-sloping portion of the Newton polygon of $\hat{B}_n^{(l)}(x)$ as follows.

Theorem 2: Let $l \in \mathbf{Z}_p$. The negative slope sides of the Newton polygon of $\hat{B}_n^{(l)}(x)$ all satisfy $\Delta y_i = -1$, for $1 \leq i \leq b$, where $\nu\left(\hat{B}_n^{(l)}(x)\right) = -b$. The corresponding $\Delta x_i = N_i$ are determined algorithmically as in the preceding theorem, so in particular $p-1|\Delta x_i$ for all i and $p|\Delta x_i$ for all $i > 1$ (and also for $i = 1$ if $p|n$).

Before deducing the Carlitz bound for the coefficients, recall the basic fact about p-divisibility of binomial coefficients [8].

$$\text{(Lucas's Theorem)} \qquad p \nmid \binom{n}{m} \text{ iff } n_i \geq m_i \text{ for all the base } p \text{ digits.} \tag{7}$$

Since $\binom{-n}{m} = (-1)^m \binom{n+m-1}{m}$, we deduce that $p \nmid \binom{-n}{m}$ iff the base p sum of $n-1$ and m has no carries. Thus if $p-1|N$, then $N + N/(p-1) = Np/(p-1)$, so $p|\binom{-N-1}{N/(p-1)}$. Also, if a_i and b_i are the lowest digits of N and $N/(p-1)$ respectively (which occur in the same place), then $a_i = p - b_i$.

Theorem 3: Let $l \in \mathbf{N}$, and suppose that the base p expansion of l has r non-zero digits. Then $\nu\left(\hat{B}_n^{(l)}(x)\right) \geq -r$, i.e., all coefficients a_i satisfy $\nu(a_i) \geq -r$.

Proof: Since $\hat{B}_n^{(n+1)}(x)$ is p-integral by (3), we have only the two cases $0 \leq l \leq n$ and $l > n+1$ to consider.

Case 1: $0 \leq l \leq n$. With the notations of Theorem 1 and by the preceding remarks, since $p \nmid \binom{l-n-1}{N_1/(p-1)}$, l must have a digit in the lowest place of N_1 or in a lower place, causing a carry from $n-l$ to that place. We prove inductively that l has at least i digits up to the bottom digit of N_i. This is true because, as above, if l has no digit in the lowest place of N_i, then there must be a carry from below for $n-l$. Thus l has a digit between the bottom places of N_i and N_{i-1} or n has a succession of digits all $p-1$ below N_i, and l has an extra digit below the succession, which causes the carry for $n-l$ to the lowest place of N_i.

Case 2: $l > n+1$. In this case, l must have a digit in the lowest place of N_1 or a digit in a lower place to prevent a carry for $l-n-1$ to that place. We have a similar proof to case 1, except that the digits of l are now required to prevent carries.

Thus $b \leq r$ in both cases, i.e., the pole has order at most r. □

Corollary 1: With the same notations, $\nu\left(\hat{B}_n^{(l)}\right) \geq -r$.

We can verify that the Carlitz bound is best possible, e.g., if $l = \sum_{i=1}^{r} a_i p^{s_i}$ with $s_1 < s_2 < \cdots < s_r$ and $1 \leq a_i \leq p-1$, then if $n = (p-1)\sum_{i=1}^{r} p^{s_i}$, taking $N_i = (p-1)p^{s_i}$ shows that $\nu\left(\hat{B}_n^{(l)}\right) = -r$, hence also $\nu\left(\hat{B}_n^{(l)}(x)\right) = -r$.

Note that our proof of Carlitz's bound gives us a new proof for the classical case, as well as for all specializations where the c_i are p-adic integers, and the bound is then attained as long as c_{p-1} is a p-adic unit.

If $l = 0$, the preceding theorem shows that for every prime p, the coefficients of $\hat{B}_n^{(0)}(x)$ are p-integral. Thus we deduce

Corollary 2: The coefficients of $\hat{B}_n^{(0)}(x)$ are in $\mathbf{Z}[c_1, c_2, \ldots, c_n]$.

Finally we turn to the case $l = 1$, i.e., the ordinary universal Bernoulli numbers $\hat{B}_n$ and polynomials $\hat{B}_n(x)$. In this case, as in [3, Corollary 2.3]

$$(n-1)!\tau_u(-n) = (-1)^d (n+d-1)!/(u!\Lambda^u) \tag{8}$$

where $u! = u_1! u_2! \ldots u_n!$ and Λ^u is as before. Hence

$$\frac{\hat{B}_n(x)}{n} = \sum_{i=0}^{n} (-1)^i c_i \binom{x-1}{i} \sum_{w(u)=n-i} (-1)^d \frac{(n+d-1)! c^u}{u! \Lambda^u}. \tag{9}$$

The following corollary follows immediately from Theorem 3, taking $l = 1$.

Corollary 3: The coefficients of $\hat{B}_n(x)$ have square-free denominators, so in particular if $a \in \mathbf{Z}$, $\hat{B}_n(a)$ has square-free denominator.

The proofs of [3, Lemma 3.1 and Theorem 3.2] give the following result.

Theorem 4: Let $a \in \mathbf{Z}_p$. Then

(1) If $p-1 \nmid n$, then $\hat{B}_n(a)/n \in \mathbf{Z}_p[c_1, c_2, \ldots, c_n]$.

(2) If $n \not\equiv 0,\ 1 (\mathrm{mod}\ p-1)$, then $\hat{B}_{n+p-1}(a)/(n+p-1) \equiv \hat{B}_n(a)c_{p-1}/n$ (mod p), where mod p is an abbreviation for mod $p\mathbf{Z}_p[c_1, c_2, \ldots]$.

The first assertion with the preceding corollary gives a modest generalization of the von Staudt result to values of universal ordinary Bernoulli polynomials, while the second generalizes the Kummer congruences. As noted in [3], the hypotheses are essential, even for the special case $a = 0$ of the ordinary universal Bernoulli numbers. See [4] for the extension of the case $a = 0$ to $n \equiv 1$ (mod $p-1$), where we have found an explicit formula for $\hat{B}_n/n$ mod p.

ACKNOWLEDGMENT

The author would like to thank the anonymous referee for several useful suggestions.

REFERENCES

[1] Adelberg, A. "On the Degrees of Irreducible Factors of Higher Order Bernoulli polynomials." *Acta Arith.*, Vol. *62* (1992): pp. 329-342.

[2] Adelberg, A. "Higher Order Bernoulli Polynomials and Newton Polygons." Applications of Fibonacci Numbers, Volume 7. Edited by G.E. Bergum, et. al. Kluwer Academic Publishers (1998): pp. 1-8.

[3] Adelberg, A. "Universal Higher Order Bernoulli Numbers and Kummer and Related Congruences." *J. Number Theory*, Vol. *84* (2000): pp. 119-135.

[4] Adelberg, A. "Kummer Congruences for Universal Bernoulli Numbers and Related Congruences for Poly-Bernoulli Numbers." *Intern. Math. Journal*, Vol. *1.1* (2002): pp. 53-63.

[5] Carlitz, L. "Some Properties of Hurwitz Series." *Duke Math. J.*, Vol. *16* (1949): pp. 285-295.

[6] Carlitz, L. "A Note on Bernoulli Numbers of Higher Order." *Scripta Mathematica*, Vol. *22* (1956): pp. 217-221.

[7] Clark, F. "The Universal von Staudt Theorem." *Trans. Amer. Math. Soc.*, Vol. *315* (1989): 591-603.

[8] Ireland, K. and Rosen, M. A Classical Introduction to Modern Number Theory, 2nd ed., Springer-Verlag, NY 1990.

[9] Nörlund, N.E. Vorlesungen über Differenzenrechnung. Chelsea Publ. Co., NY 1954 (First ed. 1923).

[10] Ray, N. "Extension of Umbral Calculus I: Penumbral Coalgebras and Generalized Bernoulli Numbers." *Adv. in Math*, Vol. *61* (1986): pp. 41-100.

[11] Young, P.T. "Congruences for Bernoulli, Euler and Stirling Numbers." *J. Number Theory*, Vol. *78* (1999): pp. 204-227.

AMS Classification Numbers: 11B63, 11B68

A GENERALIZATION OF DURRMEYER-TYPE POLYNOMIALS AND THEIR APPROXIMATION PROPERTIES

Octavian Agratini

1. INTRODUCTION

The Bernstein polynomial approximation process of discrete type defined for every function f belonging to the space $C([0,1])$ by $(B_n f)(x) = \sum_{k=0}^{n} p_{n,k}(x) f(k/n)$, where

$$p_{n,k}(x) = \binom{n}{k} x^k (1-x)^{n-k}, \; x \in [0,1], \tag{1}$$

has been the object of many investigations serving as a guide for theorems that can be proved for a large class of positive linear approximation processes on a bounded interval.

Simultaneously, the Bernstein polynomial basis $\mathcal{B}_n = (p_{n,k})_{k=\overline{0,n}}$ is a treasure of nice properties.

In order to obtain an approximation process in spaces of integrable functions, J.L. Durrmeyer [5] defined the following integral modification of the Bernstein polynomial:

$$(D_n f)(x) = (n+1) \sum_{k=0}^{n} p_{n,k}(x) \int_0^1 p_{n,k}(t) f(t) dt, \quad f \in L_1([0,1]), \; x \in [0,1], \tag{2}$$

which can be used for restoring f if its moments $\int_0^1 f(t) t^k dt$ are given. These polynomials were extensively studied by Marie Madeleine Derriennic [4]. The sequence $(D_n f)_{n \geq 1}$ appears

This paper is in final form and no version of it will be submitted for publication elsewhere.

F.T. Howard (ed.), Applications of Fibonacci Numbers, Volume 9: Proceedings of the Tenth International Research Conference on Fibonacci Numbers and their Applications, 9–18.

more complicated and maybe more difficult to compute but it possesses some desirable properties, of which the most notable are commutativity, self-adjointness and simple expansions by Legendre polynomials. It was also shown that $D_n f$, $n \in \mathbb{N}$, are positive contractions in $L_p([0,1])$, $p \geq 1$, spaces. It is the above mentioned properties that make $D_n f$ simpler than the Bernstein polynomial approximation. Therefore, we are able to prove for (2) approximation results that we are not able to prove for Bernstein polynomials from which $D_n f$ originate.

In a recent paper Michele Campiti and Giorgio Metafune [3] replaced in the polynomials $B_n f$ the binomial coefficents by general ones satisfying similar recursive properties, more precisely they replaced the sequences of constant value 1 at the sides of Pascal's triangle with arbitrary ones and defined the coefficients of their polynomials using the same rule of binomial coefficients. The new sequence does not converge to the identity operator but to an operator multiplied by an analytic function, say φ, depending on the sequences of the sides of Pascal's triangle. The authors studied the uniform convergence of these operators together with some quantitative estimates and regularity properties.

Motivated by all the above researches we propose a new general class of polynomials. The next section is devoted to construct this class. In Section 3 we investigate the convergence of the operators giving general estimates in terms of the modulus of smoothness. In the last section we focus our attention on establishing concrete examples of φ function by manipulating the numerical sequences mentioned above. Also some further ideas are presented.

2. CONSTRUCTION OF THE POLYNOMIALS $M_n f$

At first step we consider two sequences of real positive numbers $a = (a_n)_{n\geq 1}$, $b = (b_n)_{n\geq 1}$. For every $n \in \mathbb{N}$ and $k \in \{0, 1, \ldots, n\}$ we define the polynomials

$$q_{n,k}(x) = c_{n,k} x^k (1-x)^{n-k}, \quad x \in [0,1], \tag{3}$$

where the coefficients satisfy the following recursive formulas

$$c_{n+1,k} = c_{n,k} + c_{n,k-1}, \quad k = 1, \ldots, n, \quad c_{n,0} = a_n, \quad c_{n,n} = b_n. \tag{4}$$

We shall consider polynomials having the form

$$(M_n f)(x) = (n+1) \sum_{k=0}^{n} q_{n,k}(x) \int_0^1 p_{n,k}(t) f(t) dt, \quad x \in [0,1], \tag{5}$$

where $f \in L_1([0,1])$.

Actually, the polynomials defined by (3) belong to the space

$$\mathcal{P}_n^+ := \left\{ s \in \mathcal{P}_n : s(x) = \sum_{i=0}^{n} \alpha_i x^i (1-x)^{n-i}, \ \alpha_i \geq 0, \ i = 0, \ldots, n \right\},$$

where $\mathcal{P}_n$ represents the set of all algebraic polynomials of degree less than or equal to n. Best of our knowledge, such polynomials were firstly studied by Jurkat and Lorentz [6] who were concerned with density and degree of approximation questions.

As regards the relations (4), if $a_j = b_j = 1$ for every $j = 1, \ldots, n$, we have $c_{n,k} = \binom{n}{k}$ for every $k = 0, \ldots, n$. Hence, in this case $M_n f$ becomes $D_n f$ defined by (2).

It is clear that the polynomial $M_n f$ is determined uniquely by the two sequences a and b. We can point out this fact by using a more precise notation named $M_n^{\langle a,b\rangle} f$. Throughout the paper, we will use one or another of the notations as required by the context.

Remarks: (i) For every $n \in \mathbb{N}$, the operator M_n is linear positive and maps the space $L_1([0,1])$ into $\mathcal{P}_n$.

(ii) If the sequences $a^{(j)} = (a_n^{(j)})_{n\geq 1}$, $b^{(j)} = (b_n^{(j)})_{n\geq 1}$, $j \in \{1,2\}$, satisfy the following conditions $a_n^{(1)} \leq a_n^{(2)}, b_n^{(1)} \leq b_n^{(2)}$ for every $n \geq 1$, then it is easy to check that

$$M_n^{\langle a^{(1)},b^{(1)}\rangle} f \leq M_n^{\langle a^{(2)},b^{(2)}\rangle} f, \quad n \in \mathbb{N},$$

for every positive function $f \in L_1([0,1])$. In particular, if $\overline{M}$ is an upper bound for the sequences a and b, then

$$M_n^{\langle a,b\rangle} f \leq M_n^{\langle \overline{M},\overline{M}\rangle} f = \overline{M} D_n f. \tag{6}$$

(iii) By using (5) and (4) we observe that $M_n^{\langle a,b\rangle}$ depends linearly on the given sequences a and b.

Further on, we are going to investigate the sequence $(M_n f)_{n\geq 1}$.

3. PROPERTIES OF THE POLYNOMIALS $M_n f$

We will emphasize the convergence of our sequence and we will also give estimates of the rate of convergence preceded by the presentation of the following property.

Theorem 1: *Let the operator M_n be defined by (5). If f and g belong to $L_1([0,1])$ then the following identity*

$$\int_0^1 (M_n f)(x) g(x) dx = \int_0^1 f(t)(M_n g)(t) dt$$

holds. Particularly, M_n is a self-adjoint operator on the space $L_2([0,1])$.

Proof: We can write successively

$$\int_0^1 (M_n f)(x) g(x) dx = (n+1) \sum_{k=0}^{n} \int_0^1 q_{n,k}(x) g(x) dx \int_0^1 p_{n,k}(t) f(t) dt$$

$$= \int_0^1 f(t) \left\{ (n+1) \sum_{k=0}^{n} c_{n,k} t^k (1-t)^{n-k} \int_0^1 \binom{n}{k} x^k (1-x)^{n-k} g(x) dx \right\} dt$$

$$= \int_0^1 f(t)(M_n g)(t)dt.$$

If f and g belong to the complex Hilbert space $L_2([0,1])$ then the above relation implies

$$\langle M_n f, g\rangle_{L_2([0,1])} = \langle f, M_n g\rangle_{L_2([0,1])},$$

where $\langle\cdot,\cdot\rangle_{L_2([0,1])}$ stands for the inner product. We recall: $\langle f,g\rangle_{L_2([0,1])} = \int_0^1 f(x)\overline{g(x)}dx$, where $f,g \in L_2([0,1])$. This completes the proof. □

In what follows e_j stands for the j-th monomial, $e_j(t) = t^j$, $t \in [0,1]$, $j \in \mathbb{N}_0 := \mathbb{N} \cup \{0\}$. We need the following result.

Lemma 1: *If M_n is defined by (5) then the following identity*

$$(M_n e_0)(x) = \sum_{m=1}^{n-1} (a_m x(1-x)^m + b_m x^m(1-x)) + a_n(1-x)^n + b_n x^n, \tag{7}$$

$x \in [0,1]$, *holds true.*

Proof: By using the Beta function $B(\cdot,\cdot)$ we obtain

$$\int_0^1 p_{n,k}(t)dt = \binom{n}{k} B(k+1, n-k+1) = \frac{1}{n+1}, \quad k = 0,1,\ldots,n,$$

and consequently $(M_n e_0)(x) = \sum_{k=0}^n c_{n,k} x^k(1-x)^{n-k}$. Based on Remark (iii) we shall find the operators $A_{m,n}$, $B_{m,n}$, $m = \overline{1,n}$ which are associated to the sequences a, b and verify the identity

$$M_n \equiv M_n^{\langle a,b\rangle} = \sum_{m=1}^n a_m A_{m,n} + \sum_{m=1}^n b_m B_{m,n}. \tag{8}$$

Following the same technique as in [3], firstly we choose $b = 0$ and $a = \delta_m$, where $\delta_m = (\delta_{m,n})_{n\geq 1}$, $\delta_{m,n}$ being the symbol of Kronecker. We obtain $A_{m,n}e_0 = M_n^{\langle \delta_m, 0\rangle} e_0$. Taking into account the relations (4) we get

$$(A_{m,n}e_0)(x) = \sum_{k=1}^{n-m} \binom{n-m-1}{k-1} x^k(1-x)^{n-k} = x(1-x)^m, \text{ if } m = \overline{1, n-1},$$

and $(A_{n,n}e_0)(x) = (1-x)^n$.

Secondly we choose $a = 0$, $b = \delta_m$ and this leads to the identity $B_{m,n}e_0 = M_n^{\langle 0,\delta_m \rangle} e_0$. The same relations (4) imply

$$(B_{m,n}e_0)(x) = \sum_{k=m}^{n-1} \binom{n-m-1}{k-m} x^k (1-x)^{n-k} = (1-x)x^m, \text{ if } m = \overline{1, n-1},$$

and $(B_{n,n}e_0)(x) = x^n$.

Substituting the above expressions of the function $A_{m,n}e_0, B_{m,n}e_1$, $m = \overline{1,n}$, in the identity (8) we obtain the claimed result. □

Remark: From (7) we deduce $(M_n e_0)(0) = a_n$ and $(M_n e_0)(1) = b_n$. This means that the convergence of $(M_n)_{n\geq 1}$ implies the convergence of the sequences a and b. In what follows we assume that these sequences converge and set

$$\lim_{n\to\infty} a_n = l_a, \quad \lim_{n\to\infty} b_n = l_b. \tag{9}$$

Because of the above assumption we can define the functions σ, τ, φ belonging to $\mathbb{R}^{[0,1]}$ as follows

$$\sigma(x) = \begin{cases} l_a, & x = 0, \\ \sum_{m=1}^{\infty} a_m x(1-x)^m, & 0 < x \leq 1 \end{cases} \qquad \tau(x) = \begin{cases} \sum_{m=1}^{\infty} b_m x^m (1-x), & 0 \leq x < 1, \\ l_b, & x = 1, \end{cases}$$

and

$$\varphi = \sigma + \tau. \tag{10}$$

These definitions were suggested by the formula (7). Also, we mention that the boundedness of a and b guarantee that the power series which appear in the definition of σ and τ have radii of convergence greater than or equal to 1.

Moreover, $|\sigma(x)| \leq (1-x)\sup_{m\geq 1}|a_m|$ and $|\tau(x)| \leq x \sup_{m\geq 1}|b_m|$ for every $x \in [0,1]$.

In order to estimate the degree of convergence we involve ω_f, the first modulus of smoothness corresponding to a bounded function $f : [0,1] \to \mathbb{R}$,

$$\omega_f(\delta) = \sup\{|f(x) - f(y)| : x, y \in [0,1],\ |x-y| \leq \delta\}, \quad 0 \leq \delta \leq 1.$$

Among its remarkable properties we recall that for every $\delta > 0$

$$|f(t) - f(x)| \leq (1 + \delta^{-2}(t-x)^2)\omega_f(\delta), \quad (t,x) \in [0,1] \times [0,1], \tag{11}$$

see e.g. [2; Chapter 5, §. 1].

On the other hand we need the following identity due to Derriennic [4; page 327]:

$$\sum_{k=0}^{n} p_{n,k}(x) \int_0^1 p_{n,k}(t)(t-x)^2 dt = \frac{2nx(1-x) - 6x(1-x) + 2}{(n+1)(n+2)(n+3)}. \tag{12}$$

Theorem 2: *Let M_n be defined by (5). For every $x \in [0,1]$ one has*

$$|(M_nf)(x) - f(x)(M_ne_0)(x)| \le \mu(n)(1+\lambda_n(x))\omega_f(1/\sqrt{n+2}),$$

where

$$\mu(n) := \max_{m\le n}\{a_m, b_m\} \text{ and } \lambda_n(x) := 2((n-3)x(1-x)+1)/(n+3). \tag{13}$$

Proof: By using Remark (ii), for every natural n we get

$$|(M_nh)(x)| \le (n+1)\sum_{k=0}^{n} c_{n,k}x^k(1-x)^{n-k}\int_0^1 p_{n,k}(t)|h(t)|dt \le \mu(n)(D_n|h|)(x).$$

Further on, choosing $h = f - f(x)e_0$ and knowing that M_n is linear, the relations (11) and (12) allow us to write

$$|(M_nf)(x) - f(x)(M_ne_0)(x)| \le \mu(n)(n+1)\sum_{k=0}^{n} p_{n,k}(x)\int_0^1 p_{n,k}(t)|f(t)-f(x)|dt$$

$$\le \mu(n)(n+1)\sum_{k=0}^{n} p_{n,k}(x)\left(\frac{1}{n+1} + \frac{1}{\delta^2}\int_0^1 p_{n,k}(t)(t-x)^2dt\right)\omega_f(\delta)$$

$$= \mu(n)\left(1 + \frac{2}{\delta^2}\frac{(n-3)x(1-x)+1}{(n+2)(n+3)}\right)\omega_f(\delta).$$

By taking $\delta = 1/\sqrt{n+2}$ the conclusion of our theorem follows. □

Theorem 3: *Let M_n be defined by (5) such that the sequences a and b converge. For every $x \in [0,1]$ one has*

$$|(M_nf)(x) - \varphi(x)f(x)| \le \mu(n)(1+\lambda_n(x))\omega_f(1/\sqrt{n+2}) + ((1-x)^n + x^n)\nu(n)|f(x)|,$$

where $\varphi, \mu(n), \lambda_n(x)$ are defined by (10) respectively (13) and $\nu(n)$ is given by

$$\nu(n) := \sup_{j\ge n}\max\{|a_j - a_n|, |b_j - b_n|\}. \tag{14}$$

Proof: We can write

$$|(M_nf)(x) - \varphi(x)f(x)| \le |(M_nf)(x) - f(x)(M_ne_0)(x)| + |f(x)|\,|(M_ne_0)(x) - \varphi(x)|.$$

For the first quantity we apply Theorem 2. As regards the second quantity we use both Lemma 1 and the expression of the function φ, see (10). For every $0 < x < 1$ we have

$$|(M_n e_0)(x) - \varphi(x)|$$

$$= \left| a_n(1-x)^n + b_n x^n - \sum_{m=n}^{\infty} a_m x(1-x)^m - \sum_{m=n}^{\infty} b_n x^m (1-x) \right|$$

$$= \left| (1-x)^n \sum_{k=0}^{\infty} (a_n - a_{n+k}) x (1-x)^k + x^n \sum_{k=0}^{\infty} (b_n - b_{n+k}) x^k (1-x) \right|$$

$$\leq ((1-x)^n + x^n)\nu(n).$$

If $x = 0$ or $x = 1$ it is easy to see that the previous inequality holds true.

Combining the above statements we obtain the desired result. □

By a straightforward calculation we obtain

$$\max_{x\in[0,1]}((1-x)^n + x^n) = 1 \text{ and } \int_0^1 \lambda_n(x)dx = \frac{1}{3}$$

for every positive integer n. Also, for every $x \in [0,1]$, we deduce $\lambda_1(x) \leq 1/2$, $\lambda_2(x) \leq 2/5$ and for $n \geq 3$, $\lambda_n(x) \leq \lambda_n(1/2) = (n-1)/(n+3)$ consequently we can state $\lambda_n(x) < 1$. Taking into account these facts, Theorem 3 leads to the following

Corollary: *Let M_n be defined by (5) such that the sequences a and b converge.*

(i) If $f \in C([0,1])$ then $\| M_n f - \varphi f \| \leq 2\mu(n)\omega_f(1/\sqrt{n+2}) + \nu(n) \| f \|$, where $\| \cdot \|$ is the sup-norm defined by $\| h \| = \sup_{t\in[0,1]} |h(t)|$.

(ii) If $f \in L_1([0,1])$ then $\| M_n f - \varphi f \|_1 \leq \frac{4}{3}\mu(n)\omega_f(1/\sqrt{n+2}) + \nu(n) \| f \|_1$, where $\| \cdot \|_1$ is the usual norm of this space defined by $\| h \|_1 = \int_0^1 |h(t)|dt$.

Using the Remark of this section and Theorem 3 we can completely describe the convergence of $(M_n)_{n\geq 1}$.

Theorem 4: *Let M_n be defined by (5) and X be a normed linear space, where $X = C([0,1])$ or $X = L_1([0,1])$. The sequence $(M_n)_{n\geq 1}$ converges on X if and only if the sequences a and b converge.*

In this case, if φ denotes the function defined by (10), we have

$$\lim_{n\to\infty} M_n f = \varphi f$$

in the norm of the space X, for every $f \in X$.

In addition we point out the behaviour of $M_n f$ when f belongs to any Lebesgue space $L_p([0,1])$, $1 \le p \le \infty$, endowed with the usual norm $\|\cdot\|_{L_p([0,1])}$.

Theorem 5: *Let M_n be defined by (5) such that the sequences a and b admit an upper bound $\overline{M}$ less or equal to 1. Then $M_n f$ is a contraction in $L_p([0,1])$ for every $f \in L_p([0,1])$, where $1 \le p \le \infty$.*

Proof: At first we recall that these classes of functions are nested as follows: $C([0,1]) \subset L_\infty([0,1]) \subset L_p([0,1]) \subset L_1([0,1])$, for any $p \in (1,\infty)$.

The proof is simple and runs taking into account that it is sufficient to prove the result for $p = \infty$ and $p = 1$ as we can use the Riesz-Thorin theorem to obtain the result for $1 < p < \infty$ from these special cases.

In fact, $\| f \|_{L_\infty([0,1])} = ess\sup_{x\in[0,1]} |f(x)|$ and by using (6) we have

$$\| M_n f \|_{L_\infty([0,1])} \le \| f \|_{L_\infty([0,1])} M_n^{\langle \overline{M},\overline{M}\rangle} e_0 = \overline{M} \| f \|_{L_\infty([0,1])} D_n e_0 \le \| f \|_{L_\infty([0,1])},$$

since $D_n e_0 = e_0$. At the same time, for $p = 1$ choosing in the proof of Theorem 1 $g = e_0$, we easily obtain

$$\| M_n f \|_{L_1([0,1])} \le \int_0^1 |f(t)|(M_n e_0)(t)dt \le \overline{M} \int_0^1 |f(t)|(D_n e_0)(t)dt \le \| f \|_{L_1([0,1])} .$$

This way, the announced result is proved. □

4. SPECIAL CASES

It has become clear that the function φ is strongly dependent on the sequences a and b. The aim of this Section is to point out the numerous possibilities regarding the structure of the function φ when we choose the mentioned sequences.

Firstly, we keep in mind the construction of φ as a polynomial of degree less or equal to $q(\varphi \in \mathcal{P}_q)$. We consider the sequences a and b such that beginning with the rank q they are definitively constant, which means $a_n = \overline{a}$ and $b_n = \overline{b}$ for every $n \ge q$. After some manipulations, the relation (10) implies

$$\varphi(x) = \overline{a}(1-x)^q + \overline{b}x^q + x(1-x)\sum_{m=1}^{q-1}(a_m(1-x)^{m-1} + b_m x^{m-1}), \quad x \in [0,1]. \tag{15}$$

The relations (13) and (14) will have a new look, that is

$$\mu(n) = \mu(q) \text{ and } \nu(n) = 0, \quad n = q, q+1, q+2, \ldots,$$

and consequently, for every $f \in L_1([0,1])$, one has

$$\| M_n f - \varphi f \|_1 \le \frac{4}{3}\mu(q)\omega_1(1/\sqrt{n+2}), \quad n = q, q+1, q+2, \ldots .$$

Conversely, it is not difficult to prove that every polynomial φ belonging to the space $\mathcal{P}_q$ can be written as in (15) from which we can obtain the corresponding sequences a and b. In this respect we make the first step indicating the constants $\overline{a}$ and $\overline{b}$; one has $\overline{a} = \varphi(0)$, $\overline{b} = \varphi(1)$.

Secondly, let's consider a and b non-decreasing sequences. Putting $d_n := a_n - a_{n-1}$, $d'_n := b_n - b_{n-1}$ for every $n \geq 1$ with the convention $a_0 = b_0 = 0$, one has $a_m = \sum_{k=1}^{m} d_k$, $b_m = \sum_{k=1}^{m} d'_k$ and from (10), after a few calculations, we obtain

$$\varphi(x) = \sum_{m=1}^{\infty} d_m(1-x)^m + \sum_{m=1}^{\infty} d'_m x^m, \quad x \in [0,1]. \tag{16}$$

By a suitable selection of $(d_m)_{m\geq 1}$ and $(d'_m)_{m\geq 1}$ we can create a function φ with exponential growth.

For example, choosing $d_m = 0$ and $d'_m = 1/m!$, $m \geq 1$, one obtains $\varphi(x) = e^x - 1$. Alternatively if $d'_m = L_m/m!$, where (L_m) is the well-known Lucas sequence, then one arrives at the exponential generating function of this sequence, more exactly $\varphi(x) = e^{\alpha x} + e^{\beta x} - 2$ with $\alpha = (1+\sqrt{5})/2$ and $\beta = (1-\sqrt{5})/2$.

In the same relation (16) we put $d_m = 0$ and $d'_m = \frac{(\alpha)_m}{(\beta)_m m!}$, $m \geq 1$, where α and β are positive fixed numbers. We recall $(\alpha)_0 = 1$ and $(\alpha)_k = \alpha(\alpha+1)\ldots(\alpha+k-1)$ for $k \geq 1$. This choice leads us to the *confluent hypergeometric* function

$$\varphi(x) = {}_1F_1(\alpha,\beta;x) - 1, \quad x \in [0,1]. \tag{17}$$

However, this is a convergent series for all values of x and by using Kummer's equation, we get

$$x\frac{d^2\varphi}{dx^2} + (\beta - x)\frac{d\varphi}{dx} - \alpha\varphi = \alpha.$$

We have free hands to give α and β various values obtaining in (17) functions with a great personality, as reflected in [1, § 13.6, p. 509].

Final remarks: We consider the sequence $(M_n f)_{n\geq 1}$ a fertile field of investigation. Practically, we keep in mind the following directions: the study of the iterates, asymptotic properties as Voronovskaja-type formula, some qualitative properties of the function φ studying the sequences a and b, results concerning the convergence of derivatives of $M_n f$ for a differentiable function f, and also linear combinations of our operators.

REFERENCES

[1] Abramowitz, M., Stegun, I.A. "Handbook of Mathematical Functions with Formulas, Graphs and Mathematical Tables" *National Bureau of Standards Applied Mathematics Series*, Vol. *55*, Issued June, 1964.

[2] Altomare, F., Campiti, M. "Korovkin-type Approximation Theory and its Applications", *de Gruyter Studies in Mathematics*, Vol. *17* de Gruyter, Berlin, New-York, (1994).

[3] Campiti, M., Metafune, G. "Approximation Properties of Recursively Defined Bernstein-Type Operators." *J. Approx. Theory*, Vol *87* (1996): pp. 243-269.
[4] Derriennic, M.M. "Sur l'approximation de fonctions intégrables sur [0,1] par des polynômes de Bernstein modifies." *J. Approx. Theory*, Vol. *31* (1981): pp. 325-343.
[5] Durrmeyer, J.L. "Une formule d'inversion de la transformée de Laplace: Applications à la théorie des moments." Thése de 3e cycle, Faculté des Sciences de l'Université de Paris, 1967.
[6] Jurkat, W.B., Lorentz, G.G. "Uniform approximation by Polynomials with Positive Coefficients." *Duke Math. J.*, Vol. *28* (1961): pp. 463-474.

AMS Classification Numbers: 41A10, 41A35, 26D15

FIBINOMIAL IDENTITIES

Arthur T. Benjamin, Jennifer J. Quinn and Jeremy A. Rouse

Problem $A-6$ from the 1990 Putnam exam states:

If X is a finite set, let $|X|$ denote the number of elements in X. Call an ordered pair (S,T) of subsets of $\{1,2,\ldots,n\}$ *admissible* if $s > |T|$ for each $s \in S$, and $t > |S|$ for each $t \in T$. How many admissible ordered pairs of subsets of $\{1,2,\ldots,10\}$ are there? Prove your answer.

It is no coincidence that the solution, 17711, is the 21st Fibonacci number. The number of admissible ordered pairs of subsets of $\{1,2,\ldots,n\}$ with $|S| = a$ and $|T| = b$ is $\binom{n-b}{a}\binom{n-a}{b}$. As we shall show, summing over all values of a and b leads to

Identity 1:

$$\sum_{a=0}^{n}\sum_{b=0}^{n}\binom{n-b}{a}\binom{n-a}{b} = f_{2n+1}.$$

where $f_0 = 1$, $f_1 = 1$ and for $n \geq 2$, $f_n = f_{n-1} + f_{n-2}$. The published solutions [6, 7] use "convoluted" algebraic methods. Yet the presence of both Fibonacci numbers and binomial coefficients demands a combinatorial explanation. Beginning with our proof of Identity 1, we provide simple, combinatorial arguments for many *fibinomial identities* – identities that combine (generalized) Fibonacci numbers and binomial coefficients.

Fibonacci numbers can be combinatorially interpreted in many ways [10]. The primary tool used in this note will be tilings of $1 \times n$ boards with tiles of varying lengths. The identities presented are viewed as counting questions, answered in two different ways. To begin with, Identity 1 is easily seen by answering

This paper is in final form and no version of it will be submitted for publication elsewhere.

F.T. Howard (ed.), *Applications of Fibonacci Numbers, Volume 9: Proceedings of the Tenth International Research Conference on Fibonacci Numbers and their Applications*, 19–24.

Question: How many ways can a board of length $2n+1$ be tiled using (length 1) squares and (length 2) dominoes?

Answer 1: A length m board can be tiled in f_m ways, which can be seen by conditioning on whether the last tile is a square or a domino. Consequently, a board of length $2n+1$ can be tiled f_{2n+1} ways.

Answer 2: Condition on the number of dominoes on each side of the *median* square.

Any tiling of a $(2n+1)$-board must contain an odd number of squares. Thus one square, which we call the median square, contains an equal number of squares to the left and right of it. For example, the 13-tiling in Figure 1 has 5 squares. The median square, the third square, is located in cell 9.

How many tilings contain exactly a dominoes to the left of the median square and exactly b dominoes to the right of the median square? Such a tiling has $(a+b)$ dominoes and therefore $(2n+1)-2(a+b)$ squares. Hence the median square has $n-a-b$ squares on each side of it. Since the left side has $(n-a-b)+a = n-b$ tiles, of which a are dominoes, there are $\binom{n-b}{a}$ ways to tile to the left of the median square. Similarly, there are $\binom{n-a}{b}$ ways to tile to the right of the median square. Hence there are $\binom{n-b}{a}\binom{n-a}{b}$ tilings altogether.

Varying a and b over all feasible values, we obtain the total number of $(2n+1)$ tilings as $\sum_{a=0}^{n}\sum_{b=0}^{n}\binom{n-b}{a}\binom{n-a}{b}$.

Figure 1: Every square-domino tiling of odd length must have a median square. The 13-tiling above has 3 dominoes left of the median square and 1 domino to the right of the median square. The number of such tilings is $\binom{5}{3}\binom{3}{1}$.

We can extend this identity by utilizing the 3-bonacci numbers, defined by $\theta_n = 0$ for $n < 0$, $\theta_0 = 1$ and for $n \geq 1$, $\theta_n = \theta_{n-1} + \theta_{n-3}$.

Identity 2:

$$\sum_{a=0}^{n}\sum_{b=0}^{n}\sum_{c=0}^{n}\binom{n-b-c}{a}\binom{n-a-c}{b}\binom{n-a-b}{c} = \theta_{3n+2}.$$

Question: How many ways can a board of length $3n+2$ be tiled using squares and trominoes?

Answer 1: It is easy to see that θ_n counts the number of ways to tile a board of length n with squares and (length 3) trominoes. Hence there are θ_{3n+2} such tilings of a board of length $3n+2$.

Answer 2: The number of squares in any tiling of a $(3n+2)$-board must be 2 greater than a multiple of 3. Hence there will exist two *goalpost* squares, say located at cells x and y, such that there are an equal number of squares to the left of x, between x and y, and to the right of y. We condition on the number of trominoes in the three regions defined by the goalposts. If the number of trominoes in each region is, from left to right, a, b, c, then there are a total of $a+b+c$ trominoes and $(3n+2)-3(a+b+c)$ squares, including the two goalpost squares.

Thus each region has $n-(a+b+c)$ squares. The leftmost region has $n-b-c$ tiles, a of which are trominoes, and there are $\binom{n-b-c}{a}$ ways to arrange them. Likewise the tiles of the second and third region can be arranged $\binom{n-a-c}{b}$ ways and $\binom{n-a-b}{c}$ ways, respectively.

As a, b, and c vary, the total number of $(3n+2)$-tilings is the left side of our identity.

Applying the same logic, using only tiles of length 1 and k, we immediately obtain the following k-bonacci generalization.

Identity 3: *Let κ_n be the k-bonacci number defined by $\kappa_n = 0$ for $n < 0$, $\kappa_0 = 1$, and for $n \geq 1$, $\kappa_n = \kappa_{n-1} + \kappa_{n-k}$. Then for $n \geq 0$, $\kappa_{kn+(k-1)}$ equals*

$$\sum_{a_1=0}^{n}\sum_{a_2=0}^{n}\cdots\sum_{a_k=0}^{n}$$

$$\binom{n-(a_2+a_3+\cdots+a_k)}{a_1}\binom{n-(a_1+a_3+\cdots+a_k)}{a_2}\cdots\binom{n-(a_1+a_2+\cdots+a_{k-1})}{a_k} \quad (1)$$

Another generalization of Fibonacci numbers are the kth order Fibonacci numbers defined by $g_n = 0$ for $n < 0$, $g_0 = 1$, and for $n \geq 1$, $g_n = g_{n-1} + g_{n-2} + \cdots + g_{n-k}$. The next identity is proved by non-trivial algebraic methods in [8] and [9], but when viewed combinatorially, as done in [2], it is practically obvious.

Identity 4: *For all $n \geq 0$,*

$$\sum_{n_1}\sum_{n_2}\cdots\sum_{n_k}\binom{n_1+n_2+\cdots+n_k}{n_1, n_2, \cdots, n_k} = g_n,$$

the kth order Fibonacci number, where the summation is over all non negative integers $n_1, n_2, \ldots, n_k$ such that $n_1 + 2n_2 + \cdots + kn_k = n$.

Question: In how many ways can we tile a board of length n using tiles with lengths at most k?

Answer 1: By its definition, it is combinatorially clear that g_n counts this quantity.

Answer 2: Condition on the number of tiles of each length. If for $1 \leq i \leq k$ there are n_i tiles of length i, then we must have $n_1 + 2n_2 + \cdots + kn_k = n$. The number of ways to permute these tiles is given by the multinomial coefficient $\binom{n_1+n_2+\cdots+n_k}{n_1,n_2,\cdots,n_k}$.

More colorfully, for nonnegative integers $c_1, \ldots, c_k$ we define the generalized k^{th} order Fibonacci number by $h_n = 0$ for $n < 0$, $h_0 = 1$, and for $n \geq 1$, $h_n = c_1 \cdot h_{n-1} + c_2 \cdot h_{n-2} + \cdots + c_k \cdot h_{n-k}$. It is easy to see that h_n counts the number of ways to tile a board of length n with *colored* tiles of length at most k, where for $1 \leq i \leq k$, a tile of length i may be assigned any one of c_i colors. The previous identity and argument immediately generalizes to the following identity.

Identity 5: *For all* $n \geq 0$,

$$\sum_{n_1}\sum_{n_2}\cdots\sum_{n_k}\binom{n_1+n_2+\cdots+n_k}{n_1,n_2,\cdots,n_k}c_1^{n_1}c_2^{n_2}\cdots c_k^{n_k}=h_n,$$

the generalized k^{th} *order Fibonacci number, where the summation is over all non-negative integers* $n_1, n_2, \cdots, n_k$ *such that* $n_1 + 2n_2 + \cdots + kn_k = n$.

Our tiling approach also succeeds in proving even more complex fibinomial identities.

Identity 6:

$$\sum_{a_1=0}^{n}\sum_{a_2=0}^{n}\cdots\sum_{a_k=0}^{n}\binom{n}{a_1}\binom{n-a_1}{a_2}\cdots\binom{n-a_{k-1}}{a_k}=f_{k+1}^n.$$

Question: In how many ways can we simultaneously tile n distinguishable boards of length $k+1$ with squares and dominoes?

Answer 1: Since each board can be tiled f_{k+1} ways, there are f_{k+1}^n such tilings.

Answer 2: Condition on the number of dominoes covering each consecutive pair of cells. We claim there are $\binom{n}{a_1}\binom{n-a_1}{a_2}\cdots\binom{n-a_{k-1}}{a_k}$ ways to create n tilings of length $k+1$ where a_1 of them begin with dominoes, a_2 have dominoes covering cells 2 and 3, and generally for $1 \leq i \leq k$, a_i of them have dominoes covering cells i and $i+1$. To see this, notice there are $\binom{n}{a_1}$ ways to decide which of the n tilings begin with a domino (the rest begin with a square). Once these are selected, then among those $n - a_1$ tilings that do not begin with a domino there are $\binom{n-a_1}{a_2}$ ways to determine which of those will have a domino in cells 2 and 3. (The other $n - a_1 - a_2$ tilings will have a square in cell 2.) Continuing in this fashion, we see that once the tilings with dominoes covering cells $i-1$ and i are determined, there are $\binom{n-a_{i-1}}{a_i}$ ways to determine which tilings have dominoes covering cells i and $i+1$.

More generally, by tiling n distinguishable boards of length $k+1$ with squares and dominoes where the first c of them must begin with a square, the same reasoning establishes:

Identity 7: *For* $0 \leq c \leq n$,

$$\sum_{a_1=0}^{n}\sum_{a_2=0}^{n}\cdots\sum_{a_k=0}^{n}\binom{n-c}{a_1}\binom{n-a_1}{a_2}\cdots\binom{n-a_{k-1}}{a_k}=f_k^c f_{k+1}^{n-c}.$$

Identities 6 and 7 can be extended to *Gibonacci* numbers, G_n, defined by initial conditions G_0, G_1, and the Fibonacci recurrence $G_n = G_{n-1} + G_{n-2}$ for $n \geq 2$. For non-negative integers G_0 and G_1, G_n can be combinatorially defined [3] as the number of ways to tile a length n board with squares and dominoes subject to the *initial conditions* that the first tile is given a *phase*, where there are G_0 choices for the phase of a domino and G_1 choices for the phase of a square. The extension of Identity 6 to Gibonacci numbers is

Identity 8:

$$\sum_{a_1=0}^{n}\sum_{a_2=0}^{n}\cdots\sum_{a_k=0}^{n}\binom{n}{a_1}\binom{n-a_1}{a_2}\cdots\binom{n-a_{k-1}}{a_k}G_0^{a_1}G_1^{n-a_1}=G_{k+1}^n.$$

Question: In how many ways can we simultaneously tile n distinguishable boards of length $k+1$ with squares and dominoes, where an initial domino is assigned one of G_0 phases and an intial square is assigned one of G_1 phases?

Answer 1: Since each board can be tiled G_{k+1} ways, there are G_{k+1}^n such tilings.

Answer 2: As in the proof of Identity 6, we condition on the number of dominoes covering each consecutive pair of cells. The difference here is that initially choosing a_1 of the tilings to begin with a domino contributes $\binom{n}{a_1}G_0^{a_1}G_1^{n-a_1}$ to the product since we must assign domino phases to a_1 of the boards and square phases to the remaining $n-a_1$ boards.

As an immediate corollary, we have

Identity 9:

$$\sum_{a_1=0}^{n}\sum_{a_2=0}^{n}\cdots\sum_{a_k=0}^{n}\binom{n}{a_1}\binom{n-a_1}{a_2}\cdots\binom{n-a_{k-1}}{a_k}2^{a_1}=L_{k+1}^n,$$

where L_n is the nth Lucas number.

Likewise, if we require that the first c of the n boards begin with phased squares, then the same reasoning establishes the following Gibonacci extension of Identity 7.

Identity 10: *For* $0 \le c \le n$,

$$\sum_{a_1=0}^{n}\sum_{a_2=0}^{n}\cdots\sum_{a_k=0}^{n}\binom{n-c}{a_1}\binom{n-a_1}{a_2}\cdots\binom{n-a_{k-1}}{a_k}G_0^{a_1}G_1^{n-a_1}=G_1^c f_k^c G_{k+1}^{n-c}.$$

As with Identity 5, these identities can be further generalized by colorizing them. For more combinatorial proofs of fibinomial identities, see [4].

We end with an open question. The *Fibonomial Numbers* are defined like binomial coefficients with F's on top. That is, Fibonimial $\binom{n}{k}_{\mathcal{F}}=\frac{F_nF_{n-1}\cdots F_{n-k+1}}{F_1F_2\cdots F_k}$, where $F_n=f_{n-1}$ is the traditional n^{th} Fibonacci number. Amazingly $\binom{n}{k}_{\mathcal{F}}$ is always an integer [1]. We challenge the reader to find a combinatorial proof of this fact.

ACKNOWLEDGMENT

We thank the referee for many helpful suggestions. This research was supported by The Reed Institute for Decision Science and the Beckman Research Foundation.

REFERENCES

[1] Alexanderson, G.L. and Klosinski, L.F. "A Fibonacci Analogue of Gaussian Binomial coefficients." *The Fibonacci Quarterly*, Vol. *12* (1974): pp. 129-132.

[2] Benjamin, A.T., Hanusa, C.R.H. and Su, F.E. "Linear Recurrences Through Tilings and Markov Chains." *Utilitas Mathematica*, Vol. *64* (2003): pp. 3-17.

[3] Benjamin, A.T., Quinn, J.J. and Su, F.E. "Generalized Fibonacci Identities through Phased Tilings." *The Fibonacci Quarterly*, Vol. *38.3* (2000): pp. 282-288.

[4] Benjamin, A.T. and Rouse, J.A. "Recounting Binomial Fibonacci Identities." Applications of Fibonacci Numbers. Volume 10. Edited by Fredric T. Howard. Kluwer Academic Publishers, Dordrecht, The Netherlands, 2003: pp. 25-28.

[5] Carlitz, L. "The Characteristic Polynomial of a Certain Matrix of Binomial Coefficients." *The Fibonacci Quarterly*, Vol. *3.2* (1965): pp. 81-89.

[6] Klosinski, L.F. Alexanderson, G.L. and Larson, L.C. "The Fifty-First William Lowell Putnam Mathematical Competition." *American Mathematical Monthly*, Vol. *98.8* (1991): pp. 719-727.

[7] Larson, L. "The Fifty-First William Lowell Putnam Mathematical Competition." *Mathematics Magazine*, Vol. *64.2* (1991): pp. 141-144.

[8] Philippou, A.N. "A Note on the Fibonacci Sequence of Order K and the Multinomial Coefficients." *The Fibonacci Quarterly*, Vol. *21.2* (1983): pp. 82-86.

[9] Philippou, A.N and Muwafi, A.A. "Waiting for the K-th Consecutive Success and the Fibonacci Sequence of Order K." *The Fibonacci Quarterly*, Vol. *20.1* (1982): pp. 28-32.

[10] Stanley, R.P. Enumerative Combinatorics. Volume 1 *Wadsworth and Brooks/Cole* Chapter 1 problem 14a-i.

AMS Classification Numbers: 05A19, 11B39

RECOUNTING BINOMIAL FIBONACCI IDENTITIES

Arthur T. Benjamin and Jeremy A. Rouse

In [4], Carlitz demonstrates

$$F_L \sum_{x_1=0}^{n} \sum_{x_2=0}^{n} \cdots \sum_{x_L=0}^{n} \binom{n-x_L}{x_1}\binom{n-x_1}{x_2}\cdots\binom{n-x_{L-1}}{x_L} = F_{(n+1)L}, \tag{1}$$

using sophisticated matrix methods and Binet's formula. Nevertheless, the presence of binomial coefficients suggests that an elementary combinatorial proof should be possible. In this paper, we present such a proof, leading to other Fibonacci identities.

Proof: Recall that for $m \geq 1$, F_m counts the ways to tile a length $m-1$ board with squares and dominoes (see [1], [2], [3]). Hence the right side of equation (1) counts the tilings of a board with length $(n+1)L-1$.

Before explaining the left side of equation (1), we first demonstrate that any such tiling can be created in a unique way using $n+1$ *supertiles* of length L. Given a tiled board of length $(n+1)L-1$, with *cells* numbered 1 through $(n+1)L-1$, we break the tiling into $n+1$ supertiles $S_1, S_2, \cdots, S_{n+1}$ by cutting the board after cells $L, 2L, 3L, \ldots, nL$. See Figure 1.

Notice that a supertile might begin or end with a *half-domino*. For instance, if a domino covers cells L and $L+1$, then S_1 ends with a half-domino, and S_2 begins with a half-domino. A supertile that begins with a half-domino is called *open* on the left; otherwise it is *closed* on

This paper is in final form and no version of it will be submitted for publication elsewhere.

F.T. Howard (ed.), Applications of Fibonacci Numbers, Volume 9: Proceedings of the Tenth International Research Conference on Fibonacci Numbers and their Applications, 25–28.

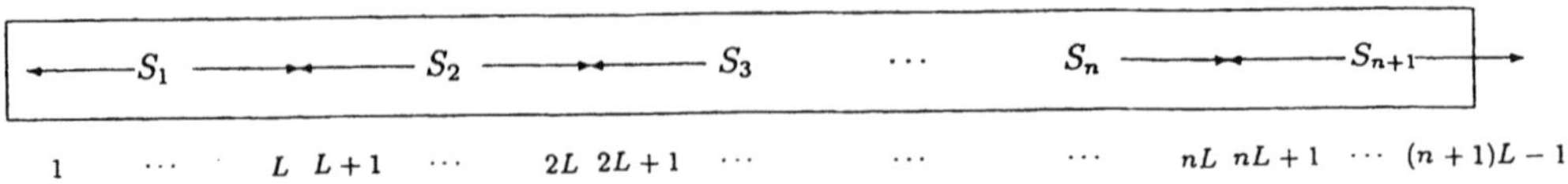

FIGURE 1. A board of length $(n+1)L-1$ (with a half-domino attached) can be split into $n+1$ supertiles of length L.

the left. Likewise a supertile is either open or closed on the right. Naturally, S_1 must be closed on the left.

For convenience, we append a half-domino to the last supertile so that S_{n+1} has length L, like all the other supertiles, and is open on the right. Notice that $S_1, \ldots, S_{n+1}$ must obey the following "following" rule:

For $1 \leq i \leq n$, S_i is open on the right iff S_{i+1} is open on the left.

Given supertiles $S_1, \ldots, S_{n+1}$, we can extract subsequences $O_1, \ldots, O_t$ and $C_1, \ldots, C_{n+1-t}$ for some $0 \leq t \leq n$, where $O_1, \ldots, O_t$ are open on the left, and $C_1, \ldots, C_{n+1-t}$ are closed on the left. By the "following" rule, there are exactly $t+1$ supertiles that are open on the right, necessarily including C_{n+1-t}. Conversely, given $0 \leq t \leq n$ and $O_1, \ldots, O_t$, $C_1, \ldots, C_{n+1-t}$ there is a unique way to reconstruct the sequence $S_1, \ldots, S_{n+1}$ that preserves the relative order of the O's and C's. Specifically, we must have $S_1 = C_1$, and for $1 \leq i \leq n$, if S_i is closed on the right then S_{i+1} is the lowest numbered unused C_j; else S_{i+1} is the lowest numbered unused O_j.

To summarize, $F_{(n+1)L}$ counts the ways to create, for all $0 \leq t \leq n$, length L supertiles $O_1, \ldots, O_t$, open on the left, and length L supertiles $C_1, \ldots, C_{n+1-t}$ closed on the left, where C_{n+1-t} is open on the right and exactly t of the other supertiles are open on the right. It remains to show that the left side of equation (1) counts the ways that such a collection of supertiles can be constructed.

Given $0 \leq t \leq n$, we begin by tiling C_{n+1-t}. Since it must end with a half-domino and has $L-1$ *free* cells, it can be tiled F_L ways. Now for any non-negative integers $x_1, \ldots, x_{L-1}$, we prove that the remaining supertiles can be created $\binom{n-x_L}{x_1}\binom{n-x_1}{x_2}\cdots\binom{n-x_{L-1}}{x_L}$ ways, where $x_L = t$ and for $1 \leq i \leq L-1$, exactly x_i, of these n supertiles have a domino beginning at its ith cell.

Since t of the supertiles (excluding C_{n+1-t}) must be open on the right, $x_L = t$ of these n supertiles have half-dominoes beginning at their L^{th} cells. Now there are $\binom{n-t}{x_1} = \binom{n-x_L}{x_1}$ ways to pick x_1 supertiles among $\{C_1, \ldots, C_{n-t}\}$ to begin with a domino. (The remaining $n-t-x_1$ C_j's (other than C_{n+1-t}) begin with a square and all of the O_j's begin with a half-domino.) Next there are $\binom{n-x_1}{x_2}$ ways to pick x_2 supertiles to have a domino covering the second and third cell among those not chosen in the last step to have a domino covering the first and second cell. The unchosen $n-x_1-x_2$ supertiles have a square on the second cell. Continuing in this fashion, there are $\binom{n-x_{i-1}}{x_i}$ ways to pick which supertiles have a domino beginning at the ith cell for $1 \leq i \leq L$. Hence $O_1, \ldots, O_t$ and $C_1, \ldots, C_{n-t}, C_{n+1-t}$ can be created in exactly $F_L\binom{n-x_L}{x_1}\binom{n-x_1}{x_2}\cdots\binom{n-x_{L-1}}{x_L}$ ways. Summing over all values of x_i gives us the left side of equation (1). □

By counting our tilings in a slightly different way, we combinatorially obtain another identity presented in [4]:

$$\sum_{i\geq 0}\sum_{j\geq 0}\binom{i+j}{i}\binom{n-j-i}{j}F_{L-1}^{i}F_{L}^{2j+1}F_{L+1}^{n-2j-i} = F_{(n+1)L}. \tag{2}$$

Proof: $F_{(n+1)L}$ counts the ways to create supertiles $S_1,\ldots,S_{n+1}$ subject to the same conditions as before. This time, we classify supertiles in four ways, depending on whether they are closed on the left only, right only, both, or neither. If, for some $0 \leq j \leq \lfloor\frac{n}{2}\rfloor$, $S_1,\ldots,S_{n+1}$ contains exactly j supertiles $R_1,\ldots,R_j$ closed on the right only, then there must be exactly $j+1$ supertiles $L_1,\ldots,L_{j+1}$ closed on the left only. Subsequently, $S_1,\ldots,S_{n+1}$ has subsequence

$$L_1, R_1, L_2, R_2, \ldots, L_j, R_j, L_{j+1}.$$

For example, see Figure 2. Since each of the supertiles above has length L with one half-domino and $L-1$ free cells, this subsequence can be tiled $(F_L)^{2j+1}$ ways.

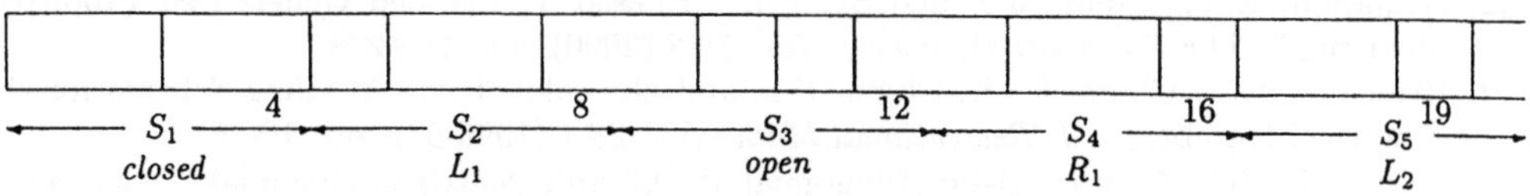

FIGURE 2. When this length 19 board (plus half-domino) is split after every 4 cells, we create 5 supertiles that are closed, respectively, on both sides, left side, neither side, right side, and left side.

Now suppose $S_1,\ldots,S_{n+1}$ is to have exactly i supertiles that are open at both ends, where $0 \leq i \leq n-2j$. We first place these supertiles, like i identical balls to be placed in $j+1$ distinct buckets, between any L_k and R_k or after L_{j+1}. Since there are $\binom{a+b-1}{a}$ ways to place a identical balls into b distinct buckets, there are $\binom{i+j}{i}$ ways to do this. Once placed, since each has $L-2$ free cells, they can be tiled $(F_{L-1})^i$ ways.

Finally, the remaining $n-2j-i$ supertiles that are closed on both ends can be placed into $j+1$ different buckets (before L_1 or between any R_k and L_{k+1}) in $\binom{n-j-i}{n-2j-i} = \binom{n-j-i}{j}$ ways. Once placed, they can be tiled $(F_{L+1})^{n-2j-i}$ ways.

Consequently, the number of legal ways to choose supertiles $S_1,\ldots,S_{n+1}$ with exactly j supertiles closed on the right only and i supertiles open on both ends is $\binom{i+j}{i}\binom{n-j-i}{j}F_{L-1}^{i}F_{L}^{2j+1}F_{L+1}^{n-2j-i}$. (Notice that the second binomial coefficient causes this quantity to be zero whenever $n-j-i<j$, i.e., when $2j+i>n$.) Summing over all i and j proves equation (2). □

Notice that both equations (1) and (2) imply that for all $n \geq 1$, F_L divides F_{nL}. However, a more direct combinatorial proof is possible, without invoking supertiles. Specifically, we have:

$$F_L \sum_{j=1}^{n} (F_{L-1})^{j-1} F_{(n-j)L+1} = F_{nL}. \tag{3}$$

Proof: The right side counts the ways to tile a board of length $nL-1$. The left side of (3) counts this by conditioning on the first j, $1 \leq j \leq n$, for which the tiling has a square or domino ending at cell $jL-1$. Such a tiling consists of $j-1$ tilings of length $L-2$, each followed by a domino. This is followed by a tiling of the next $L-1$ cells (cells $(j-1)L+1$ through $jL-1$), followed by a tiling of the remaining $nL-jL$ cells. This can be accomplished $(F_{L-1})^{j-1} F_L F_{(n-j)L+1}$ ways, and the identity follows. □

The authors gratefully acknowledge the assistance of Jennifer J. Quinn. This research was suppoted by The Reed Institute for Decision Science and the Beckman Research Foundation.

REFERENCES

[1] Benjamin, A.T. and Quinn, J.J. "Recounting Fibonacci and Lucas Identities." *College Mathematics Journal*, Vol. *30.5* (1999): pp. 359-366.

[2] Benjamin, A.T., Quinn, J.J. and Su, F.E. "Phased Tilings and Generalized Fibonacci Identities." *The Fibonacci Quarterly*, Vol. *38.3* (2000): pp. 282-288.

[3] Brigham, R.C., Caron, R.M., Chinn, P.Z. and Grimaldi, R.P. "A Tiling Scheme for the Fibonacci Numbers." *J. Recreational Math*, Vol. *28.1* (1996-97): pp. 10-16.

[4] Carlitz, L. "The Characteristic Polynomial of a Certain Matrix of Binomial Coefficients." *The Fibonacci Quarterly*, Vol. *3.2* (1965): pp. 81-89.

AMS Classification Numbers: 05A19, 11B39

THE FIBONACCI DIATOMIC ARRAY APPLIED TO FIBONACCI REPRESENTATIONS

Marjorie Bicknell-Johnson

1. INTRODUCTION

The Fibonacci diatomic array of this paper is a self-generating array which begins with two rows, each containing two 1's; each row begins and ends with two 1's. Each interior element in each row is in row $(n-1)$ or else is the sum of two adjacent elements in row $(n-2)$, according as the column number is of the form a_p or b_p, where (a_p, b_p) is a Wythoff pair. The Fibonacci diatomic array counts the number of Fibonacci representations $R(N)$ of non-negative integers N. It is also possible to generate the Fibonacci diatomic array from a single row of two 1's, or to use a single row of two 1's to generate either the odd rows or the even rows of the array. Both the array of Fibonacci representations and the Fibonacci diatomic array of this paper illustrate many known identities relating to Fibonacci representations while suggesting new identities.

Let $R(N)$ denote the number of Fibonacci representations of the non-negative integer N; that is, the number of representations of N using distinct Fibonacci numbers $F_k, k \geq 2$, written in descending order throughout this paper. We define $R(0) = R(F_1) = 1$. The Zeckendorf representation of N, denoted Zeck N, is the unique representation of N using only non-consecutive Fibonacci numbers $F_k, k \geq 2$. If F_n is the largest Fibonacci number contained in N, we will say that Zeck N begins with F_n; similarly, if F_t is the smallest Fibonacci number used in Zeck N, we will write Zeck N ends with F_t, or Zeck $N = F_n + \cdots + F_t$. For example, $N = 58$ has Zeck $N = F_{10} + F_4$, and $R(58) = 7$ since 58 can be represented as $55+3, 55+2+1, 34+21+3, 34+21+2+1, 34+13+8+3, 34+13+8+2+1$, and $34+13+5+3+2+1$. Carlitz [2] and Klarner [6] prove that $R(F_n - 1) = 1$ and the $R(F_n) = \lfloor n/2 \rfloor$, where $\lfloor x \rfloor$ denotes the floor function. Many basic properties of $R(N)$ appear in [1], [2] and [6]. We will use the subscripts appearing in Zeck N to write $R(N)$.

This paper is in final form and no version of it will be submitted for publication elsewhere.

F.T. Howard (ed.), *Applications of Fibonacci Numbers, Volume 9: Proceedings of the Tenth International Research Conference on Fibonacci Numbers and their Applications*, 29–38.

There are several ways to compute $R(N)$; for example,

$$\prod_{j=2}^{\infty}(1+x^{F_j})=\sum_{N=0}^{\infty}R(N)x^N \tag{1.1}$$

$R(N)=\{1,1,1,2,1,2,2,1,3,2,2,3,1,3,3,2,4,2,3,3,1,\ldots,\}, N=0,1,2,\ldots,$ is sequence $A000119$ in [8]. Here, Table 1 lists $R(N)$ by interval, where the nth row contains successive values for $R(N)$ for $F_n-1\leq N\leq F_{n+1}-1, n\geq 1$.

1:	1																												
2:	1	1																											
3:	1	1																											
4:	1	2	1																										
5:	1	2	2	1																									
6:	1	3	2	2	3	1																							
7:	1	3	3	2	4	2	3	3	1																				
8:	1	4	3	3	5	2	4	4	2	5	3	3	4	1															
9:	1	4	4	3	6	3	5	5	2	6	4	4	6	2	5	5	3	6	3	4	4	1							
10:	1	5	4	4	7	3	6	6	3	8	5	5	7	2	6	6	4	8	4	6	6	2	7	5	5	8	3	6	6
11:	1	5	5	4	8	4	7	7	3	9	6	6	9	3	8	8	5	10	5	7	7	2	8	6	6	10	4	8	8
12:	1	6	5	5	9	4	8	8	4	11	7	7	10	3	9	9	6	12	6	9	9	3	11	8	8	13	5	10	10
13:	1	6	6	5	10	5	9	9	4	12	8	8	12	4	11	11	7	14	7	10	10	3	12	9	9	15	6	12	12
14:	1	7	6	6	11	5	10	10	5	14	9	9	13	4	12	12	8	16	8	12	12	4	15	11	11	18	7	14	14
15:	1	7	7	6	12	6	11	11	5	15	10	10	15	5	14	14	9	18	9	13	13	4	16	12	12	20	8	16	16
16:	1	8	7	7	13	6	12	12	6	17	11	11	16	5	15	15	10	20	10	15	15	5	19	14	14	23	9	18	18
17:	1	8	8	7	14	7	13	13	6	18	12	12	18	6	17	17	11	22	11	16	16	5	20	15	15	25	10	20	20
18:	1	9	8	8	15	7	14	14	7	20	13	13	19	6	18	18	12	24	12	18	18	6	23	17	17	28	11	22	22
19:	1	9	9	8	16	8	15	15	7	21	14	14	21	7	20	20	13	26	13	19	19	6	24	18	18	30	12	24	24
20:	1	10	9	9	17	8	16	16	8	23	15	15	22	7	21	21	14	28	14	21	21	7	27	20	20	33	13	26	26
21:	1	10	10	9	18	9	17	17	8	24	16	16	24	8	23	23	15	30	15	22	22	7	28	21	21	35	14	28	28
	a	b	a	a	b	a	b	a	a	b	a	a	b	a	b	a	a	b	a	b	a	a	b	a	a	b	a	b	a
			C	C	D	C	D	D	C	E	D	D	F	C	E	E	D	2D	D	F	F	C	G	E	E	H	D	2D	2D

Table 1
The Array of Fibonacci Representations

We will use basic properties of Wythoff pairs in writing the Fibonacci diatomic array. Among many others, references [3], [4], [5], and [6] discuss Wythoff pairs with applications to Fibonacci numbers. While there are many references to Fibonacci representations in the literature, the arrays of this paper provide a new way to study their properties, as well as giving a visually pleasing display.

2. THE ARRAY OF FIBONACCI REPRESENTATIONS

Many basic formulas for Fibonacci representations are known. We restate some basic properties of $R(N)$ from [1] in notation consistent with this paper.

Lemma 2.1: If sequence $\{g_n\}$ satisfies the Fibonacci recurrence $g_{n+2}=g_{n+1}+g_n$, then $R(g_n-1)=R(g_k-1), n\geq k$, for k sufficiently large. In particular, $R(F_n-1)=1, n\geq 1$.

Lemma 2.2: If Zeck N ends with F_m+F_t, then $R(N)$ satisfies the following:

(i) $R(N)=R(N-1)$ when m and t have opposite parity, for $t=2$ or $t=3$;

(ii) $R(N)=R(N-1)+R(N+1)$, t even, $t\geq 4$;

(iii) $R(N)=R(N-1)+R(N+2)$, t odd, $t\geq 5$.

Lemma 2.3: $R(F_{2c}) = R(F_{2c+1}) = c$, and $R(F_n) = \lfloor n/2 \rfloor$.

Lemma 2.4: Let Zeck $N = F_n + F_p + \cdots + F_t = F_n + K$. Then

(i) $R(N) = cR(F_p + \cdots + F_t) = cR(K)$ if $n = p + 2c - 1$;

(ii) $R(N) = cR(K) + R(F_{p+1} - K - 2)$ if $n = p + 2c$;

(iii) $R(N) = R(F_n + K) = R(F_{n+1} - K - 2)$.

Lemma 2.5: Let Zeck $N = F_n + F_p + \cdots + F_t = F_n + K$. Then,

(i) $R(N) = R(F_{n-p+1})R(K) = \lfloor (n-p+1)/2 \rfloor R(k)$, n and p opposite parity;

(ii) $R(N) = R(F_{n-p})R(K) = \lfloor (n-p)/2 \rfloor R(K) + R(F_{p+1} - K - 2)$, n and p the same parity.

We now turn to Table 1. The columns of Table 1 illustrate Lemmas 2.1 - 2.5 and other properties of $R(N)$. The rows of the Fibonacci representation array contain values for $R(N)$ for consecutive integers N. If $m_{n,k}$ is the kth element in the nth row,

$$m_{n,k} = R(N), N = F_n + k - 2, F_n - 1 \le N \le F_{n+1} - 1, 1 \le k \le F_{n-1} + 1, n = 1, 2, \ldots,. \quad (2.1)$$

Column 1 illustrates $R(F_n - 1) = R(F_{n+1} - 1) = 1$ from Lemma 2.1. Column 2 shows $R(F_{2c}) = R(F_{2c+1}) = c$ from Lemma 2.3, since $m_{n,2} = R(F_n), n \ge 1$.

In Table 1, the columns marked (C) show

$$R(F_n + F_p - 1) = \lfloor (n-p+2)/2 \rfloor, p \ge 2, n \ge p+1, \quad (2.2)$$

which can be proved from Lemmas 2.1 and 2.3, since $R(F_n + F_p - 1) = R(F_{n-p+2} + F_2 - 1) = R(F_{n-p+2})$. If n and p have opposite parity, and if Zeck $N = F_n + F_p + \cdots = F_n + K$, then a multiple of the pth row appears within the nth row,

$$R(F_n + K) = \lfloor (n-p+2)/2 \rfloor R(K), p \ge 2, n \ge p+1, F_p - 1 \le K \le F_{p+1} - 1. \quad (2.3)$$

For example, within row 10, 3 times row 5 appears as terms 6 through 9. There are numerous placeds in Table 1 where a multiple of an entire previous row appears within a row. To prove Eq. (2.3), combine Lemmas 2.4(i) and 2.5(i) with Eq. (2.2).

The columns (D) in Table 1 illustrate

$$R(F_n + F_p + F_{p-3} - 1) = n - p + 1 = R(F_n + F_p + F_{p-2} - 1), \quad (2.4)$$

which can be proved from Lemma 2.1 and 2.5. Using Lemma 2.1, shift subscripts down $(p-5)$ to write $R(N) = R(F_n + F_p + F_{p-3} - 1) = R(F_{n-p+5} + F_5 + F_2 - 1) = R(F_{n-p+5} + F_5)$. If n and p have opposite parity, then $R(N) = \lfloor (n-p+1)/2 \rfloor R(F_5) = n - p + 1$ from Lemma 2.5. If n and p have the same parity, then $R(N) = \lfloor (n-p)/2 \rfloor R(F_5) + R(F_6 - F_5 - 2) = \lfloor (n-p)/2 \rfloor \cdot (2) + 1 = n - p + 1$. The second half of (2.4) has a similar derivation.

The elements in each row of Table 1 are symmetric about the center, illustrating Lemma 2.4(iii). The elements within the nth row are governed by Lemmas 2.1 and 2.2; in fact the nth row of the array of Fibonacci representations can be generated from the previous two rows. For example, row 7 can be formed from rows 5 and 6:

Row 5 :	1	2	2	1					
Row 6 :	1	3	2	2	3	1			
Row 7 :	$\underline{1}$	3	$\underline{3}$	$\underline{2}$	4	$\underline{2}$	3	$\underline{3}$	$\underline{1}$
	a	b	a	a	b	a	b	a	a
N, Row 7 :	12	13	14	15	16	17	18	19	20

The underlined terms of row 7 are copies from row 6, and the non-underlined terms are the sum of a pair of adjacent terms from row 5, taken in order: $3 = 1 + 2, 4 = 2 + 2, 3 = 2 + 1$. Row 7 contains $R(N)$ for $12 \leq N \leq 20$. The terms in columns marked "a" are copies from row 6 and have Zeck N ending in 1 or 2; terms in columns marked "b" are sums from row 5 and have Zeck N ending in $F_t, t \geq 4$.

The nth row in Table 1 contains all of the terms of row $(n-1)$, interspersed with sums of two consecutive terms from row $(n-2)$. Any term in row n in a column marked "a" equals a term in row $(n-1)$; any term in a column marked "b" is the sum of adjacent terms in row $(n-2)$. Theorem 2.1 describes the generation of rows in the array of Fibonacci representations.

Theorem 2.1: Let Zeck $N = F_n + F_p + \cdots + F_w + F_t$, so that $R(N)$ appears in the nth row of the array of Fibonacci representations. If Zeck N ends with 1 or 2, then $R(N) = R(N^*)$, where $R(N^*)$ appears in row $(n-1)$ and is the first term in row $(n-1)$ not yet used in row n. If Zeck N ends with $F_t \geq 3$, then $R(N) = R(N^{**}) + R(N^{**}+1)$, where $R(N^{**})$ and $R(N^{**}+1)$ are adjacent terms in row $(n-2)$, and $R(N^{**})$ is the first term in row $(n-2)$ not yet used in forming row n.

2 :	1	1																											
4 :	1	2	1																										
6 :	1	3	2	2	3	1																							
8 :	1	4	3	3	5	2	4	4	2	5	3	3	4	1															
10 :	1	5	4	4	7	3	6	6	3	8	5	5	7	2	6	6	4	8	4	6	6	2	7	5	5	8	3	6	6
12 :	1	6	5	5	9	4	8	8	4	11	7	7	10	3	9	9	6	12	6	9	9	3	11	8	8	13	5	10	10
14 :	1	7	6	6	11	5	10	10	5	14	9	9	13	4	12	12	8	16	8	12	12	4	15	11	11	18	7	14	14
16 :	1	8	7	7	13	6	12	12	6	17	11	11	16	5	15	15	10	20	10	15	15	5	19	14	14	23	9	18	18
18 :	1	9	8	8	15	7	14	14	7	20	13	13	19	6	18	18	12	24	12	18	18	6	23	17	17	28	11	22	22
20 :	1	10	9	9	17	8	16	16	8	23	15	15	22	7	21	21	14	28	14	21	21	7	27	20	20	33	13	26	26

Table 1a
The Array of Fibonacci Representations: Even Rows

1 :	1																												
3 :	1	1																											
5 :	1	2	2	1																									
7 :	1	3	3	2	4	2	3	3	1																				
9 :	1	4	4	3	6	3	5	5	2	6	4	4	6	2	5	5	3	6	3	4	4	1							
11 :	1	5	5	4	8	4	7	7	3	9	6	6	9	3	8	8	5	10	5	7	7	2	8	6	6	10	4	8	8
13 :	1	6	6	5	10	5	9	9	4	12	8	8	12	4	11	11	7	14	7	10	10	3	12	9	9	15	6	12	12
15 :	1	7	7	6	12	6	11	11	5	15	10	10	15	5	14	14	9	18	9	13	13	4	16	12	12	20	8	16	16
17 :	1	8	8	7	14	7	13	13	6	18	12	12	18	6	17	17	11	22	11	16	16	5	20	15	15	25	10	20	20
19 :	1	9	9	8	16	8	15	15	7	21	14	14	21	7	20	20	13	26	13	19	19	6	24	18	18	30	12	24	24
21 :	1	10	10	9	18	9	17	17	8	24	16	16	24	8	23	23	15	30	15	22	22	7	28	21	21	35	14	28	28
			C	C	D	C	D	D	C	E	D	D	F	C	E	E	D	2D	D	F	F	C	G	E	E	H	D	2D	2D

Table 1b
The Array of Fibonacci Representations: Odd Rows

Proof: If Zeck N ends in 1, then Zeck $N = F_n + F_w + \cdots + F_x + F_2$, where $x \geq 4$ because no consecutive Fibonacci numbers are used in Zeck N. Applying Lemma 2.1 to shift subscripts down 1,

$$\begin{aligned} R(N) &= R(F_n + F_w + \cdots + F_x + 1) = R(F_n + F_w + \cdots + F_x + F_3 - 1) \\ &= R(F_{n-1} + F_{w-1} + \cdots + F_{x-1} + F_2 - 1) = R(F_{n-1} + F_{w-1} + \cdots + F_{x-1}), \end{aligned} \tag{2.5}$$

an element in row $(n-1)$. Similarly, if Zeck N ends in 2, $t = 3$ and $x \geq 5$. Shifting subscripts down 1, $R(N) = R(F_n + F_w + \cdots + F_x + F_4 - 1) = R(F_{n-1} + F_{w-1} + \cdots + F_{x-1} + 1)$, a term in row $(n-1)$.

If Zeck N ends in F_t, $t \geq 4$, there are two cases. If t is even, then $R(N) = R(N-1) + R(N+1)$ from Lemma 2.2(iii). Applying Lemma 2.1 to shift subscripts down 2,

$$\begin{aligned} R(N) &= R(F_n + F_w + \cdots + F_{2c} - 1) + R(F_n + F_w + \cdots + F_{2c} + F_3 - 1) \\ &= R(F_{n-2} + F_{w-2} + \cdots + F_{2c-2} - 1) + R(F_{n-2} + F_{w-2} + \cdots + F_{2c-2}), \end{aligned} \tag{2.6}$$

the sum of two consecutive terms from row $(n-2)$. If t is odd, the demonstration is similar, using $R(N) = R(N-1) + R(N+2)$ from Lemma 2.2(iv). In both cases, $R(N)$ is the sum of a pair of consecutive terms from row $(n-2)$.

If $R(N)$ in row n is computed sequentially for N increasing throughout the interval $F_n - 1 \leq N \leq F_{n+1} - 1$, the terms used from rows $(n-1)$ and $(n-2)$ must also be taken sequentially. □

In Table 1, the columns marked (E), (F), (G) and (H) seem mysterious until we divide the array into even rows only, or odd rows only, as in Tables 1a and 1b. Then observe the many columns which contain arithmetic progressions. For example, terms of column (E) in Table 1a have the form $5 + 3w$; in Table 1b, 3w. Column (E) satisfies

$$\begin{aligned} &R(F_{2c} + F_6) = [(2c-6)/2]R(F_6) + R(F_7 - F_6 - 2) = (c-3)(3) + 2, c = 4, 5, \ldots, n = 2c \\ &R(F_{2c+1} + F_6) = [((2c+1) - 6 + 1)/2]R(F_6) = (c-2)(3), c = 4, 5, \ldots, n = 2c+1, \end{aligned}$$

from Lemma 2.5. Columns (F), (G), and (H) display similar properties.

3. THE FIBONACCI DIATOMIC ARRAY

The Fibonacci Diatomic Array is a self-generating array which begins with two rows, each containing two 1's, numbered row 2 and row 3 for convenience. Each succeeding row $n \geq 4$ is formed from the preceding two rows, interspersing the elements of row $(n-1)$ with sums of pairs of adjacent elements from row $(n-2)$. We use Wythoff pairs (a_p, b_p) to make an explicit correspondence of elements in row n to elements in rows $(n-1)$ and $(n-2)$.

The Wythoff pairs (a_p, b_p) are self-generating. Let $a_1 = 1, b_p = a_p + p, p \geq 1$, and let a_p be the least integer not yet used. The first ten pairs are

p:	1	2	3	4	5	6	7	8	9	10
a_p:	1	3	4	6	8	9	11	12	14	16
b_p:	2	5	7	10	13	15	18	20	23	26

The Wythoff numbers are sequences A000201 and A001950 in [8], also given by

$$a_p = \lfloor \alpha p \rfloor \text{ and } b_p = \lfloor \alpha^2 p \rfloor = a_p + p, p \geq 1, \tag{3.1}$$

where $\alpha = (1 + \sqrt{5})/2$. We will need some properties of Wythoff pairs from [3], [4], [5], [7], which are restated as Lemma 3.1.

```
1                   1
1                   1
1         2         1
1       2   2       1
1   3   2   2   3   1
1 3   3 2 4 2 3   3 1

1                               1
1                               1
1               2               1
1           2       2           1
1     3     2       2     3     1
1   3   3   2   4   2   3   3   1
1 4 3   3 5 2 4   4 2 5 3   3 4 1

1                                                   1
1                                                   1
1                         2                         1
1                   2           2                   1
1         3         2           2         3         1
1       3   3       2     4     2       3   3       1
1   4   3   3   5   2   4   4   2   5   3   3   4   1
1 4   4 3 6 3 5   5 2 6 4   4 6 2 5   5 3 6 3 4   4 1

1                                                                                   1
1                                                                                   1
1                                         2                                         1
1                               2                   2                               1
1               3               2                   2               3               1
1           3       3           2         4         2           3       3           1
1     4     3       3     5     2       4   4       2     5     3       3     4     1
1   4   4   3   6   3   5   5   2   6   4   4   6   2   5   5   3   6   3   4   4   1
1 5 4   4 7 3 6   6 3 8 5   5 7 2 6   6 4 8 4 6   6 2 7 5   5 8 3 6   6 3 7 4   4 5 1
```

Table 2
The Fibonacci Diatomic Array

Lemma 3.1: Any positive integer M is either of the form a_p or b_p but not both, where a_p and b_p are defined by (3.1).

(i) If Zeck M ends with F_w, then $M = a_p$ for some $p \geq 1$ if w is even; $M = b_p$ for w odd.

(ii) The interval $F_n - 1 \leq M \leq F_{n+1} - 1$ contains $(F_{n-1}+1)$ integers M, of which $(F_{n-2}+1)$ have the form a_p and (F_{n-3}) have the form b_p. □

The Fibonacci Diatomic Array is defined as follows. Let $m_{n,k}$ be the kth term in the nth row of the Fibonacci Diatomic Array, $n \geq 2, 1 \leq k \leq F_{n-1}+1$. Then $m_{2,1} = m_{2,2} = 1, m_{3,1} = m_{3,2} = 1$, and for $n \geq 4$,

$$m_{n,k} = m_{n-1,p} \text{ if } k = a_p, \text{ and } m_{n,k} = m_{n-2,p} + m_{n-2,p+1} \text{ if } k = b_p, \tag{3.2}$$

where (a_p, b_p) form a Wythoff pair as given by (3.1).

Table 2 contains the Fibonacci Diatomic Array, written to illustrate columns of constants. We generate row 7 as an example.

Row 5 :	1				2		2				1
Row 6 :	1		3		2		2		3		1
Row 7 :	1	3		3	2	4	2	3		3	1
Term k :	1	2		3	4	5	6	7		8	9

Write $m_{7,k}$ from definition (3.2), taking $k = 1, \ldots, 9$:

$$\begin{array}{llll}
k = 1 = a_1, & m_{7,1} = m_{6,1} & k = 2 = b_1, & m_{7,2} = m_{5,1} + m_{5,2} \\
k = 3 = a_2, & m_{7,3} = m_{6,2} & & \\
k = 4 = a_3, & m_{7,4} = m_{6,3} & k = 5 = b_2, & m_{7,5} = m_{5,2} + m_{5,3} \\
k = 6 = a_4, & m_{7,6} = m_{6,4} & k = 7 = b_3, & m_{7,7} = m_{5,3} + m_{5,4} \\
k = 8 = a_5, & m_{7,8} = m_{6,5} & & \\
k = 9 = a_6, & m_{7,9} = m_{6,6}. & &
\end{array}$$

In writing the rows for Table 2, an extra space is allowed between equal terms in a row. Here, extra spaces appear between both pairs of 3's in row 7, so that the 3's in row 6 can be centered above them.

Except for being written to show columns of constants, the Fibonacci diatomic array looks suspiciously like the array of Fibonacci representations. Indeed, they are the same for $n \geq 2$, as we prove in Theorem 3.1.

Theorem 3.1: The Fibonacci diatomic array contains the same elements as the array of Fibonacci representations, except for row 1. Let $m_{n,k}$ be the kth term in the nth row of the Fibonacci diatomic array, as defined in (3.2), and define $R(0) = 1$. Then, $m_{2,1} = m_{2,2} = 1, m_{3,1} = m_{3,2} = 1$, and for $n \geq 4$,

$$m_{n,k} = R(N) \text{ where } N = F_n + k - 2, F_n - 1 \leq N \leq F_{n+1} - 1, 1 \leq k \leq F_{n-1} + 1. \tag{3.3}$$

Proof: From (3.2), $m_{n,k} = m_{n-1,p}$ if $k = a_p$. If $k = a_p$, then $N = F_n + a_p - 2 = F_n + (\cdots + F_w + F_{2c}) - 2$ for some $c \geq 1$. If $c = 1$, then $N = F_n + \cdots + F_w - 1$; Zeck N ends in 1 or 2, according as w is odd or even. If $c = 2$, Zeck N ends with $F_4 - 2 = 1$. If $c > 2$, then Zeck N ends with $5+1$, since $N = F_n + \cdots + F_w + F_{2c} - 2 = F_n + \cdots + F_w + (F_{2c-1} + F_{2c-3} + \cdots + 13 + 5 + 2 + 1) - 2$. From Theorem 2.1, $R(N) = R(N^*)$, the first term not yet used in the preceding row; that is, the pth term in row $(n-1)$ when $k = a_p$.

Also from (3.2), $m_{n,k} = m_{n-2,p} + m_{n-2,p+1}$ if $k = b_p$. If $k = b_p$, then $N = F_n + b_p - 2 = F_n + (\cdots + F_w + F_{2c+1}) - 2$ for some $c \geq 1$. If $c = 1$, Zeck N ends in $F_w + F_3 - 2 = F_w$, $w \geq 5$. If $c > 1$,

$$N = F_n + \cdots + F_w + (F_{2c+1} - 1) - 1 = F_n + \cdots + F_w + (F_{2c} + F_{2c-2} + \cdots + 8 + 3 + 1) - 1$$

so that Zeck N ends in F_4. In both cases, Zeck N ends in $F_t, t \geq 4$. By Theorem 2.1, $R(N)$ is the sum of the first pair of adjacent elements from row $(n-2)$ not yet used. Thus, $m_{n,k} = R(N)$. □

Corollary 3.1.1: The Fibonacci Diatomic Array can be generated from a single row, containing two 1's. If $m_{n,k}$ is the kth term in the nth row, $n \geq 2, 1 \leq k \leq F_{n-1} + 1$, then $m_{2,1} = m_{2,2} = 1$, and for $n \geq 3$, if

$$k = a_p, m_{n,a_p} = m_{n-1,p}; \text{ if } k = b_p, m_{n,b_p} + m_{n-1,a_{p+1}}. \tag{3.4}$$

Proof: Row 3 can be generated from row 2 as follows. Since $a_1 = 1, m_{3,1} = m_{2,1} = 1$, while $m_{3,2} = m_{2,1} + m_{2,3}$ since $b_1 = 2$ and $a_2 = 3$. But $m_{2,3}$ doesn't exist; thus, $m_{3,2} = m_{2,1} = 1$. Row 3 has $F_3 - 1 \leq N \leq F_4 - 1$ so that $N = 1, 2$, and row 3 contains $R(1) = 1, R(2) = 1$. Row n can be generated from $(n-1), n \geq 3$. From (3.2), $m_{n,k} = m_{n-1,p}$ if $k = a_p$; if $k = b_p, m_{n,k} = m_{n-2,p} + m_{n-2,p+1} = m_{n-1,a_p} + m_{n-1,a_{p+1}}$. □

Corollary 3.1.2: The even rows of the Fibonacci Diatomic Array can be generated from a single row containing two 1's. If $m_{n,k}$ is the kth term in the nth row, $n \geq 2, 1 \leq k \leq F_{n-1} + 1$, then $m_{2,1} = m_{2,2} = 1$, and for $n \geq 4$, n even,

$$m_{n,k} = m_{n-2,p} + m_{n-2,p+1} \text{ if } k = b_p;$$
$$m_{n,k} = m_{n-2,w} \text{ if } k = a_{a_w} \text{ and } m_{n,k} = m_{n-2,a_w} + m_{n-2,a_{w+1}} \text{if } a_{b_w}. \tag{3.5}$$

Proof: The case $k = b_p$ is given in (3.2). The case $k = a_{a_w}$ follows from $m_{n,k} = m_{n-1,a_w} = m_{n-2,w}$. The case $k = a_{b_w}$ gives $m_{n,k} = m_{n-1,b_w} = m_{n-3,w} + m_{n-3,w+1} = m_{n-2,a_w} + m_{n-2,a_{w+1}}$. Thus, row n can be generated from row $(n-2)$ only. □

Corollary 3.1.3: The odd rows of the Fibonacci Diatomic Array can be generated from a single row containing two 1's, using the recursion of Equation (3.5) but taking n odd, starting with $m_{3,1} = m_{3,2} = 1$.

Proof: The proof is similar to that of Corollary 3.1.2. □

The columns of constants in the Fibonacci Diatomic Array are an important feature, since each such column corresponds to a Fibonacci sequence. For example, the columns of 2's correspond to $R(3), R(5), R(9), R(15), R(25), \ldots, R(N-1), \ldots$, and $R(3), R(6), R(10), R(17), R(28), \ldots, R(N^*-1), \ldots$, where $\{N\} = \{4, 6, 10, 16, 26, \ldots\}$ and $\{N^*\} = \{4, 7, 11, 18, 29, \ldots\}$ are Fibonacci sequences.

Theorem 3.2: Whenever a column of constant values for $R(N-1)$ appears in the Fibonacci Diatomic Array, the corresponding values of N form a Fibonacci sequence.

Proof: Theorem 3.2 illustrates Lemma 2.1. Each column of constants branches off from $R(N_1)$ for Zeck $N_1 = F_n + F_p + \cdots + F_t, t \geq 4$, where $R(N_1)$ appears in row n. The related values for N_j form two Fibonacci sequences, each containing N_1. In row $(n+1), R(N_2) = R(F_{n+1} + F_{p+1} + \cdots + F_{t+1} + F_2 - 1) = R(N_1)$ and $R(N_2^*) = R(F_{n+1} + F_{p+1} + \cdots + F_{t+1} + F_3 - 1) = R(N_1)$, where $N_2 = N_2^* - 1$. In row $(n+2)$, $R(N_3) = R(F_{n+2} + F_{p+2} + \cdots + F_{t+2} + F_3 - 1) = R(N_1)$ and $R(N_3^*) = R(F_{n+2} + F_{p+2} + \cdots + F_{t+2} + F_4 - 1) = R(N_1)$, where $N_3 = N_3^* - 1$. Notice that $R(M-1) = R(M)$ as in Lemma 2.2 (i), in rows $(n+1)$ and $(n+2)$. We also have two Fibonacci sequences, $N_1, N_2, N_3, \ldots$, and $N_1^*, N_2^*, N_3^*, \ldots$, each containing N_1, as required. $\square$

```
1
1 1
1 1
1 2 1
1 2 2 1
1 3 2 2 3 1
1 3 3 2 4 2 3 3 1
1 4 3 3 5 2 4 4 2 5 3 3 4 1
1 4 4 3 6 3 5 5 2 6 4 4 6 2 5 5 3 6 3 4 4 1
1 5 4 4 7 3 6 6 3 8 5 5 7 2 6 6 4 8 4 6 6 2 7 5 5 8 3 6 6 3 7 4 4 5 1
... 2 8 6 6 10 4 8 8 4 10 6 6 8 2 ...
... 2 8 8 6 12 6 10 10 4 12 8 8 12 4 10 10 6 12 6 8 8 2 ...
... 2 10 8 8 14 6 12 12 6 16 10 10 14 4 12 12 8 16 8 12 12 4 14 10 10 16 6 12 12 6 14 8 8 10 2 ...
... 4 16 12 12 20 8 16 16 8 20 12 12 16 4 ...
... 4 16 16 12 24 12 20 20 8 24 16 16 24 8 20 20 12 24 12 16 16 4 ...
... 4 20 16 16 28 12 24 24 12 32 20 20 28 8 24 24 16 32 16 24 24 8 28 20 20 32 12 24 24 12 28 16 16 20 4 ...
... 8 32 24 24 40 16 32 32 16 40 24 24 32 8 ...
... 8 32 32 24 48 24 40 40 16 48 32 32 48 16 40 40 24 48 24 32 32 8 ...
... 8 40 32 32 56 24 48 48 24 64 40 40 56 16 48 48 32 64 32 48 48 16 56 40 40 64 24 48 48 24 56 32 32 40 8 ...
```

Table 3
The Central Elements
The Array of Fibonacci Representations

4. THE CENTRAL ELEMENTS

If we center-justify Table 1 to form Table 3, we find many powers of 2, and the central section of row n contains twice the entire row $(n-3)$. If the first three terms of a column are t, u, and v, the column can be split into three interweaved sequences: $\{t, 2t, 4t, 8t \ldots\}$; $\{u, 2u, 4u, 8u, \ldots\}$; and $\{v, 2v, 4v, 8v, \ldots\}$.

Theorem 4.1: In the nth row of the Fibonacci diatomic array, $n = 3i+1$, the central term is $2^i, i = 1, 2, \ldots$.

Proof: By inspection, the central terms of rows 4 and 7, are 2 and 4. Row $(3i+1)$ has $F_{3i+1-1} + 1 = F_{3i} + 1$ terms, with central term number $k = ((F_{3i}+1)+1)/2 = F_{3i}/2 + 1$, where we recall that F_{3i} is even. By induction, one could show that $k = F_{3i}/2 + 1 = F_{3i-2} + F_{3i-5} + \cdots + F_7 + F_4 + 2$. Then, the term number k is of the form b_p, and the central term of row $n = (3i+1)$ is the sum of the two central terms from row $(n-2)$, namely, $2^{i-1} + 2^{i-1}$. The rows $(n-1)$ and $(n-2)$ each have a pair of central terms; each of these central terms has a term number of form a_p and thus equals a term from the preceding row; row $(n-3) = (3i-2)$ has central term 2^{i-1}. Alternately, taking $c = 2$ in Lemma 1.4(i) yields $R(F_{m+3} + F_m + \ldots) = 2R(F_m + \ldots)$, which can be applied repeatedly to $N = F_{3i+1} + k - 2$. Then, $R(N) = R(F_{3i+1} + F_{3i-2} + F_{3i-5} + \cdots + F_7 + F_4) = 2R(F_{3i-2} + F_{3i-5} + \cdots + F_7 + F_4) = \cdots = 2^{i-1}R(F_4) = 2^{i-1} \cdot = 2^i$. $\square$

REFERENCES

[1] Bicknell-Johnson, M. and Fielder, D.C. "The Number of Representations of N Using Distinct Fibonacci Numbers, Counted by Recursive Formulas." *The Fibonacci Quarterly*, Vol. *37.1* (1999): pp. 47-60.
[2] Carlitz, L. "Fibonacci Representations." *The Fibonacci Quarterly*, Vol. *6.4* (1968): pp. 193-220.
[3] Hoggatt, V.E. Jr. and Bicknell, M. "A Generalization of Wythoff's Game." *The Fibonacci Quarterly*, Vol. *17.3* (1979): pp. 198-211.
[4] Horadam, A.F. "Wythoff Pairs." *The Fibonacci Quarterly*, Vol. *16.2* (1978): 147-151.
[5] Kimberling, C. "The Zeckendorf Array Equals the Wythoff Array." *The Fibonacci Quarterly*, Vol. *33.1* (1995): pp. 3-8.
[6] Klarner, D.A. "Partitions of N into Distinct Fibonacci Numbers." *The Fibonacci Quarterly*, Vol. *6.4* (1968): pp. 235-44.
[7] Lang, W. "The Wythoff and the Zeckendorf Representations of Numbers are Equivalent." Applications of Fibonacci Numbers. Volume 6. Dordrecht: Kluwer, 1996, pp. 321-37.
[8] Sloane, N.J.A. "On-Line Encyclopedia of Integer Sequences." http://www.research.att.com/~njas/sequences/.

AMS Classification Numbers: 11B39, 11B37, 11Y55

ON PURPLE PARROTS, FIBONACCI NUMBERS, AND COLOR THEORY

Marjorie Bicknell-Johnson

Imitating colors from nature, with many different greens and earth-tones, makes pleasing color combinations, or at least gives us what we are used to seeing. But what colors will we choose for dyes for yarns to weave a plaid, or which tubes of paint do we want to open to plan an abstract painting? What ink proportions do we want to print computer graphics? Suprisingly, the Fibonacci numbers 1, 2, 3, 5, 8, 13, 21, form proportional progressions of pleasing colors, something like choosing chord progressions in music. But, first, a little color history.

Color science has been deeply studied. In art, two major systems of color notation give numerical designations for hues, tints, and shades of color. Both the system devised by Munsell [5] and that by Oswald [6] are widely used in science, technology and industry; Oswald's system is more popular in Europe than in America. (As an aside, Oswald received the Nobel Prize for Chemistry in 1909.) Computer graphics programs such as Adobe ®Illustrator ® employ a CYMK system for choosing inks for the printer, a RGB system for colors for the screen and web pages, and standardized Pantone ® colors. The present paper does not study any of that; the 24-color wheel uses ideas from Oswald's system. The Fibonacci colors are chosen from a color wheel with approximately equal steps of perceived hue difference around the wheel. The color wheel in this paper is built using several easy-to-describe colors in CMY space and linearly interpolating other colors in the gaps. Such an idealized 24-color wheel exists only in the imagination; in fact, monitor calibration, scanning, printing, copying, preparing overhead transparencies, or using an opaque projector all change the perceived colors. This paper reports colors that made a pleasing display on an inkjet printer, and Adobe ®Illustrator ® was used because the same drawing can quickly be colored using many color schemes. The colors are not reproduced here; who knows how they would turn out?

This paper is in final form and no version of it will be submitted for publication elsewhere.

F.T. Howard (ed.), Applications of Fibonacci Numbers, Volume 9: Proceedings of the Tenth International Research Conference on Fibonacci Numbers and their Applications, 39–42.

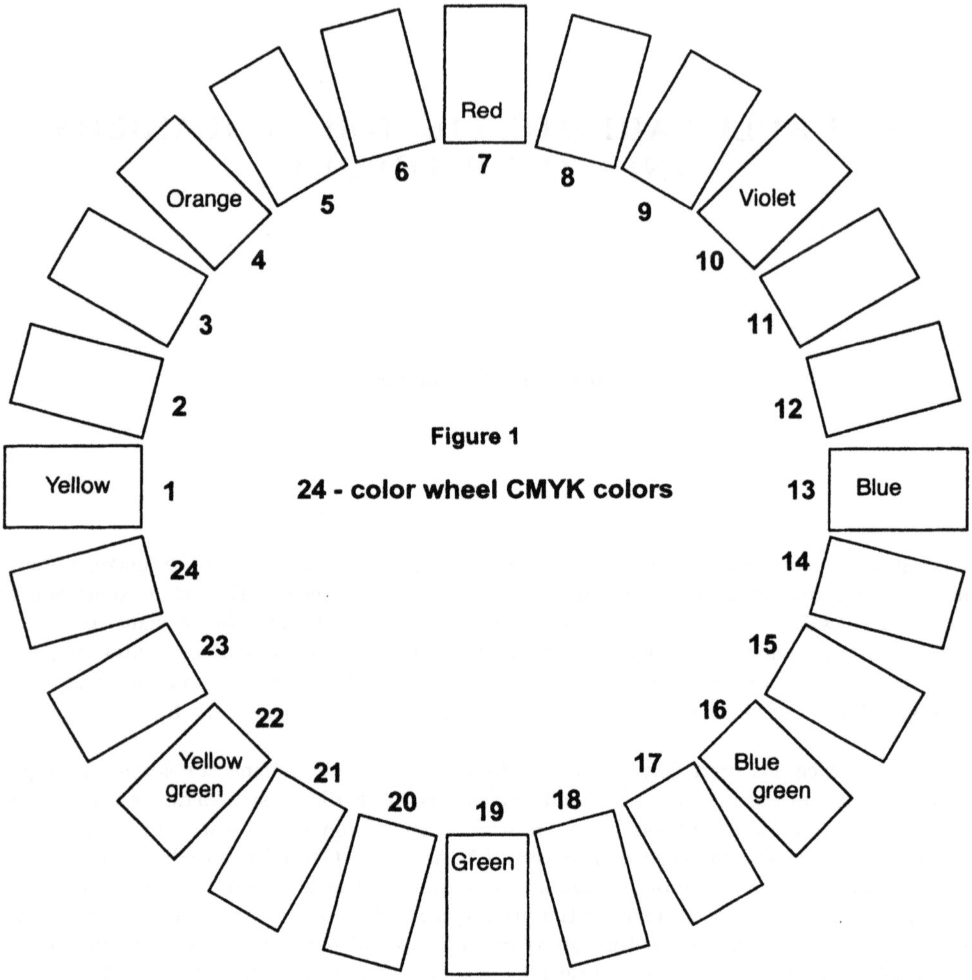

Figure 1

24 - color wheel CMYK colors

To find Fibonacci color schemes, build a color wheel (Figure 1) with 24 divisions, with yellow, red, blue, and green spaced equally, the colors between linearly graduating in the CMY coordinates, where cyan $= C = (100\ 0\ 0)$, magenta $= M = (0\ 100\ 0)$, yellow $= Y = (0\ 0\ 100)$; in this system, black $= (100\ 100\ 100)$. Place yellow at 1, red $=$ magenta at 7, blue $=$ cyan at 13, and green at 19. The CMYK values used with Adobe Illustrator and an inkjet printer appear in Table 1, taking $K = 0$. The resulting color wheel is not reproduced here. From a lay person's perspective, magenta ink leans towards rose-pink red and cyan is a wimpy sky-blue, but no matter. Also, colors of ink are not linearly related to visual color perception. None of this affects the result, as long as the hues appear to progress in "equal steps." But, back to the promised Fibonacci colors.

Color	#	C	M	Y	Color	#	C	M	Y	Color	#	C	M	Y	Color	#	C	M	Y
Yellow	1	0	0	100	Red	7	0	100	0	Blue	13	100	0	0	Green	19	100	0	100
	2	0	34	100		8	34	100	0		14	100	0	17		20	83	0	100
	3	0	67	100		9	67	100	0		15	100	0	34		21	67	0	100
Orange	4	0	100	100	Violet	10	100	100	0	Blu.Gr	16	100	0	50	Yel.Gr	22	50	0	100
	5	0	100	67		11	100	67	0		17	100	0	67		23	34	0	100
	6	0	100	34		12	100	34	0		18	100	0	83		24	17	0	100
Red	7	0	100	0	Blue	13	100	0	0	Green	19	100	0	100	Yellow	1	0	0	100

Table 1: 24 Color Wheel CMYK Colors

The Fibonacci numbers provide a way to choose a pleasing palette of colors, forming a proportional progression of color around the 24-color wheel; for example, colors 1, 2, 3, 5, 8, 13, 21 on the original wheel, where colors 1, 2, 3 will predominate, with color 21 as an accent color. The first number can be taken at any position on the 24-color wheel. Copy Figure 1 on white paper, removing the rectangles numbers 1, 2, 3, 5, 8, 13, 21. Each placement of the cut out wheel over the perceptually uniform 24-color wheel makes a pleasing color combination.

Figure 2
Image from Adobe Illustrator Classroom in a Book [1]

For example, start at 10 which is violet, and choose colors 10, 11, 12, 14, 17, 22, and 6. Actually, the color combinations taken have more to do with color theory than with Fibonacci numbers. Colors 1, 2, 3 are analogous colors; colors 1 and 13 are complementary; colors 5, 13, 21 form a triad (vertices of an equilateral triangle). Other pleasing schemes come from taking the vertices of squares, or from multiples of 3; Ostwald [6], [7] describes many color schemes, all based upon divisors of 24. The Fibonacci colors can be illustrated by coloring the same picture several ways, such as painting purple parrots, blue-green parrots, yellow-green parrots from Figure 2.

The application of Fibonacci numbers to a color wheel was part of a color theory course taught by Fritz Faiss in the 1970's at California State University, Northridge, as reported in [2]. According to Faiss, who studied art at the Bauhaus in the 1930's, since the exact colors cannot be purchased, the color wheel has to be adjusted by eye. The "proportional progress" is one of twenty color groupings discussed in [4], where Faiss relates color and philosophy, using some ideas taken from Goethe's [3] subjective theory of colors. The first proportional progress scheme, beginning with yellow, gives an impression of intricate sensuality and emphasizes the warm colors. With blue in first place, the emphasis is on the cool colors, but yellow and purple create an extreme counter-balance, or a harmony in diversity. Faiss related harmony in color to harmony in music. For Faiss, blue-green in first place is an exceptional chord; the warm, positive sound of orange becomes lost in the complementary blue-green, which expands into green and yellow-green. When violet is number one, blue comes into action, and loses to yellow in yellow-green; it is a strange, unusual sound. Starting with yellow-green, everything seems off-beat, different, but humble and mild. According to Faiss, of all the color harmonies taken from the 24-color wheel, the proportional progress appears supreme, superior, and the most challenging.

Fritz Faiss said that everyone studying art at the Bauhaus in the 1930's knew the 24-color wheel. Ostwald [7] used number sequences to describe many color combinations in terms of human sensation, but Fibonacci numbers were not included. Munsell [5] standardized colors with a hue circle, but did not mention Fibonacci numbers. We are not trying to develop color theory, only to show an entertaining application of Fibonacci numbers, a clever motivational device when studying a color wheel.

REFERENCES

[1] Adobe Image Library. Adobe ® Illustrator ® Classrom in a Book. San Jose, CA: Adobe Systems, Inc., 1998.

[2] Bicknell-Johnson, M. "Fibonacci Chromotology or How to Paint Your Rabbit." *The Fibonacci Quarterly*, Vol. *16.5* (1978): pp. 426-428.

[3] Gothe, Johann W. Theory of Colors. Cambridge, MA: M.I.T. Press, 2000.

[4] Faiss, Fritz Concerning the Way of Color: An Artist's Approach. Valencia Hills, California: Green Hut Press, 1972.

[5] Munsell, Albert H. A Color Notation. 11th Edition. Baltimore: Munsell Color Company, 1946.

[6] Ostwald, Wilhelm Color Science. Translated from the German by J. Scott Taylor. London: Winsor and Newton, 1933.

[7] Ostwald, Wilhelm Harmonthek. Housed in the rare book collection, University of Munich, Munich, Germany.

AMS Classification Numbers: 11B39

FINDING FIBONACCI IN A FRACTAL

Nathan C. Blecke, Kirsten Fleming and George William Grossman

The focus of this paper is to further investigate properties of a two-dimensional fractal that involves counting and Fibonacci numbers. We determine the fractal dimension using the Box Counting Theorem and also the concept of similitude. We find affine transformations that generate some of the set of points that are in the fractal, which have the form $A\mathbf{x}+b$ for a pair of two matrices A_1 and A_2 and some vectors $\mathbf{x}, b_1$ and b_2. We denote these transformations S_1 and S_2. We find examples of the limit points generated, by taking repeated applications of the operators on some starting points (which are vertices of triangles) in some prescribed order. The fractal, denoted G, is the countable intersection of the countable union of a set of triangles. The fractal is shown to be a totally disconnected set.

1. INTRODUCTION

Weierstrass discovered what was thought to be the first nowhere differentiable continuous function. This discovery was later shown to be incorrect, see Hairer and Wanner [9, p. 262 and p. 365]. Despite the lack of success other mathematicians continued working on the subject. Montel in 1903-04 did this by describing the set of limit points as "attracting orbits" and formulated the theory of normal families, which describes the behavior of iterates, see Alexander [1, p. 125]. "Montel's theory of normal families is ... a way to characterize the behavior of the iteratres of a given function $\phi(z)$ for arbitrary points in the complex plane .... If for all $z_0 \in D$, the sequence $\{\phi^n(z_0)\}$ converges to a fixed point x, then the family $\{\phi^n : n \in Z_0^+\}$, where Z_0^+ denotes the set of non-negative integers, was shown to be normal on D, and in fact converges uniformly to the constant function $G(z) \equiv x$", see Alexander [1, p. 125].

In 1906 Koch, Fatou, and Julia began developing these concepts. Koch had presented the "Koch Snowflake" as an example of a non-differentiable continuous curve, which encloses

This paper is in final form and no version of it will be submitted for publication elsewhere.

F.T. Howard (ed.), Applications of Fibonacci Numbers, Volume 9: Proceedings of the Tenth International Research Conference on Fibonacci Numbers and their Applications, 43–62.

a finite area, see Alexander [1, p. 128]. Fatou and Julia focused their efforts on the complex plane and looked at the idea of normality. By definition "A set $\Im \subset C(G,\Omega)$, where $C(G,\Omega)$ is the set of continuous functions from G to Ω, is normal if each sequence in $\Im$ has a subsequence which converges to a function $f \in C(G,\Omega)$", see Conway [5, p. 146]. Fatou studied the normal set of functions and Julia was studying the non-normal set. These two sets are referred to as Fatou and Julia sets and these sets are the complements of each other, see Alexander [1, p. 126].

In studying these sets, Fatou altered mathematical sets and developed functions that produce sets that are highly related to the original set. One such function is a perturbation of the Cantor function. The original set, Cantor's dust, is transformed to a new set referred to as the "real Fatou dust", see Mandelbrot [11, p. 182]. It has limit points of zero and one, see Mandelbrot [11, p. 192]. Fatou observed that the old set and the new set are of the category "totally disconnected perfect" set, see Alexander [1, p. 129] where totally disconnected is described as having no overlap, see Mandelbrot [11, p. 115] and perfect is defined to be the set of all limit points of the set, see Barnsley [3, p. 19].

Mandelbrot, upon the advice of an uncle read the papers by Fatou and Julia and developed an interest in these types of functions and in 1975 he coined the term "Fractal" and developed one of the first definitions of fractal, see Lesmoir-Gordon [10, p. 7]. It reads as follows: "A fractal is any set where the Hausdorff-Besicovitch dimension (or fractal dimension) is greater than the topological dimension, $D > D_T$," see Mandelbrot [11, p. 15]. Dimension is discussed in the next section.

Many fractals have been developed such as the Cantor set or the Sierpinski Gasket, see Mandelbrot [11, p. 80, p. 143]. Dr. George Grossman of Central Michigan University developed one such fractal, which is the focus of this paper, see Grossman [8]. His idea was actually the result of a joint paper on orthogonal projections, see Angelos [2]. The next person to study this fractal was Mr. Rowley who focused solely on the dimension of the set, see Rowley [12].

2. DIMENSION AND SIMILITUDE

Fractal dimension of a set describes how densely the set occupies the metric space in which it lies, see Barnsley [3, p. 3]. One such method for determining the fractal dimension is the **Box-Counting Theorem** (Although it may not always be possible to analytically calculate the dimension except for sets that are easily described for example, similarity transformations).

Theorem 1: *Box Counting Theorem*, see Crownover [6, p. 110]: Let $A \in H(X)$ where H is a Hausdorff space and (x, d) is a metric space (where metric space (x, d) is a nonempty set X of elements together with a metric d). For each $\varepsilon > 0$ let $N(\varepsilon)$ denote the smallest number of closed balls of radius $\varepsilon > 0$ needed to cover A. If

$$d = -\lim_{\varepsilon \to 0} \frac{\log N(\varepsilon)}{\log \varepsilon} \tag{1}$$

exists, then d is called the *box counting dimension* of A.

Proposition 1: The box counting dimension of a finite set of disjoint points is 0.

Proof: Clearly the numerator in (1) is finite, however the denominator is unbounded as $\varepsilon \to 0$. □

Proposition 2: Suppose $N(\varepsilon)$, which is the number of balls of radius $\varepsilon > 0$ needed to make a covering is equal to

$$F_{n+1} = \frac{1}{\sqrt{5}}\tau^{n+1}\left(1-\left(-\frac{1}{\tau+1}\right)\right)^{n+1}, \text{ where } \tau = \frac{1+\sqrt{5}}{2} \text{ with} \tag{2}$$

$$\tau^2 = \tau + 1,\ \tau > 1 \text{ and } \frac{-1}{\tau} = \frac{1-\sqrt{5}}{2} = -(\tau - 1).$$

Then the box counting dimension d is approximately $1.38848\ldots$.

Proof: Let $\varepsilon > 0$ and let radius $r = \frac{1}{\sqrt{2^k}}$. Select k such that $\varepsilon \geq r = \frac{1}{\sqrt{2^k}}$ hence $k > \frac{-2\log\varepsilon}{\log 2}$.

We use for k the smallest integer satisfying the above inequality. So $k = \left\lfloor \frac{-2\log\varepsilon}{\log 2} \right\rfloor + 1$ with the floor function $\lfloor \bullet \rfloor$.

Using this and equation (1) and (2) we get with $n = n(\varepsilon) = \frac{-2\log\varepsilon}{\log 2} + 1$

$$-\lim_{\varepsilon\to 0} \frac{\log\left(\frac{1}{\sqrt{5}}\tau^{n(\varepsilon)+1}\left(1-\left(\frac{-1}{\tau^2}\right)^{n(\varepsilon)+1}\right)\right)}{\log\varepsilon}$$

$$= \lim_{\varepsilon\to 0}\left(\frac{\log\sqrt{5}}{\log\varepsilon} - \frac{\log\tau^{n(\varepsilon)+1}}{\log\varepsilon} - \frac{\log\left(1-\left(\frac{-1}{\tau+1}\right)^{n(\varepsilon)+1}\right)}{\log\varepsilon}\right)$$

$$= \lim_{\varepsilon\to 0}\frac{\log\tau^{n(\varepsilon)+1}}{\log\varepsilon}$$

because the first and third terms $\underset{\varepsilon\to 0}{\to} 0$, since $\left(\frac{-1}{\tau+1}\right)^{n(\varepsilon)+1} \underset{\varepsilon\to 0}{\to} 0$ due to

$$n(\varepsilon) \underset{\varepsilon\to 0}{\to} +\infty = -\lim_{\varepsilon\to 0}\frac{(n(\varepsilon)+1)\log\tau}{\log\varepsilon} = -\lim_{\varepsilon\to 0}\frac{2\left(1-\frac{\log\varepsilon}{\log 2}\right)\log\tau}{\log\varepsilon}$$

$$= +\lim_{\varepsilon\to 0}\frac{\log\tau}{\log 2} = \frac{\log\tau}{\log 2} \approx 1.3884\ldots.$$ □

Note:

$$\lim_{\varepsilon\to 0}\left(\frac{1-\sqrt{5}}{2}\right)^{2-\frac{2\log\varepsilon}{\log 2}} = \lim_{\varepsilon\to 0}\left(\frac{-1}{\tau}\right)^{2-\frac{2\log\varepsilon}{\log 2}} = \left(\frac{-1}{\tau}\right)^{2}\lim_{\varepsilon\to 0}(-\tau)^{2\frac{\log\varepsilon}{\log 2}}$$

$$= \left(\frac{-1}{\tau}\right)^{2}\lim_{\varepsilon\to 0}(\tau^2)^{\frac{\log\varepsilon}{\log 2}} = \frac{1}{\tau^2}\lim_{\varepsilon\to 0}\left(\frac{1}{\tau^2}\right)^{-\frac{\log\varepsilon}{\log 2}} \to 0$$

when $0 < \frac{1}{\tau^2} < 1$. A similar proof using an alternate form of the Box Counting Theorem was done in 1996, see Rowley [12, p. 25].

We now introduce a definition, see Crownover [6, p. 67].

Definition 1: *The transformaiton $S : \mathfrak{R}^n \to \mathfrak{R}^n$ is a similitude (with similarity ratio $r > 0$) if*

$$\| S(x) - S(y) \|_2 = r \| x - y \|_2, \; x, \; y \in \mathfrak{R}^n. \tag{3}$$

where $\| \bullet \|_2$ denotes the Euclidean norm in $\mathfrak{R}^2$.

Definition 2: A square real matrix Q is orthogonal if the inner product of any two unique rows or columns is zero and any row or column has unit length.

Next, we have the following theorem, see Crownover [6, p. 67].

Theorem 2: *Let $S : \mathfrak{R}^n \to \mathfrak{R}^n$ be transformation that is a similitude (with similarity ratio $r > 0$) S is affine and has the form*

$$S(x) = rQ\mathbf{x} + b \tag{4}$$

where Q is an orthogonal matrix and b is a vector.

The following pair of equations gives the transformations in $\mathfrak{R}^2$ that we consider in this paper

$$A_1\begin{pmatrix} x \\ y \end{pmatrix} + \frac{1}{2}\begin{pmatrix} 1 \\ 1 \end{pmatrix}, \text{ where } A_1 = \frac{1}{2}\begin{pmatrix} -1 & -1 \\ 1 & -1 \end{pmatrix}, \tag{5}$$

whence $r_1 = \frac{1}{\sqrt{2}}$ and $Q_1 = \sqrt{2}A_1$.

$$A_2\begin{pmatrix} x \\ y \end{pmatrix} + \frac{1}{2}\begin{pmatrix} -1 \\ 0 \end{pmatrix}, \text{ where } A_2 = \frac{1}{2}\begin{pmatrix} 1 & 0 \\ 0 & 1 \end{pmatrix}, \tag{6}$$

whence $r_2 = \frac{1}{2}$ and $Q_2 = 2A_2 = I$.

We observe that the matrices Q_1 and Q_2 are orthogonal by definition 2 and that these matrices generate all eight orientations of the triangles seen in levels $L1 - L6$. Also, it follows from the definition 1 that Q_1 and Q_2 are both similitudes with ratios $1/\sqrt{2}$ and $1/2$ respectively and a angle of $\alpha = \frac{3\pi}{4}$ in the matrix $\begin{pmatrix} \cos\alpha & -\sin\alpha \\ \sin\alpha & \cos\alpha \end{pmatrix}$. In other words Q_1 rotates the figure $\frac{3\pi}{4}$

in the positive sense, and shrinks by a factor of $\frac{1}{\sqrt{2}}$ and translates with a vector $\left(\frac{1}{2},\frac{1}{2}\right)$. Q_2 reduces the original triangle by a factor of $\frac{1}{2}$ and translates with a vector of $\left(\frac{-1}{2},0\right)$.

When it is known that the affine transformations are similitudes, it is possible to determine the dimension of the set a second way, see Crownover [6, p. 113].

Definition 3: Suppose A is a compact set (where a compact set is both closed and bounded). A is self-similar if there are similitudes $S_1, S_2, \ldots, S_N$ and that

$$A = S_1(A) \cup S_2(A) \cup \cdots \cup S_N(A) \tag{7}$$

and the sets $S_i(A)$ with similarity ratios $r_i, i = 1, 2, \ldots, N$, do not have significant overlap which means that the dimension of the set of the union is zero if it is countable, see Crownover [6, p. 115].

We now introduce a theorem, see Crownover [6, p. 113].

Theorem 3: *Suppose A is self-similar (as in (7)). Suppose as well that the set $S_i(A)$ are disjoint and that d is the unique real solution to*

$$r_1^d + r_2^d + \cdots + r_N^d = 1. \tag{8}$$

Then, if $B_d(A) > 0$, (where $B_d(A)$ is an open ball with radius r and dimension d) the box dimension of A is

$$\dim_B(A) = d. \tag{9}$$

Proposition 3: Suppose A has affine transformations that are similitudes and meet the requirements of theorem 3 with scaling factors $r_1 = \frac{1}{\sqrt{2}}$, and $r_2 = \frac{1}{2}$, then $d = \frac{2\log\tau}{\log 2} \approx 1.3884\ldots.$

Proof: By (8) $\left(\frac{1}{\sqrt{2}}\right)^d + \left(\frac{1}{2}\right)^d = 1$ i.e. $x^2 = -x + 1$ with $x = \left(\frac{1}{2}\right)^{\frac{d}{2}}$ and solutions $x = -\tau$ and $x = \frac{1}{\tau}$. To get positive d one uses $x = \frac{1}{\tau}$ to obtain $d = \frac{2\log\tau}{\log 2} \approx 1.3884\ldots.$ □

3. MAPPINGS

3.1 Algorithm for Generating the Fractal:

The goal of this chapter is provide an explanation of how the fractal is generated as well as show that the number of triangles in the fractal follows a *Fibonacci* sequence. Once this is established then the dimension will follow from Chapter II. First we need to see how each one of the vertices is generated and this will be done through example. Let us consider the application of S_1 and S_2 to the points of the right triangle with vertices $(-1,0)$, $(1,0)$ and $(0,1)$, (see Fig. 1, p. 7). We will define this triangle t as level $L = 1$. We get the following points

$$S_1\begin{pmatrix}-1\\0\end{pmatrix} = \begin{pmatrix}1\\0\end{pmatrix},\ S_1\begin{pmatrix}1\\0\end{pmatrix} = \begin{pmatrix}0\\1\end{pmatrix},\ S_1\begin{pmatrix}0\\1\end{pmatrix} = \begin{pmatrix}0\\0\end{pmatrix}.$$

These are the vertices of a triagle similar to the original one with sides scaled by $1/\sqrt{2}$ and rotated 135 degrees counter-clockwise, (see Fig. 2, p. 8). Let us do the same for S_2. We get

$$S_2\begin{pmatrix}-1\\0\end{pmatrix}=\begin{pmatrix}-1\\0\end{pmatrix},\ S_2\begin{pmatrix}0\\1\end{pmatrix}=\begin{pmatrix}-1/2\\1/2\end{pmatrix},\ S_2\begin{pmatrix}1\\0\end{pmatrix}=\begin{pmatrix}0\\0\end{pmatrix}.$$

This defines a triangle similar to the original one with sides scaled by 1/2 and moved to the left quadrant (see Fig. 2, p. 8). This figure will be referred to as level $L=2$ (Each of the following shapes will be called "level L" where $L=1,2,\ldots$, etc.). We will repeat this process by applying S_1 to the vertices of level $L=2$ and S_2 to the vertices of level $L=1$. We get

$$S_1\begin{pmatrix}-1\\0\end{pmatrix}=\begin{pmatrix}1\\0\end{pmatrix},\ S_1\begin{pmatrix}0\\1\end{pmatrix}=\begin{pmatrix}0\\0\end{pmatrix},\ S_1\begin{pmatrix}1\\0\end{pmatrix}=\begin{pmatrix}0\\1\end{pmatrix},\ S_1\begin{pmatrix}0\\0\end{pmatrix}=\begin{pmatrix}1/2\\1/2\end{pmatrix},\ S_1\begin{pmatrix}-1/2\\1/2\end{pmatrix}=\begin{pmatrix}1/2\\0\end{pmatrix},$$

$$S_2\begin{pmatrix}-1\\0\end{pmatrix}=\begin{pmatrix}-1\\0\end{pmatrix},\ S_2\begin{pmatrix}0\\1\end{pmatrix}=\begin{pmatrix}-1/2\\1/2\end{pmatrix},\ S_2\begin{pmatrix}1\\0\end{pmatrix}=\begin{pmatrix}0\\0\end{pmatrix}.$$

The right half of the figure is similar to the previous level with sides scaled $1/\sqrt{2}$ and rotated 135 degrees counter-clockwise. The left half is the original triangle scaled by 1/2 and moved to the left quadrant (see Fig. 3, p. 8).

In order to generate level $L=4$ of the fractal we will apply S_1 to all the vertices of level $L=3$ and S_2 to all the vertices of level $L=2$. We get

$$S_1\begin{pmatrix}0\\1\end{pmatrix}=\begin{pmatrix}0\\0\end{pmatrix},\ S_1\begin{pmatrix}0\\0\end{pmatrix}=\begin{pmatrix}1/2\\1/2\end{pmatrix},\ S_1\begin{pmatrix}-1/2\\1/2\end{pmatrix}=\begin{pmatrix}1/2\\0\end{pmatrix},S_1\begin{pmatrix}-1\\0\end{pmatrix}=\begin{pmatrix}1\\0\end{pmatrix},$$

$$S_1\begin{pmatrix}1\\0\end{pmatrix}=\begin{pmatrix}0\\1\end{pmatrix},\ S_1\begin{pmatrix}1/2\\0\end{pmatrix}=\begin{pmatrix}1/4\\3/4\end{pmatrix},\ S_1\begin{pmatrix}1/2\\1/2\end{pmatrix}=\begin{pmatrix}0\\1/2\end{pmatrix},$$

and

$$S_2\begin{pmatrix}0\\0\end{pmatrix}=\begin{pmatrix}-1/2\\0\end{pmatrix},\ S_2\begin{pmatrix}-1/2\\1/2\end{pmatrix}=\begin{pmatrix}-3/4\\1/4\end{pmatrix},\ S_2\begin{pmatrix}-1\\0\end{pmatrix}=\begin{pmatrix}-1\\0\end{pmatrix},$$

$$S_2\begin{pmatrix}0\\1\end{pmatrix}=\begin{pmatrix}-1/2\\1/2\end{pmatrix},S_2\begin{pmatrix}1\\0\end{pmatrix}=\begin{pmatrix}0\\0\end{pmatrix}.$$

As was stated earlier, the right half of level $L=4$ is level $L=3$ scaled by $1/\sqrt{2}$ and rotated 135 degrees counter-clockwise, and the left side is level $L=2$ scaled by 1/2 and placed in the left half.

To continue the generation of the fractal we must continue to use these transformations S_1 and S_2 as follows: To obtain level $L \geq 2$

1. Apply S_1 to every vertex in level $L-1$

2. If $L=2$ apply S_2 to every vertex in level 1 (until level $L=3$) and if $L \geq 3$ apply S_2 to every vertex in level $L-2$ (see Fig. 3, 4, 5, and 6, p. 8, 9, 10).

3.2 Where Each Point will be Mapped:

Lemma 1: If $(p,q) \in \mathcal{R}^2$ then applying S_1 yields

$$(x,y) := S_1(p,q) = \left(\frac{-1}{2}(p+q)+\frac{1}{2}, \frac{1}{2}(p-q)+\frac{1}{2}\right).$$

Proof: Put $x=p$ and $y=q$ in (5)

Lemma 2: If (p,q) are selected so that $0 \leq p \leq 1$, $0 \leq q \leq 1$, and $q \leq -p+1$ then $(0 \leq x \leq y \leq -x+1 \leq 1.)$

Proof: S_1 is a rotation of 135° in the positive sense with a scale factor of $1/\sqrt{2}$ and a shift horizontally of 1/2 and vertically of 1/2.

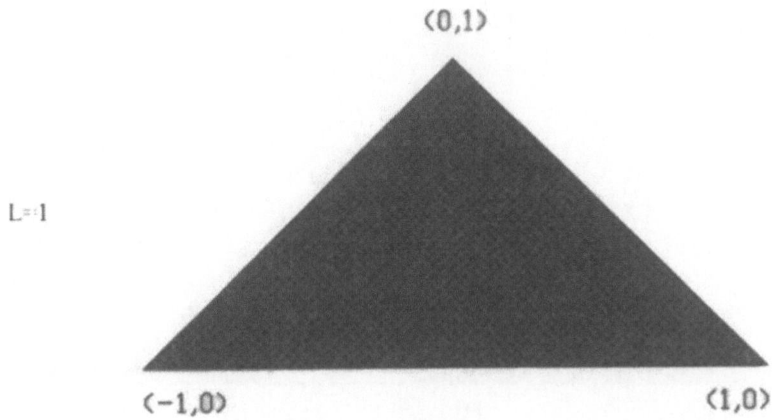

Figure 1

L=2

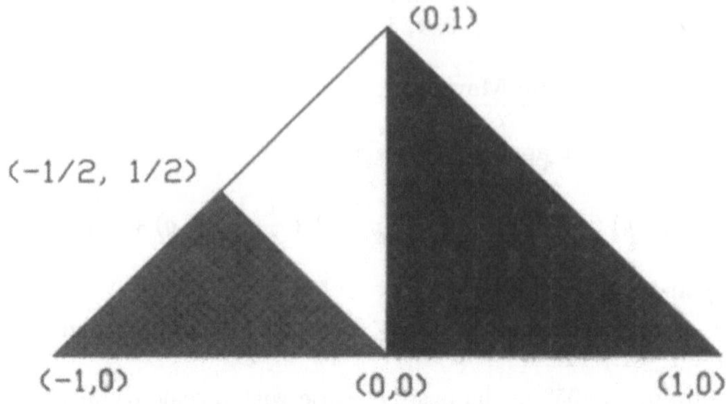

Figure 2

L=3

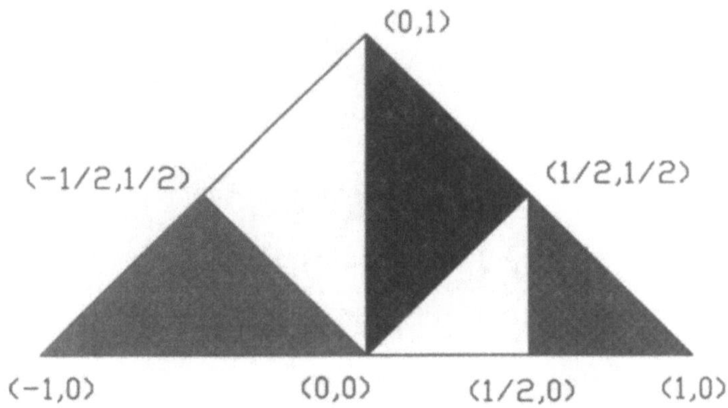

Figure 3

L=4

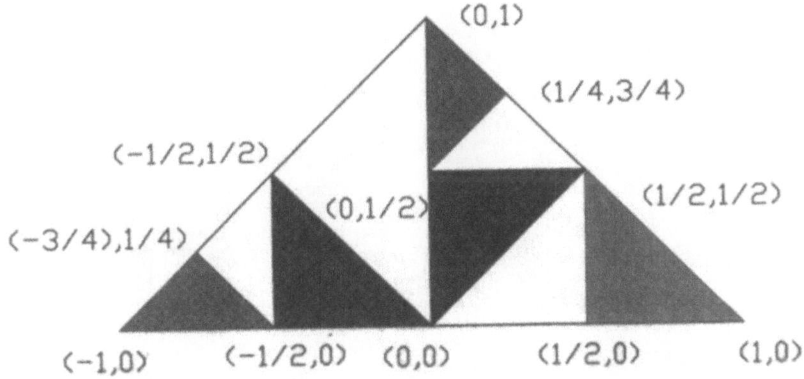

Figure 4

L=5

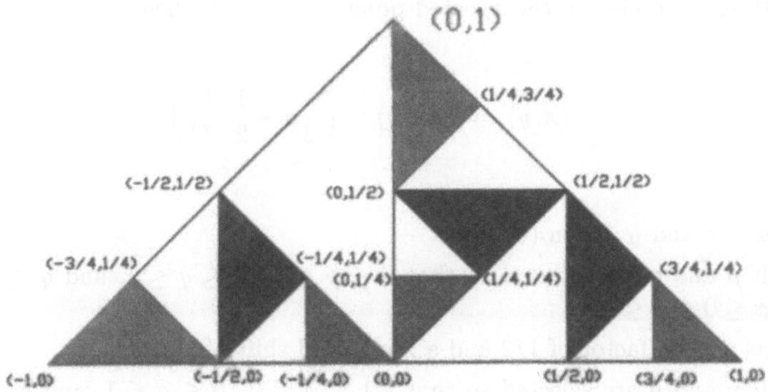

Figure 5

L=6

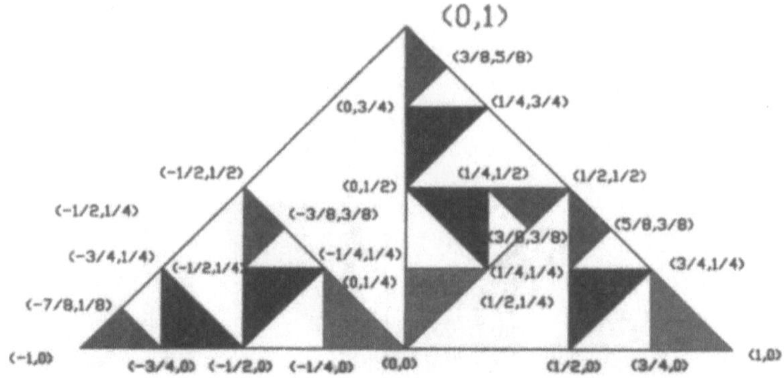

Figure 6

Lemma 3: If $-1 \leq p \leq 0$, $0 \leq q \leq 1$, and $q \leq p+1$ then $0 \leq y \leq x \leq -y+1$.

Proof: See proof of lemma 2.

Lemma 4: If S_2 is applied to the selected point $(p, q) \in \Re^2$, then

$$(x, y) := S_2(p, q) = \left(\frac{1}{2}p - \frac{1}{2}, \frac{1}{2}q\right).$$

Proof: Use $x = p$ and $y = q$ in (6).

Lemma 5: If p and q are selected so that $0 \leq p \leq 1$, $0 \leq q \leq 1$, and $q \leq -p+1$ then S_2 yields $\frac{-1}{2} \leq x \leq 0 \leq y \leq -x$.

Proof: S_2 has a scale factor of 1/2 and a horizontal shift of -1/2.

Lemma 6: If p and q are selected so that $-1 \leq p \leq 0$, $0 \leq q \leq 1$, and $q \leq p+1$ then S_2 yields $-1 \leq x \leq -\frac{1}{2}$ with $0 \leq y \leq x+1$.

Proof: See proof of lemma 5.

3.3 Finding the Fibonacci Sequence:

Lemma 7: If we count the number of triangles at level L, as we generate the fractal according to the algoirthm stated earlier, we find a Fibonacci number F_{L+1}.

Proof: At level one or $L = 1$ we have the original triangle (See Fig. 1). In order to generate level two or $L = 2$ we apply the algorithm described in chapter 3.1 (See Fig. 2).

Let $L = 3$ we generate the right hand triangles from S_1 applied to the level $L-1$ and the left hand triangle from S_2 applied to the level $L-2$. The number of triangles present in the third level is a Fibonacci number $3 = F_4 = F_3 + F_2$. Assuming this is true for the Kth level L, we prove this for the level $L = K + 1$. By Assumption $F_K = F_{K-1} + F_{K-2}$ and applying the algorithm described earlier we find the number of triangles $F_{K+1} = F_K + F_{K-1}$, where F_K is the number of triangles generated by S_1 and the number of triangles F_{K-1} is generated by S_2. □

4. HOW TO GENERATE POINTS

Recall equations (5) and (6), both of the transformations have matrices, $A_i, i = 1, 2, \ldots$. In hopes of generating a general formula for all points in the set we consider compositions of the form (however this is not accomplished in this paper)

$$S_i^{m_1} \circ S_j^{m_2} \ldots S_i^{m_{n-1}} \circ S_j^{m_n}\mathbf{x}, \qquad \mathbf{x} = \begin{pmatrix} x \\ y \end{pmatrix}. \tag{10}$$

where $m_1, m_2, \ldots, m_n \in \mathcal{N}$, (where $\mathcal{N}$ is natural numbers) and $i, j \in \{1, 2\}$ where either $i = 1$ and $j = 2$ or $i = 2$ and $j = 1$. We remark that $S_i^m \circ S_i^n = (S_i \circ S_i)^{m+n}$, however, $(S_i \circ S_j)^m \neq S_i^m \circ S_j^m$ if $i \neq j$.

The first thing that must be calculated is the repeated compositions of the matrices A_1 and A_2.

Lemma 8:

$$a) \quad A_1^n = \left(\frac{1}{\sqrt{2}}\right)^n \begin{pmatrix} \cos\tau_n & -\sin\tau_n \\ \sin\tau_n & \cos\tau_n \end{pmatrix}, \tau_n := \frac{3}{4}n\pi, \; n \in \mathcal{N}_0, \tag{11}$$

$$b) \quad A_2^n = \frac{1}{2^n}\begin{pmatrix} 1 & 0 \\ 0 & 1 \end{pmatrix}. \tag{12}$$

Proof: a) By induction on n, using addition theorems for trigonometric functions.

b) Obvious. □

Note for the readers, the previous approach and following generation was given to us by an unknown referee:

A_1^n is found from the fact that Q_1 is a 2×2 matrix with determinant of $Q_1 = 1$ and $x \equiv \frac{1}{2} trQ_1 = \frac{-1}{\sqrt{2}} = \cos\left(\frac{3\pi}{4}\right)$. Use the Cayley-Hamilton theorem: $Q_1^2 = 2xQ_1 - I = -\sqrt{2}Q_1 - I$

(Where I is the identity matrix). Therefore $Q_1^n = U_{n-1}(x)Q_1 - U_{n-2}(x)I$ with Chebyshev's

polynomials of the first kind: $U_n(x) := \frac{\sin((n+1)\cos^{-1}x)}{\sin(\cos^{-1}x)}, x = -\frac{1}{\sqrt{2}}$ with $\cos^{-1} - \left(\frac{1}{\sqrt{2}}\right) = \frac{3}{4}\pi$:

$$U_{n-1}(x)=\left(\frac{\sin(\frac{3}{4}n\pi)}{\sin(\frac{3}{4}\pi)}\right)=\sqrt{2}\sin\tau_n,\ \tau_n=\tfrac{3}{4}n\pi.$$

$$U_{n-2}\left(x=\frac{-1}{\sqrt{2}}\right)=-(\sin\tau_n+\cos\tau_n)\left(\text{addition theorem: }\sin\left(\frac{3}{4}n\pi-\frac{3}{4}\pi\right)=\dots\right)$$

$$\Longrightarrow Q_1^n=\begin{pmatrix}\cos\tau_n & -\sin\tau_n\\ \sin\tau_n & \cos\tau_n\end{pmatrix},\text{ or } A_1^n=\left(\frac{1}{\sqrt{2}}\right)^n Q_1^n.$$

Corollary 1:

$$\lim_{n\to\infty}A_1^n=\begin{pmatrix}0&0\\0&0\end{pmatrix},\ \lim_{n\to\infty}A_2^n=\begin{pmatrix}0&0\\0&0\end{pmatrix}.$$

Proof: This follows easily from Lemma 8, because $|\cos\tau_n|$ and $|\sin\tau_n|$ are bounded by 1. □

Now we consider compositions of the form $S_1^p v,\ S_2^p v$.

Lemma 9: For integers $p\geq 1$

$$S_1^p v=A_1^p v+A_1^{p-1}t_1+\cdots+A_1t_1+t_1=A_1^p v+(I-A_1)^{-1}(I-A_1^p)t_1, \tag{13}$$

$$S_2^p v=A_2^p v+A_2^{p-1}t_2+\cdots+A_2t_2+t_2=A_2^p v+(I-A_2)^{-1}(I-A_2^p)t_2. \tag{14}$$

Proof: It suffices to prove (13), since (13) and (14) are the same except for subscript. We proceed inductively. $p=1$ is true with $A_i^0\equiv I,\ i=1,2$.

Assume (14) and we obtain

$$S_1^{p+1}v=S_1S_1^p v=A_1(A_1^p v+A_1^{p-1}t_1+\cdots+A_1t_1+t_1)+t_1=A_1^{p+1}v+A_1^p t_1+\cdots+A_1t_1+t_1$$

because $1+A_1+A_1^2+\cdots+A^{p-1}=(I-A_1)^{-1}(I-A_1^p)(\text{if }\det(I-A_1)\neq 0)$. □

Corollary 2: We have independence of v

$$\lim_{p\to\infty}S_1^p v=\sum_{k=0}^{\infty}A_1^k t_1=(I-A_1)^{-1}t_1,\ \lim_{p\to\infty}S_2^p v=\sum_{k=0}^{\infty}A_2^k t_2=(I-A_2)^{-1}t_2. \tag{15}$$

Proof: This follows from Corollary 1 and Lemma 9. □

Lemma 10: *For any vector v in $\Re^2$,*

$$\lim_{p\to\infty}S_2^p v=(I-A_2)^{-1}=\begin{pmatrix}-1\\0\end{pmatrix}. \tag{16}$$

Proof:

$$(I - A_2)^{-1} = 2I \text{ and } t_2 = \frac{1}{2}\begin{pmatrix}-1\\0\end{pmatrix} \qquad \square \tag{17}$$

Lemma 11: *For any vector v in $\Re^2$*

$$\lim_{p\to\infty} S_1^p v = (I - A_1)^{-1} = \frac{1}{5}\begin{pmatrix}1\\1\end{pmatrix}.$$

Proof: $(I - A_1)^{-1} = \frac{1}{5}\begin{pmatrix}3 & 1\\ -1 & 3\end{pmatrix}$ and $t_1 = \frac{1}{2}\binom{1}{1}$. $\square$

Lemma 12:

$$(S_1 \circ S_2)^n v = (A_1A_2)^n v + \sum_{k=0}^{n-1}(A_1A_2)^k(A_1t_2 + t_1), \tag{18}$$

$$(S_2 \circ S_1)^n v = (A_2A_1)^n v + \sum_{k=0}^{n-1}(A_2A_1)^k(A_2t_1 + t_2). \tag{19}$$

Proof: We see that (19) follow from (18) by interchanging the subscripts 1 and 2 so it suffices to prove (18). We proceed by induction. By definition of S_2v and S_1v we have that

$$(S_1 \circ S_2)v = S_1 \circ (A_2v + t_2) = A_1(A_2v + t_2) + t_1 = A_1A_2v + A_1t_2 + t_1. \tag{20}$$

Thus, by assuming (18) we get

$$(S_1 \circ S_2)^{n+1}v = (S_1 \circ S_2)\left((A_1A_2)^n v + \sum_{k=0}^{n-1}(A_1A_2)^k(A_1t_2 + t_1)\right)$$

$$= (A_1A_2)^{n+1}v + \sum_{k=0}^{n-1}(A_1A_2)^{k+1}(A_1t_2 + t_1) + A_1t_2 + t_1$$

$$= (A_1A_2)^{n+1}v + \sum_{k=0}^{n}(A_1A_2)^k(A_1t_2 + t_1). \qquad \square$$

Corollary 3:

$$(A_1A_2)^n = \frac{1}{2^n}A_1^n = \left(\frac{1}{2\sqrt{2}}\right)^n \begin{pmatrix}\cos\tau_n & -\sin\tau_n\\ \sin\tau_n & \cos\tau_n\end{pmatrix}, \tau_n := \frac{3}{4}n\pi,\ n \in \mathcal{N}_0,$$

this is an adaptation of lemma 8(a) where $A_1A_2 = \frac{1}{2}A_1$. If we are interested in $(S_1 \circ S_2)v$ for $n \to \infty$ then we look to the following $\sum_{k=0}^{n-1}(A_1A_2)^k = (1 - A_1A_2)^{-1}(1 - (A_1A_2)^n)$, as long as $(\det(I - A_1A_2) \neq 0)$ and note that $\lim_{n\to\infty}(A_1A_2) = 0$ by lemma 8(a).

Corollary 4:

Keeping the aforementioned in mind

$$(I - A_1A_2)^{-1} = \left(I - \frac{1}{2}A_1\right)^{-1} = \left(\frac{1}{4}\begin{pmatrix} 5 & 1 \\ -1 & 5 \end{pmatrix}\right)^{-1} = \frac{2}{13}\begin{pmatrix} 5 & -1 \\ 1 & 5 \end{pmatrix} \text{ and } A_1t_2 + t_1 = \frac{1}{4}\begin{pmatrix} 3 \\ 1 \end{pmatrix}.$$

And $A_2t_1 + t_2 = \frac{1}{4}\binom{-1}{1}$.

We now consider compositions of the form $S_1 \circ S_2 \circ \cdots \circ S_1 \circ S_2 \circ S_1 v$ where j is the number of S's present. (Note that when j is odd, $j \geq 3$, and that S_1 occurs $(j+1)/2$ times and S_2 occurs $(j-1)/2$ times. We are able to get, for exchanging lables 1 and 2, $S_2 \circ S_1 \circ \cdots \circ S_2 \circ S_1 \circ S_2 v$. Now for j compositions we have,

Lemma 13:

$$\underbrace{(S_1 \circ S_2 \circ \cdots \circ S_2 \circ S_1)v}_{\#\text{ of } S\text{'s}=j}$$

$$= (A_1A_2)^p A_1 v + (1 - A_1A_2)^{-1}\left(1 - (A_1A_2)^{p+1}\right)t_1 + (1 - A_1A_2)^{-1}\left(1 - (A_1A_2)^p\right)A_1t_2$$

$$\text{where } p = \frac{j-1}{2} \text{ as long as } (\det(I - A_1A_2) \neq 0). \tag{21}$$

Proof: This can be proven by induction on j.

First we let $j = 3$.

$$(S_1 \circ S_2 \circ S)v = (A_1A_2A_1)v + (A_1A_2)t_1 + A_1t_2 + t_1$$

$$= (A_1A_2A_1)v + \left(\sum_{k=0}^{1}(A_1A_2)^{1-k}\right)t_1 + \left(\sum_{r=1}^{1}(A_1A_2)^{1-r}\right)A_1t_2$$

$$= (A_1A_2)A_1v + (1 - A_1A_2)^{-1}\left(1 - (A_1A_2)^2\right)t_1 + (1 - A_1A_2)^{-1}\left(1 - (A_1A_2)^1\right)A_1t_2$$

We assume true for arbitrary j,

$$(S_1 \circ S_2)(S_1 \circ S_2 \circ \cdots \circ S_1 \circ S_2 \circ S_1)v$$

$$= (S_1 \circ S_2)\left((A_1A_2)^p A_1 v + \left(\sum_{k=0}^{p}(A_1A_2)^{p-k}\right)t_1 + \left(\sum_{r=1}^{p}(A_1A_2)^{p-r}\right)A_1t_2\right)$$

$$= (A_1 \circ A_2)\left((A_1A_2)^p A_1 v + \left(\sum_{k=0}^{p}(A_1A_2)^{p-k}\right)t_1 + \left(\sum_{r=1}^{p}(A_1A_2)^{p-r}\right)A_1t_2\right)$$

$$+ A_1t_2 + t_1$$

$$= \left((A_1A_2)^{p+1} A_1 v + \left(\sum_{k=0}^{p+1}(A_1A_2)^{p+1-k}\right)t_1 + \left(\sum_{r=1}^{p+1}(A_1A_2)^{p+1-r}\right)A_1t_2\right)$$

$$= (A_1A_2)^{p+1}A_1v + (1-(A_1A_2)^{-1}\left(1-(A_1A_2)^{p+2}\right)t_1$$

$$+ (1 - A_1A_2)^{-1}\left(1-(A_1A_2)^{p+1}\right)A_1t_2$$

$$\underbrace{(S_1 \circ S_2 \circ \cdots \circ S_1 \circ S_2 \circ S_1)}_{\#\text{ of } S\text{'s}=j+2} \qquad \square$$

<u>**Lemma 14:**</u>

$$\underbrace{(S_1 \circ S_2 \circ \cdots \circ S_2 \circ S_1)^n v}_{\#\text{ of } S\text{'s}=j}$$

$$= ((A_1A_2)^p A_1)^n v + \left(\sum_{w=0}^{n-1}((A_1A_2)^p A_1)^w\right)\left(\left(\sum_{k=0}^{p}(A_1A_2)^{p-k}\right)t_1 + \left(\sum_{r=1}^{p}(A_1A_2)^{p-r}\right)A_1t\right)$$

$$= ((A_1A_2)^p A_1)^n v + (1 - ((A_1A_2)^p)A_1)^{-1}$$
$$(1-((A_1A_2)^pA_1)^n)\left((1-A_1A_2)^{-1}\left(1-(A_1A_2)^{p+1}\right)t_1 + (1-A_1A_2)^{-1}(1-(A_1A_2)^p)A_1t_2\right)$$

as long as $(\det(1 - ((A_1A_2)^p)A_1) \neq 0)$ (22)

By lemma 13 we can let j be arbitrary and let $n = 2$,

$$\underbrace{(S_1 \circ S_2 \circ \cdots \circ S_1 \circ S_2 \circ S_1)}_{\#\text{ of } S\text{'s}=j}\underbrace{(S_1 \circ S_2 \circ \cdots \circ S_1 \circ S_2 \circ S_1)v}_{\#\text{ of } S\text{'s}=j}$$

$$= \underbrace{(S_1 \circ S_2 \circ \cdots \circ S_1 \circ S_2 \circ S_1)}_{\#\text{ of } S\text{'s}=j}\left((A_1A_2)^pA_1v + \left(\sum_{k=0}^{p}(A_1A_2)^{p-k}\right)t_1 + \left(\sum_{r=1}^{p}(A_1A_2)^{p-r}\right)A_1t_2\right)$$

$$= \underbrace{(A_1A_2\ldots A_1A_2A_1)}_{\#\text{ of } A\text{'s}=j}\left((A_1A_2)^pA_1v + \left(\sum_{k=0}^{p}(A_1A_2)^{p-k}\right)t_1 + \left(\sum_{r=1}^{p}(A_1A_2)^{p-r}\right)A_1t_2\right)$$

$$+ \left(\sum_{k=0}^{p}(A_1A_2)^{p-k}\right)t_1 + \left(\sum_{r=1}^{p}(A_1A_2)^{p-r}\right)A_1t_2$$

$$= ((A_1A_2)^pA_1)\left((A_1A_2)^pA_1v + \left(\sum_{k=0}^{p}(A_1A_2)^{p-k}\right)t_1 + \left(\sum_{r=1}^{p}(A_1A_2)^{p-r}\right)A_1t_2\right)$$

$$+ \left(\sum_{k=0}^{p}(A_1A_2)^{p-k}\right)t_1 + \left(\sum_{r=1}^{p}(A_1A_2)^{p-r}\right)A_1t_2$$

$$= ((A_1A_2)^pA_1)^2v + \left(((A_1A_2)^pA_1)\left(\sum_{k=0}^{p}(A_1A_2)^{p-k}\right)t_1 + ((A_1A_2)^pA_1)\right.$$

$$\left.\left(\sum_{r=1}^{p}(A_1A_2)^{p-r}\right)A_1t_2\right) + \left(\sum_{k=0}^{p}(A_1A_2)^{p-k}\right)t_1 + \left(\sum_{r=1}^{p}(A_1A_2)^{p-r}\right)A_1t_2$$

$$= ((A_1A_2)^pA_1)^2v + \left(((A_1A_2)^pA_1)\left(\sum_{k=0}^{p}(A_1A_2)^{p-k}\right)t_1 + ((A_1A_2)^pA_1)\right.$$

$$\left.\left(\sum_{r=1}^{p}(A_1A_2)^{p-r}\right)A_1t_2\right) + \left(\sum_{k=0}^{p}(A_1A_2)^{p-k}\right)t_1 + \left(\sum_{r=1}^{p}(A_1A_2)^{p-r}\right)A_1t_2$$

$$= ((A_1A_2)^pA_1)^2v + ((A_1A_2)^pA_1 + 1)\left(\left(\sum_{k=0}^{p}(A_1A_2)^{p-k}\right)t_1 + \left(\sum_{r=1}^{p}(A_1A_2)^{p-r}\right)A_1t_2\right)$$

$$= ((A_1A_2)^pA_1)^2v$$

$$+ (1 - ((A_1A_2)^p) A_1)^{-1} \left(1 - ((A_1A_2)^p A_1)^2\right) \left((1 - A_1A_2)^{-1} \left(1 - (A_1A_2)^{p+1}\right) t_1\right.$$

$$\left. + (1 - A_1A_2)^{-1} (1 - (A_1A_2)^p) A_1 t_2\right)$$

$$= \underbrace{(S_1 \circ S_2 \circ \cdots \circ S_1 \circ S_2 \circ S_1)^2}_{\#\text{ of } S\text{'s}=j}$$

Assume true for arbitrary n and recall that j is arbitrary,

$$\underbrace{(S_1 \circ S_2 \circ \cdots \circ S_1 \circ S_2 \circ S_1)}_{j} \underbrace{(S_1 \circ S_2 \circ \cdots \circ S_1 \circ S_2 \circ S_1)^n}_{j}$$

$$= \underbrace{(A_1A_2 \ldots A_1A_2A_1)}_{\#\text{ of } A\text{'s}=j} \left(((A_1A_2)^p A_1)^n v\right.$$

$$+ \left(\sum_{w=0}^{n-1} ((A_1A_2)^p A_1)^w\right) \left(\left(\sum_{k=0}^{p} (A_1A_2)^{p-k}\right) t_1 + \left(\sum_{r=1}^{p} (A_1A_2)^{p-r}\right) A_1 t_2\right)\Bigg)$$

$$+ \left(\sum_{k=0}^{p} (A_1A_2)^{p-k}\right) t_1 + \left(\sum_{r=1}^{p} (A_1A_2)^{p-r}\right) A_1 t_2$$

$$= ((A_1A_2)^p A_1) \left(((A_1A_2)^p A_1)^n v\right.$$

$$+ \left(\sum_{w=0}^{n-1} ((A_1A_2)^p A_1)^w\right) \left(\left(\sum_{k=0}^{p} (A_1A_2)^{p-k}\right) t_1 + \left(\sum_{r=1}^{p} (A_1A_2)^{p-r}\right) A_1 t_2\right)\Bigg)$$

$$+ \left(\sum_{k=0}^{p} (A_1A_2)^{p-k}\right) t_1 + \left(\sum_{r=1}^{p} (A_1A_2)^{p-r}\right) A_1 t_2$$

$$= \Bigg(((A_1A_2)^p A_1)^{n+1} v + \left(\sum_{w=0}^{n} ((A_1A_2)^p A_1)^w\right)$$

$$\left(\left(\sum_{k=0}^{p} (A_1A_2)^{p-k}\right) t_1 + \left(\sum_{r=1}^{p} (A_1A_2)^{p-r}\right) A_1 t_2\right)\Bigg)$$

$$= ((A_1A_2)^p A_1)^{n+1} v + (1 - ((A_1A_2)^p) A_1)^{-1} \left(1 - ((A_1A_2)^p A_1)^{n+1}\right)$$

$$((1 - A_1A_2)^{-1} (1 - (A_1A_2)^{p+1}) t_1 + (1 - A_1A_2)^{-1} (1 - (A_1A_2)^p) A_1 t_2)$$

$$\underbrace{(S_1 \circ S_2 \circ \cdots \circ S_1 \circ S_2 \circ S_1)^{n+1}}_{j}. \qquad \square$$

Using the substitution $A_1A_2 = \frac{1}{2}A_1$ one is then able to determine the limit points for all $n \to \infty$ as long as $(\det(1 - (\frac{1}{2}A_1)^n A_1) \neq 0)$.

5. DESCRIPTION OF THE FRACTAL

The purpose of this chapter is to define the fractal formally and show that this fractal is a compact uncountable set.

Definition 4: The fractal which is generated by the algorithm in section 3.1 is the following set,

$$G = \bigcap_{i=2}^{\infty} \left(\bigcup_{k=2}^{Fi} V_k \right) \tag{23}$$

where V_k is the kth right triangle at the ith stage of the fractal generation.

We now introduce a definition, see Crownover [6, p. 46].

Definition 5: A set is bounded when its diameter is finite. Diameter of set A is

$$\delta(A) = \sup\{\| \mathbf{x}\text{-}\mathbf{y} \|_2 \colon \mathbf{x},\mathbf{y} \in A\}. \tag{24}$$

Lemma 15: The fractal G is compact.

Proof: The fractal is closed since the countable intersection of closed sets and bounded since contained with the original triangle with vertices having diameter

$$\delta(G) = \sup\{\| \mathbf{x}\text{-}\mathbf{y} \|_2\} = 2$$

since the points (-1,0) and (1,0) are elements of G. A closed and bounded set in $\Re^2$ is compact. $\square$

Theorem 5: The fractal G is uncountable and totally disconnected.

Proof: The fractal dimension is approximately $\frac{\log \tau}{\log 2} \approx 1.3884\ldots$ and exceeds the topological dimension of the real line so the set G is uncountable. Let g be any point in G, then g is the vertex of a triangle or an accumulation point since it is in the countable intersection of a subsequence of triangles $\{V_{k_i}\}_{i=1}^{\infty}$. Clearly, there is no open neighborhood of g within G. Moreover, a triangle V_{k_i} of arbitrarily small diameter can be enclosed by an open set, the boundary of which intersects the triangle only at the three vertices of the triangle and has disjoint intersection with any other triangle V_{kj} at the same level L see figures on pages 7-10. The union of all the vertices taken as sets, over every level L is a countable set and so has topological dimension 0, see Crownover [6, p. 289]. Thus, G is compact and has topological dimension 0 and so is totally disconnected, see Crownover [6, p. 287-88]. (It is conjectured that G is an irregular 1-set, see Falconer [7, p. 45].) It is noted that the components of G are all points, see Willard [15, p. 210]. □

6. CONCLUSION

We have discovered that this fractal has a *Fibonacci* sequence embedded in it by counting triangles and because of this, we were able to determine the fractal dimension. This was supported by the property of similitude. Both methods yielded a dimension of approximately 1.3884.... Also, we were able to develop an algorithm for generating some of the limit points of the set. We were finally able to show that the set is compact, uncountable and totally disconnected.

ACKNOWLEDGMENTS

We would like to acknowledge the work of two individuals for there help with this paper. First, Leela Rakesh, who is member of Central Michigan University's Mathematics Department and was a member of the ad-hoc Thesis Committee that reviewed the paper. We would also like to thank the unknown referee who helped with the development of lemma 8 and the generation thereafter.

I (Nathan Blecke) would like to also thank Mr. John Ingles and Dr. George Grossman for their encouragement during this entire process. It is greatly appreciated.

REFERENCES

[1] Alexander, Daniel S. A History of Complex Dynamics: From Schroder to Fatou and Julia. Braunschweig/Weisbaden, Germany: Aspects of Mathematics, Friedr. Vieweg & Sohn Verlagsgesellschaft mbh, 1994.

[2] Angelos, J., Grossman, G., Kaufman, E., Lenker, T., Rakesh, L. "Limit Cycles for Successive Projections onto Hyperplanes in R^n." *The Fibonacci Quarterly*, Vol. *35* (1997): pp. 206-224.

[3] Barnsley, Michael F. Fractals Everywhere. Academic Press, Boston, 1988.

[4] Berkey, Dennis D. Calculus. Saunders. College Publishing, New York, 1988.

[5] Conway, John B. Functions of One Complex Variable. Graduate Texts in Mathematics 2nd ed., Springer-Verlag, New York 1978.

[6] Crownover, Richard M. Introduction to Fractals and Chaos. Jones and Bartlett, Boston, 1995.

[7] Falconer, K. J. The Geometry of Fractal Sets. Cambridge University Press, 1985.

[8] Grossman, G. "Fractal Construction by Orthogonal Projection Using the Fibonacci Sequence." *The Fibonacci Quarterly*, Vol. *35.3* (1997): pp. 206-224.

[9] Hairer, E. and Wanner, G. Analysis by Its History. Springer-Verlag, New York, 1996.

[10] Lesmoir-Gordon, Nigel, Rood, Will and Edney, Ralph. Introducing Fractal Geometry. Icon Books, Cambridge, United Kingdom, 2000.

[11] Mandelbroit, Benoit B. The Fractal Geometry of Nature. W.H. Freeman and Co., New York, 1999.

[12] Rowley, Kevin T. Fractals and Their Dimensions. Plan B paper, Central Michigan University, 1996.

[13] Royden, H.L. Real Analysis. Macmillan Publishing Company, New York, 1988.
[14] Smart, James R. Modern Geometries. Brooks/Cole Publishing Company, Pacific Grove, California, 1988.
[15] Willard, Stephen. General Topology. Addison-Wesley Publishing Company, Inc., Reading Massachusetts, 1970.

AMS Classification Numbers: 11B39, 29A80

POSITIVE INTEGERS $(a^2+b^2)/(ab+1)$ ARE SQUARES

Jens-P. Bode and Heiko Harborth

1. INTRODUCTION

In 1988, the International Mathematical Olympiad (IMO) was held in Canberra and on day II the following Problem 6 was posed (see [1,2]):

Let a and b be positive integers such that $ab+1$ divides a^2+b^2. Show that $(a^2+b^2)/(ab+1)$ is the square of an integer.

In [1,4] proofs are given without the explicit determination of pairs (a,b) for which $(a^2+b^2)/(ab+1)$ is an integer. The purpose of this paper is to explicitely describe the set of all integer triples (a,b,x) which satisfy the diophantine equation $a^2+b^2=x(ab+1)$.

2. POSITIVE INTEGERS

We will first determine all positive integers a,b for which $x=(a^2+b^2)/(ab+1)$ is an integer. It will be shown that x is a square for all solutions (a,b,x), with a and b positive, and hence Problem 6 from the 1988 IMO is an immediate consequence.

Theorem 1: For $a \geq b \geq 1$, all solutions (a,b,x) of the diophantine equation

$$a^2+b^2=x(ab+1) \tag{1}$$

are (1,1,1) and

$$a=a_s(t)=\sum_{i=0}^{\lfloor s/2 \rfloor}(-1)^i\binom{s-i}{i}t^{2s+1-4i},$$

$$b=a_{s-1}(t), \text{ and}$$

$$x=t^2$$

for $s \geq 1,\ t \geq 2$, and with $a_0=t$.

This paper is in final form and no version of it will be submitted for publication elsewhere.

F.T. Howard (ed.), *Applications of Fibonacci Numbers, Volume 9: Proceedings of the Tenth International Research Conference on Fibonacci Numbers and their Applications*, 63–67.

Proof: Let (a, b, x) be any solution of (1). Let $a = vb + r$, with $v \geq 1$ and $0 \leq r \leq b - 1$. We distinguish the two cases $r = 0$ and $r \geq 1$.

If $r = 0$, then

$$x = \frac{v^2b^2 + b^2}{vb^2 + 1} = v + \frac{b^2 - v}{vb^2 + 1}.$$

Since $-(vb^2 + 1) < b^2 - v < vb^2 + 1$, we obtain an integral x for $v = b^2$ only. Thus

$$a = t^3, \ b = t, \ \text{and } x = v = t^2 \text{ for } t \geq 1 \tag{2}$$

are all solutions for $r = 0$.

If $r \geq 1$, that is, $b \geq 2$, then

$$x = \frac{v^2b^2 + 2vbr + r^2 + b^2}{vb^2 + br + 1} = v + \frac{b^2 + vbr - v + r^2}{vb^2 + br + 1}. \tag{3}$$

With $1 \leq r \leq b - 1$ we have

$$0 < b^2 + vbr - v + r^2 < 2(vb^2 + br + 1)$$

so that x is an integer only if

$$b^2 + vbr - v + r^2 = vb^2 + br + 1.$$

Combining this with (3) shows that $x = v + 1$. Solving for v yields

$$v = 1 + \frac{r^2 - 2}{b^2 - br + 1} = -1 + \frac{b^2 + z^2}{bz + 1} = -1 + x \tag{4}$$

if the substitution $r = b - z$, $2 \leq z \leq b - 1$, is used. Note that $z = 1$ is impossible since $v = -1 + b - (b - 1)/(b + 1)$ is not an integer for $b \geq 2$.

From equation (4) we conclude that (a, b, x) is a solution of (1) only if the smaller triple (b, z, x) solves (1). On the other hand, any solution (b, z, x) for $z > 1$ determines the larger solution (a, b, x) with

$$a = vb + r = (x - 1)b + b - z = xb - z. \tag{5}$$

Thus any solution (a, b, x) with $b \geq 2$ determines an infinite sequence (a_s, b_s, x) of solutions with constant values of x, and with $a = a_s$, $b = b_s = a_{s-1}$, and $z = z_s = b_{s-1} = a_{s-2}$. Since b_s is decreasing with s, for some s, say $s = 1$, the value of a_s is a multiple of b_s. Then the case $r = 0$ determines the smallest triple in this sequence to be $(a_1, b_1, x) = (t^3, t, t^2)$, $t \geq 2$ (see (2)), and $x = t^2$ is always a square solving Problem 6. For $b = t = 1$ only the solution (1,1,1) exists (see (2)).

From (5) we have for a_s the recursive formula

$$a_s = t^2 a_{s-1} - a_{s-2} \text{ with } a_0 = t,\ a_1 = t^3. \tag{6}$$

Using the recursion (6) the solutions asserted in Theorem 1 can be checked inductively. □

All solutions (a_s, b_s) are polynomials in t. It may be remarked that the sum of the absolute values of the coefficients of the polynomial $a_{s-1} = b_s$ equals the Fibonacci number F_s.

Equation (6) is a linear second-order difference equation with constant coefficients. By the well-known methods another form of the solution given in Theorem 1 is

$$a_s(t) = \frac{t}{\sqrt{t^4-4}}\left(\frac{t^2+\sqrt{t^4-4}}{2}\right)^{s+1} - \frac{t}{\sqrt{t^4-4}}\left(\frac{t^2-\sqrt{t^4-4}}{2}\right)^{s+1}.$$

We remark that recently Luca, Osgood, and Walsh [3] used the above reduction method to discuss the quadratic equations $x^2-(k^2-4)y^2 = 4k$, k odd.

3. INTEGERS

With regard to the symmetry of (1) in a and b, Theorem 1 gives the complete solution of (1) in integers for $ab > 0$. If $ab = 0$ then the solutions $(a, b, x) = (t, 0, t^2)$ are trivial. It remains $ab < 0$, say $a \geq -b \geq 1$. The solution of (1) in this case is similar to the solution for $ab > 0$.

Theorem 2: For $a \geq -b \geq 1$ if all solutions (a, b, x) of the diophantine equation (1) are

$$a = a_s = a_1 \sum_{i=0}^{\lfloor (s-1)/2 \rfloor} (-1)^i \binom{s-1-i}{i} 5^{s-1-2i}$$

$$- a_0 \sum_{i=0}^{\lfloor (s-2)/2 \rfloor} (-1)^i \binom{s-2-i}{i} 5^{s-2-2i}, \text{ for } s \geq 2,$$

$b = b_s = -a_{s-1}$ for $s \geq 1$ and

$x = -5$ with $a_0 = 1$ and $a_1 = 2$ or $a_1 = 3$.

Proof: With $\beta = -b$ we consider

$$a^2 + \beta^2 = x(1 - a\beta) \tag{7}$$

for $a \geq \beta \geq 1$ instead of (1). Let (a, β, x) be any solution of (7). Let $a = v\beta + r$, with $v \geq 1$ and $0 \leq r \leq \beta - 1$. We have the two cases $r = 0$ and $r \geq 1$.

If $r = 0$, then

$$x = \frac{v^2\beta^2 + \beta^2}{1 - v\beta^2} = -v - \frac{\beta^2 + v}{v\beta^2 - 1}.$$

Thus x is not an integer if $0 < \beta^2 + v < v\beta^2 - 1$, that is, if

$$2 < (\beta^2 - 1)(v - 1). \tag{8}$$

Inequality (8) is fulfilled for $v, \beta \geq 2$. It remain the two cases $v = 1$ and $\beta = 1$. For $v = 1$ we have $x = -2 - 2/(\beta^2 - 1)$ and x cannot be an integer. From $\beta = 1$ we obtain $x = -v - 1 - 2/(v - 1)$ which is an integer only if $v = 2$ or $v = 3$. This implies

$$(a, \beta, x) = (2, 1, -5) \text{ and } (a, \beta, x) = (3, 1, -5) \tag{9}$$

as solutions of (7).

If $r \geq 1$, that is, $\beta \geq 2$, then

$$x = -v - \frac{\beta^2 + r^2 + v\beta r + v}{v\beta^2 + \beta r - 1}. \tag{10}$$

With $1 \leq r \leq \beta - 1$ we have

$$0 < \beta^2 + r^2 + v\beta r + v < 2(v\beta^2 + \beta r - 1)$$

so that x is an integer only if

$$\beta^2 + r^2 + v\beta r + v = v\beta^2 + \beta r - 1.$$

Combining this with (10) shows that $x = -v - 1$. Solving for v yields

$$v = 1 + \frac{r^2 + 2}{\beta^2 - \beta r - 1} = -1 - \frac{\beta^2 + z^2}{1 - \beta z} = -1 - x \tag{11}$$

if the substitution $r = \beta - z, 1 \leq z \leq \beta - 1$, is used.

From equation (11) we conclude that (a, β, x) solves (7) only if the smaller triple (β, z, x) is a solution of (7). Also, any solution (β, z, x) determines the large solution (a, β, x) with

$$a = v\beta + r = (-1 - x)\beta + \beta - z = -x\beta - z. \tag{12}$$

Thus any solution (a, β, x) of (7) induces an infinite sequence of solutions with constant values of x, and with $a = a_s, \beta = \beta_s = a_{s-1}$, and $z = z_s = \beta_{s-1} = a_{s-2}$. Since β_s is decreasing with s, for some s, say $s = 1$, the value of a_s is a multiple of β_s. Then the case $r = 0$ determines only two such sequences of solutions starting with the two triples in (9). In both cases we have $x = -5$, not being a square.

From (12) we obtain the recursion

$$a_s = 5a_{s-1} - a_{s-2} \text{ with } a_0 = 1 \text{ and } a_1 = 2 \text{ or } a_1 = 3.$$

Again the solutions asserted in Theorem 2 can be checked inductively ($\beta_s = -b_s$) and the equivalent solutions of this difference equation are

$$a_s = c_1\left(\frac{5+\sqrt{21}}{2}\right)^s + (1-c_1)\left(\frac{5-\sqrt{21}}{2}\right)^s \text{ with } c_1 = \frac{1}{2} \pm \frac{\sqrt{21}}{42}. \qquad \square$$

Summarizing, the fraction $(a^2+b^2)/(ab+1)$ is always a square for $ab \geq 0$ and only -5 is possible for $ab < 0$.

REFERENCES

[1] Andreescu, T. and Gelca, R. Mathematical Olympiad Challenges. Birkhäuser, Basel 2000.

[2] Atanassou, E. Aufgabe 6. *Praxis der Mathematik*, Vol. *30* (1988): p. 501.

[3] Luca, F., Osgood, C.F., and Walsh, P.G. Diophantine Approximations and a Problem from the 1988 IMO. Preprint, 2002.

[4] Runge, E. and Sewerin, H. Lösung der Aufgabe 6. *Praxis der Mathematik*, Vol. *31* (1989): p. 60.

AMS Classification Numbers: 11B37, 11D09, 39A05

ON THE FIBONACCI LENGTH OF POWERS OF DIHEDRAL GROUPS

C.M. Campbell, P.P. Campbell, H. Doostie and E.F. Robertson

1. INTRODUCTION

The idea of forming a sequence of group elements based on a Fibonacci-like recurrence relation was initially introduced by Wall in [14] and later developed by other authors, see [1], [7], [16]. This idea was then refined in [3] and [5] to that of a Fibonacci orbit and Fibonacci length of a given group $G = \langle A \rangle$ where $A = \{a_1, a_2, \ldots, a_n\}$ as follows:

Definition 1.1: For the finitely generated group $G = \langle A \rangle$, where $A = \{a_1, \ldots, a_n\}$, the *Fibonacci orbit* of G with respect to the generating set A, written $F_A(G)$, is the sequence $x_1 = a_1, \ldots, x_n = a_n, x_{i+n} = \prod_{j=1}^{n} x_{i+j-1}$, $i \geq 1$. If $F_A(G)$ is periodic then the length of the period of the sequence is called the *Fibonacci length* of G with respect to the generating set A, written $LEN_A(G)$. When it is clear which generating set is being investigated we will write $LEN(G)$ for $LEN_A(G)$. It is clear that $F_A(G)$ is periodic if G is an epimorphic image of a Fibonacci group; see [13] for more information on Fibonacci groups.

As can be seen from the above definitions the Fibonacci length of a group depends in general on the chosen generating set. It was noted in [3] that, for any generating pair $\{a, b\} \in D_{2m}, LEN_{\{a,b\}}(D_{2m}) = 6$, where D_{2m} is the dihedral group of order $2m$. In this paper we intend to investigate this phenomenon further by calculating the Fibonacci length of D_{2m}^i, for any integer i, with respect to the natural generating set. We use the natural generating set for D_{2m}, as in [4], defined as satisfying $\langle a_1, b_1 | a_1^2 = 1, b_1^m = 1, (a_1 b_1)^2 = 1 \rangle$. This is extended to the direct product by using the following well known method of construction: if $G = \langle X | R \rangle$ and $N = \langle Y | S \rangle$ then $G \times N = \langle X, Y | R, S, [X, Y] \rangle$ where $[X, Y] = \{[x, y] : x \in X, y \in Y\}$, see [9]. In the above we have used relations and relators as appropriate and will continue to do so in this paper. When rewriting words we use the convention found in [11], namely the use of

This paper is in final form and no version of it will be submitted for publication elsewhere.

F.T. Howard (ed.), *Applications of Fibonacci Numbers, Volume 9: Proceedings of the Tenth International Research Conference on Fibonacci Numbers and their Applications*, 69–85.

underscores to highlight the subwords which are replaced in passing from one word to the next.

Over the years there has been interest in particular types of presentations for direct powers of groups, with one strand of this interest motivated by questions of Wiegold in [15]; see for example [2] and [4]. Other strands are discussed in [10] and [12]. In this paper we find the structure of the Fibonacci orbit of D_∞^i and use this to find the Fibonacci length of D_{2m}^i. We show that for $i > 1$ the Fibonacci length of D_{2m}^i on the natural generating set no longer remains constant but varies with i and m. We also examine the Fibonacci length of a restricted set of D_{2m}^i, for odd m, on a different generating set.

The results in this paper were suggested by data from computer programs written in the computational algebra system GAP [8].

2. THE FIBONACCI ORBIT OF D_∞^i, $i \geq 2$

D_∞^i has a presentation with $2i$ generators and $2i^2$ relations:

$$D_\infty^i = \langle a_1, b_1, a_2, b_2, \ldots, a_i, b_i | a_l^2 = (a_l b_l)^2 = 1, \\ [a_j, a_k] = [a_j, b_k] = [b_j, b_k] = 1, \\ 1 \leq l \leq i,\ 1 \leq j < k \leq i \rangle.$$

The elements of the group defined by the above presentation can be written in the normal form $a_i^{q_i} a_{i-1}^{q_{i-1}} \ldots a_1^{q_1} b_i^{r_i} b_{i-1}^{r_{i-1}} \ldots b_1^{r_1}$ where the q_l are either 0 or 1. Throughout this paper we will always reduce elements of the Fibonacci orbit to this normal form.

In order to elucidate the main proof of this section we start by considering the Fibonacci type sequence of elements of D_∞^2 as

$$x_1 = a_1, x_2 = b_1, x_3 = a_2, x_4 = b_2, x_n = x_{n-4} x_{n-3} x_{n-2} x_{n-1},\ (n \geq 5).$$

We have:

Lemma 2.1: *Every element of the Fibonacci orbit $\{x_j\}$ of $D_\infty^2 = \langle a_1, b_1, a_2, b_2 \rangle$ may be represented by:*

$$x_j = \begin{cases} a_1, & j \equiv 1, -4 \bmod 10 \\ b_1^{\pm 1}, & j \equiv 2, -3 \bmod 10 \\ a_2 b_1^{\pm 2(j-3)/5}, & j \equiv 3, -2 \bmod 10 \\ b_2^{\pm 1} b_1^{\pm 2(j-4)(j+1)/5^2}, & j \equiv 4, -1 \bmod 10 \\ a_2 a_1 b_2^{\pm 1} b_1^{\pm (2j^2 - 5^2)/5^2}, & j \equiv 5, 0 \bmod 10 \end{cases}$$

where the positive exponent is chosen for the first value of j and the negative exponent is chosen for the second value of j.

Proof: The assertion may be proved by induction on j. We first prove the anchor step of the induction. Since $a^iba^i = b^{-1}, i$ odd, we have

$$\begin{aligned}
x_1 &= a_1,\\
x_2 &= b_1,\\
x_3 &= a_2,\\
x_4 &= b_2,\\
x_5 &= a_2a_1b_2b_1,\\
x_6 &= b_1a_2b_2a_1b_1a_2b_2 = a_1,\\
x_7 &= a_2b_2a_1b_1a_2b_2a_1 = a_1b_1a_1 = b_1^{-1},\\
x_8 &= b_2a_1b_1a_2b_2a_1b_1^{-1} = b_2a_2b_2 \cdot a_1b_1a_1b_1^{-1} = a_2b_1^{-2},\\
x_9 &= a_1b_1a_2b_2a_1b_1^{-1}a_2b_1^{-2} = a_2b_2a_2 \cdot a_1b_1a_1b_1^{-3} = b_2^{-1}b_1^{-4},\\
x_{10} &= a_1b_1^{-1}a_2b_1^{-2}b_2^{-1}b_1^{-4} = a_2a_1b_2^{-1}b_1^{-7}.
\end{aligned}$$

Now let $k \equiv 0 \bmod 10$ and assume that the result holds for all values up to $k+5$, namely

$$\begin{aligned}
x_{k+1} &= a_1,\\
x_{k+2} &= b_1,\\
x_{k+3} &= a_2b_1^{2((k+3)-3)/5} = a_2b_1^{2k/5},\\
x_{k+4} &= b_2b_1^{2((k+4)-4)((k+4)+1)/25} = b_2b_1^{2k(k+5)/25},\\
x_{k+5} &= a_1a_2b_1^{(2(k+5)^2-25)/25}b_2 = a_2a_1b_2b_1^{(2k^2+20k+25)/25}.
\end{aligned}$$

We now prove that the next ten entries have the required form and thus complete the induction.

$$\begin{aligned}
x_{k+6} &= b_1a_2b_1^{2k/5}b_2b_1^{2k(k+5)/25}a_2a_1b_2b_1^{(2k^2+20k+25)/25}\\
&= b_1^{1+2k/5+2k^2/25+10k/25}a_1b_1^{(2k^2+20k+25)/25}a_2b_2a_2b_2\\
&= a_1.
\end{aligned}$$

$$\begin{aligned}
x_{k+7} &= a_2b_1^{2k/5}b_2b_1^{2k(k+5)/25}a_2a_1b_2b_1^{(2k^2+20k+25)/25}a_1\\
&= b_1^{2k/5+2k^2/25+10k/25}a_1b_1^{(2k^2+20k+25)/25}a_1a_2b_2a_2b_2\\
&= b_1^{-1}.
\end{aligned}$$

$$\begin{aligned}
x_{k+8} &= b_2 b_1^{2k(k+5)/25} a_2 a_1 b_2 b_1^{(2k^2+20k+25)/25} a_1 b_1^{-1} \\
&= b_1^{2k^2/25+10k/25} a_1 b_1^{(2k^2+20k+25)/25} a_1 b_1^{-1} b_2 a_2 b_2 \\
&= a_2 b_1^{-2((k+8)-3)/5}.
\end{aligned}$$

$$\begin{aligned}
x_{k+9} &= a_2 a_1 b_2 b_1^{(2k^2+20k+25)/25} a_1 b_1^{-1} a_2 b_1^{-2((k+8)-3)/5} \\
&= a_1 b_1^{(2k^2+20k+25)/25} a_1 b_1^{-2k/5-3} a_2 b_2 a_2 \\
&= b_2^{-1} b_1^{-2((k+9)-4)((k+9)+1)/25}.
\end{aligned}$$

$$\begin{aligned}
x_{k+10} &= a_1 b_1^{-1} a_2 b_1^{-2((k+8)-3)/5} b_2^{-1} b_1^{-2((k+9)-4)((k+9)+1)/25} \\
&= a_1 b_1^{-2k/5-7-2k^2/25-15k/25} a_2 b_2^{-1} \\
&= a_2 a_1 b_2^{-1} b_1^{-(2(k+10)^2-25)/25}.
\end{aligned}$$

$$\begin{aligned}
x_{k+11} &= b_1^{-1} a_2 b_1^{-2((k+8)-3)/5} b_2^{-1} b_1^{-2((k+9)-4)((k+9)+1)/25} a_2 a_1 b_2^{-1} b_1^{-(2k^2+40k+175)/25} \\
&= b_1^{-10k/25-7-2k^2/25-6k/5} a_1 b_1^{-2k^2/25-8k/5-7} a_2 b_2^{-1} a_2 b_2^{-1} \\
&= a_1.
\end{aligned}$$

$$\begin{aligned}
x_{k+12} &= a_2 b_1^{-2((k+8)-3)/5} b_2^{-1} b_1^{-2((k+9)-4)((k+9)+1)/25} a_2 a_1 b_2^{-1} b_1^{-(2k^2+40k+175)/25} a_1 \\
&= b_1^{-2k/5-2-2k^2/25-6k/5-4} a_1 b_1^{-2k^2/25-8k/5-7} a_1 a_2 b_2^{-1} a_2 b_2^{-1} \\
&= b_1.
\end{aligned}$$

$$\begin{aligned}
x_{k+13} &= b_2^{-1} b_1^{-2((k+9)-4)((k+9)+1)/25} a_2 a_1 b_2^{-1} b_1^{-(2k^2+40k+175)/25} a_1 b_1 \\
&= b_1^{-2k^2/25-6k/5-4} a_1 b_1^{-2k^2/25-8k/5-7} a_1 b_1 b_2^{-1} a_2 b_2^{-1} \\
&= a_2 b_1^{2((k+13)-3)/5}.
\end{aligned}$$

$$x_{k+14} = a_2 a_1 b_2^{-1} b_1^{-(2k^2+40k+175)/25} a_1 b_1 a_2 b_1^{2(k+10)/5}$$

$$= a_1 b_1^{-(2k^2+40k+175)/25} a_1 b_1^{1+2(k+10)/5} a_2 b_2^{-1} a_2$$

$$= b_2 b_1^{2((k+14)-4)((k+14)+1)/25}.$$

$$x_{k+15} = a_1 b_1 a_2 b_1^{2(k+10)/5} b_2 b_1^{2((k+14)-4)((k+14)+1)/25}$$

$$= a_1 b_1^{1+2(k+10)/5+2((k+14)-4)((k+14)+1)/25} a_2 b_2$$

$$= a_2 a_1 b_2 b_1^{(2(k+15)^2-25)/25}. \quad \square$$

In the same manner as the previous case we examine the Fibonacci sequence of D^3_∞ i.e.

$$x_1 = a_1, x_2 = b_1, x_3 = a_2, x_4 = b_2, x_5 = a_3, x_6 = b_3, x_n = \prod_{j=1}^{6} x_{n-7+j}, \quad (n \geq 7).$$

Lemma 2.2: *Every element of the Fibonacci orbit $\{x_j\}$ of $D^3_\infty = \langle a_1, b_1, a_2, b_2, a_3, b_3 \rangle$ may be represented by:*

$$x_j = \begin{cases} a_1, & j \equiv 1, -6 \bmod 14 \\ b_1^{\pm 1}, & j \equiv 2, -5 \bmod 14 \\ a_2 b_1^{\pm 2(j-3)/7}, & j = 3, -4 \bmod 14 \\ b_2^{\pm 1} b_1^{\pm 2(j-4)(j+3)/7^2}, & j \equiv 4, -3 \bmod 14 \\ a_3 b_2^{\pm 2(j-5)/7} b_1^{\pm 4(j-5)(j+2)(j+9)/(3\times 7^3)}, & j \equiv 5, -2 \bmod 14 \\ b_3^{\pm 1} b_2^{\pm 2(j+1)(j-6)/7^2} b_1^{\pm 2(j-6)(j+1)(j+8)(j+15)/(3\times 7^4)}, & j \equiv 6, -1 \bmod 14 \\ a_3 a_2 a_1 b_3^{\pm 1} b_2^{\pm(2j^2-7^2)/7^2} b_1^{\pm(2(j/7)^4+8(j/7)^3+4(j/7)^2-8(j/7)-3)/3}, & j \equiv 7, 0 \bmod 14 \end{cases}$$

where the positive exponent is chosen for the first value of j and the negative exponent is chosen for the second value of j.

Proof: As in Lemma 2.1, it is sufficient to compute $x_1, x_2, \ldots, x_{14}$ and the result follows by induction on j. $\square$

We now seek to generalize the above to obtain a normal form for the Fibonacci orbit of $D^i_\infty, i \geq 2$. We will need the following result in our calculations.

Lemma 2.3: *For $n \geq 3$ the following polynomial identity holds:*

$$2 + \sum_{j=3}^{n-1} \frac{2^{j-2}}{(j-2)!} m(m+1)(m+2)\ldots(m+j-3)$$

$$+ \sum_{j=3}^{n} \frac{2^{j-2}}{(j-2)!}(m-1)m(m+1)\ldots(m+j-4)$$

$$= \frac{2^{n-2}}{(n-2)!} m(m+1)(m+2)\ldots(m+n-3)$$

with the convention that when $n = 3$, the first term in the first summation on the left hand side is zero.

Proof: We use induction on n. When $n = 3$ the result holds since $2 + 0 + \frac{2}{1!}(m-1) = 2m$, i.e. $\frac{2}{1!}m = 2m$. When $n = 4$ the result also holds since

$$2 + \frac{2}{1!}m + \frac{2}{1!}(m-1) + \frac{2^2}{2!}(m-1)m = 2m(m+1), \text{ i.e. } \frac{2^2}{2!}m(m+1) = 2m(m+1).$$

Now assume that the result holds for all values less than n. Then

$$2 + \sum_{j=3}^{n-1} \frac{2^{j-2}}{(j-2)!} m(m+1)\ldots(m+j-3) + \sum_{j=3}^{n} \frac{2^{j-2}}{(j-2)!}(m-1)m\ldots(m+j-4)$$

$$= 2 + \sum_{j=3}^{n-2} \frac{2^{j-2}}{(j-2)!} m(m+1)\ldots(m+j-3) + \sum_{j=3}^{n-1} \frac{2^{j-2}}{(j-2)!}(m-1)m\ldots(m+j-4)$$

$$+ \frac{2^{n-3}}{(n-3)!} m(m+1)\ldots(m+n-4) + \frac{2^{n-2}}{(n-2)!}(m-1)m\ldots(m+n-4).$$

By the inductive hypothesis this is equal to

$$\frac{2^{n-3}}{(n-3)!}m(m+1)\dots(m+n-4)+\frac{2^{n-3}}{(n-3)!}m(m+1)\dots(m+n-4)$$

$$+\frac{2^{n-2}}{(n-2)!}(m-1)m\dots(m+n-4)$$

$$=\frac{2^{n-2}}{(n-3)!}m(m+1)\dots(m+n-4)+\frac{2^{n-2}}{(n-2)!}(m-1)m\dots(m+n-4)$$

$$=\frac{2^{n-2}}{(n-2)!}m(m+1)\dots(m+n-3)$$

as required. □

Now we make the following observations about the Fibonacci orbits $\{x_j\}$ of the groups $D^i_\infty = \langle a_1, b_1, a_2, b_2, \dots, a_i, b_i\rangle$ (where $i \in \mathbb{N}$).

Lemma 2.4: *The exponent of b_l in x_j is zero if $j \equiv 1, 2, \dots, 2l-1 \mod (2i+1)$, where $l \in \{1, 2, \dots, i\}$.*

Proof: This is easy to see if we look at the 'structure' of the entries of the Fibonacci orbit $\{x_j\}$ of D^i_∞. For $j \in \{1, 2, \dots, 2i\}$ we have

$$x_j = \begin{cases} a_{(j+1)/2}, & \text{if } j \text{ is odd,} \\ b_{j/2}, & \text{if } j \text{ is even,} \end{cases}$$

and

$$x_{2i+1} = \prod_{d=1}^{2i} x_d.$$

Now it is easy to see that the next term of the Fibonacci orbit to contain an a_l will be

$$x_{2l+2i} = \prod_{d=2l}^{2l+2i-1} x_d.$$

Likewise the next b_l occurs in

$$x_{2l+2i+1} = \prod_{d=2l+1}^{2l+2i} x_d.$$

The final stage of the induction follows by using an argument analogous to that above. □

Remark: It can easily be shown that the exponents of b_l can be calculated once one knows the exponents of b_1 since the exponents of b_l 'lag' behind the exponents of b_1. This holds because all b_l's initially have exponent one and from Lemma 2.4 above it can be seen when the exponents of b_l are nonzero. As an example of this 'lag' see the D^3_∞ case where, when $j \equiv 3 \bmod 14$, the power of b_1 is $2(j-3)/7$, and when $j \equiv 5 \bmod 14$, the power of b_2 is $2(j-5)/7$.

These results are best illustrated by looking at the D^3_∞ case (given separately in Lemma 2.2). It can be used as an example in the following proof to elucidate concepts. In this case the orbit is

$$\begin{aligned} Z =&(a_1, b_1, a_2, b_2, a_3, b_3, a_3a_2a_1b_3b_2b_1,\\ & a_1, b_1^{-1}, a_2b_1^{-2}, b_2^{-1}b_1^{-4}, a_3b_2^{-2}b_1^{-4}, b_3^{-1}b_2^{-4}b_1^{-8}, a_3a_2a_1b_3^{-1}b_2^{-7}b_1^{-19},\\ & a_1, b_1, \ldots). \end{aligned}$$

So the Fibonacci orbit behaves as if it is in 'layers' of length $2i+1$ where in alternate layers the exponents of b_l are all positive and all negative. Since one layer's 'structure' depends only on the previous layer, a proof by induction will only need the first layer to be proved to anchor the induction. Also we note here that it is unnecessary to know the general form of the $k(2i+1)^{th}$ entry ($k \in \mathbb{N}$) in the Fibonacci orbit since this will always be the product of all the previous terms in its layer and so this entry has a known shape, namely $a_ia_{i-1}\ldots a_1b_i^{w_i}b_{i-1}^{w_{i-1}}\ldots b_1^{w_1}$ where w_l is the sum of the exponents of b_l in the layer.

We are now ready to give the main result of this section.

Proposition 2.5: *The exponents of b_1 in the Fibonacci orbit $\{x_j\}$ of D^i_∞ are given in the table below*

$$\begin{array}{ll} 0, & j \equiv 1,\ 2i+2 \bmod (4i+2)\\ \pm 1, & j \equiv 2,\ 2i+3 \bmod (4i+2)\\ \pm A_n\Big(\prod_{d=0}^{n-3}(j-n+d(2i+1))\Big)/(2i+1)^{n-2}, & j \equiv n,\ 2i+1+n \bmod (4i+2)\\ \sum_{d=j-2i}^{j-1} z_d, & j \equiv 2i+1, 0 \bmod (4i+2) \end{array}$$

where $A_n = 2^{n-2}/(n-2)!$, z_r is the exponent of b_1 in x_r, $n \equiv j \bmod (2i+1)$ so $3 \le n \le 2i$ and the positive forms of elements in the orbit are chosen if $1 \le j \bmod (4i+2) \le 2i+1$; otherwise choose the (second) negative forms.

Proof: Let the exponent of b_1 in the x_r entry of the Fibonacci orbit be z_r. Assume that the result is true for all values of j such that $1 \le j < k$. There are several cases to examine:

Case 1. $k \equiv 1, 2i+2 \bmod (4i+2)$

In this case we have

$$\underline{b_1^{z_{k-2i}+\cdots+z_{k-2}}a_1b_1^{z_{k-2i}+\cdots+z_{k-2}}} = a_1.$$

Case 2. $k \equiv 2, 2i+3 \mod (4i+2)$

Here we have

$$\underline{b_1^{z_{k-2i}+\cdots+z_{k-3}} a_1 b_1^{z_{k-2i-2}+\cdots+z_{k-3}}} a_1 = \underline{a_1 b_1^{0+z_{k-2i-1}} a_1} = b_1^{-z_{k-2i-1}} = b_1^{\pm 1}.$$

Case 3. $k \equiv n, 2i+n+1 \mod (4i+2),\ 3 \leq n \leq 2i+1$

When we are trying to calculate the exponent of b_1 for the kth entry in the Fibonacci orbit we need only concentrate on the terms in a_1 and b_1. The exponent of b_1 and a_1 in the kth entry in the Fibonacci orbit is

$$\left(\prod_{l=k-2i}^{x-1} b_1^{z_l}\right)\left(a_1 \prod_{l=x-2i}^{x-1} b_1^{z_l}\right) a_1 \left(\prod_{l=x+1}^{k-1} b_1^{z_l}\right)$$

where $x = (2i+1)\lfloor k/(2i+1) \rfloor$. (Note the first bracket is from the layer below that of x_k, the second bracket is the last entry in the lower layer). The above sum can be simplified using the group relations as follows

$$\underline{\left(\prod_{l=k-2i}^{x-1} b_1^{z_l}\right)\left(a_1 \prod_{l=x-2i}^{x-1} b_1^{z_l}\right)} a_1 \left(\prod_{l=x+1}^{k-1} b_1^{z_l}\right) = \underline{a_1 \left(\prod_{l=x-2i}^{k-2i-1} b_1^{z_l}\right) a_1} \left(\prod_{l=x+1}^{k-1} b_1^{z_l}\right)$$

$$= \left(\prod_{l=x-2i}^{k-2i-1} b_1^{-z_l}\right)\left(\prod_{l=x+1}^{k-1} b_1^{z_l}\right).$$

Thus the exponent of b_1 is given by

$$\sum_{l=x+1}^{k-1} z_l - \sum_{l=x-2i}^{k-2i-1} z_l$$

$$= \pm\,(0+1+A_3((x+3)-3)/(2i+1)+\ldots$$

$$+ A_{k-x-1}\left[\prod_{d=0}^{k-x-4}((k-1)-(k-x-1)+d(2i+1))\right]/(2i+1)^{k-x-3})$$

$$-(\mp(0+1+A_3((x-2i+2)-3)/(2i+1)+\ldots$$

$$+ A_{k-x}\left[\prod_{d=0}^{k-x-3}((k-2i-1)-(k-x)+d(2i+1))\right]/(2i+1)^{k-x-2}))$$

So we want to show that

$$2+\sum_{l=3}^{n-1}\frac{2^{l-2}}{(l-2)!}m(m+1)\dots(m+l-3)+\sum_{l=3}^{n}\frac{2^{l-2}}{(l-2)!}(m-1)m(m+1)\dots(m+1-4)$$

$$=\frac{2^{n-2}}{(n-2)!}m(m+1)\dots(m+n-3)$$

where $n \equiv k \bmod (2i+1)$, $m=\lfloor k/(2i+1)\rfloor$ and $2<n<2i+1$ and $\sum_{l=3}^{n-1}(2^{l-2}m(m+1)\dots(m+l-3)/(l-2)!)$ is zero if $n=3$. Now the result follows by using Lemma 2.3.

Case 4. $k \equiv 2i+1,\ 0 \bmod (4i+2)$

Here there is nothing to prove. □

3. THE FIBONACCI LENGTH OF D^i_{2m}, $i \geq 2$

In order to give an expression for the Fibonacci length of D^i_{2m}, $i \geq 2$, we require the following definitions and lemmas.

Definition: In D^i_{2m} let $MinLEN(D^i_{2m}) = 2m(2i+1)/(4,m)$.

Lemma 3.1: *In D^i_{2m}, $i \geq 2$, $MinLEN(D^i_{2m})$ divides $LEN(D^i_{2m})$, and the quotient $LEN(D^i_{2m})/MinLEN(D^i_{2m})$ only involves odd prime divisors of m.*

Proof: We first note that $MinLEN(D^i_{2m})$ is the smallest possible Fibonacci length since we must have $(4i+2)|LEN(D^i_{2m})$ and, from the third entry in the Fibonacci orbit of D^i_{2m}, we also have $m|(2LEN(D^i_{2m})/(2i+1))$. Thus $MinLEN(D^i_{2m})$ divides $LEN(D^i_{2m})$.

Now let $l = MinLEN(D^i_{2m})$. By Proposition 2.5 for $3 \leq n \leq 2i$ we have

$$z_{l+n}=\frac{2^{n-2}l(l+(2i+1))\dots(l+(n-3)(2i+1))}{(n-2)!(2i+1)^{n-2}}.$$

If $\bar{l}=l/(2i+1)=2m/(4,m)$ the above becomes

$$z_{l+n}=\frac{2^{n-2}\bar{l}(\bar{l}+1)\dots(\bar{l}+(n-3))}{(n-2)!}.$$

Now z_{l+n} is obviously an integer. Since z_{l+n} is the power of b_1 in the Fibonacci orbit we require that m divides $z_{LEN(D^i_{2m})+n}$, $3 \leq n \leq 2i$. It may occur that $m \nmid z_{l+n}$ because powers of primes from the factorization of m that also occur in the numerator of z_{l+n} may be factored out by the $(n-2)!$ denominator. Let these 'missing' primes be $p_j^{\alpha_j}\dots p_r^{\alpha_r}$. Now multiplying l by a factor q less than $p_j^{\alpha_j}\dots p_r^{\alpha_r}$ will not be sufficient. For m will not divide z_{ql+n} since the denominator $(n-2)!$ will be the same as in the z_{l+n} case and the numerator will still be a product of 2^{n-2} and $n-2$ consecutive integers but this time starting at $2qm/(4,m)$. Thus $(n-2)!$ will still factor out $p_j^{\alpha_j}\dots p_r^{\alpha_r}$ and since q is less than $p_j^{\alpha_j}\dots p_r^{\alpha_r}$ we still have $m \not| z_{ql+n}$.

If we let $q = p_j^{\alpha_j}\dots p_r^{\alpha_r}$ then we will have $m|z_{ql+n}$. □

__Definition:__ Let $\pi_{m,i} \in \mathbb{N}$ be defined as satisfying the equation $LEN(D^i_{2m}) = \pi_{m,i} MinLEN(D^i_{2m})$.

We now give a property of $\pi_{m,i}$ that will be used in a later lemma.

__Lemma 3.2:__ *For any fixed m the sequence $(\pi_{m,i})_{i=2}^{\infty}$ is monotonically increasing.*

__Proof:__ Let $l = LEN(D^i_{2m}) = 2m\pi_{m,i}(2i+1)/(4,m)$ and $\bar{l} = l/(2i+1)$. From Proposition 2.5 for $3 \le n \le 2i$ we have

$$\begin{aligned} z_{l+n} &= \frac{2^{n-2}\bar{l}(\bar{l}+1)\dots(\bar{l}+(n-3))}{(n-2)!} \\ &= \frac{2^{n-2}(2m\pi_{m,i}/(4,m))((2m\pi_{m,i}/(4,m))+1)\dots((2m\pi_{m,i}/(4,m))+(n-3))}{(n-2)!}. \end{aligned}$$

As i increases, and m remains the same, the number of entries in the layers of the Fibonacci orbit increases as does $(2i-2)!$. This gives the possibility that the number and size of primes in the prime decomposition of $(2i-2)!$ will increase and as a consequence so will $\pi_{m,i}$. Note if $\pi_{m,i+1} \not> \pi_{m,i}$ then since $(2i)! > (2i-2)!$, $\pi_{m,i+1} = \pi_{m,i}$. □

Now we show a multiplicative property of $\pi_{m,i}$.

__Lemma 3.3:__ *For $m,n \in \mathbb{N}$, $(m,n)=1$ and $i \ge 2$, $\pi_{mn,i} = \pi_{m,i}\pi_{n,i}$.*

__Proof:__ This proof is similar to that of the previous lemma. Let $l = MinLEN(D^i_{2(mn)})$. As in Lemma 3.1 we look at

$$z_{l+k} = \frac{2^{k-2}l(l+(2i+1))\dots(l+(k-3)(2i+1))}{(k-2)!(2i+1)^{k-2}},$$

where $3 \le k \le 2i$. Now $\pi_{mn,i}$ is found by seeing which primes in the prime decomposition of mn are factored out by the $(k-2)!$ denominator. Any primes that are factored out by the denominator are multiplied back so that $mn|z_{l+k}$. Now since $(m,n)=1$ and the layers of the Fibonacci orbit are the same length the result follows immediately. □

__Lemma 3.4:__ *If p is an odd prime and $\alpha \in \mathbb{N}$ then $\pi_{p^\alpha,i} = \pi_{p,i}$. Also for any $\beta \in \mathbb{N}$, $\pi_{2^\beta,i} = 1$.*

__Proof:__ Here we use induction on i.

When $i=2$, letting $l = MinLEN(D^i_{2p}) = 2p5/(4,p) = 10p$ and using Proposition 2.5 we obtain

$$\begin{aligned} z_{l+3} &= 2(10p), \\ z_{l+4} &= 2(10p)(10p+1). \end{aligned}$$

Both z_{l+3} and z_{l+4} are divisible by p and so $l = MinLEN(D^{\mathbf{i}}_{2p}) = LEN(D^{\mathbf{i}}_{2p})$ and $\pi_{p,2} = 1$. Using an analogous argument we see that $\pi_{p^\alpha,2} = 1$.

Now assume that $\pi_{p^\alpha,i} = \pi_{p,i}$ for all $i \le \mu$. We examine the case $i = \mu + 1$. By Lemma 3.2 there are three possibilities:

Case 1. $\pi_{p^\alpha,\mu+1} = \pi_{p^\alpha,\mu}$ and $\pi_{p,\mu+1} = \pi_{p,\mu}$

Here there is nothing to prove.

Case 2. $\pi_{p^\alpha,\mu+1} > \pi_{p^\alpha,\mu}$

Let $l = LEN(D^{\mu}_{2p^\alpha})$, $3 \le n \le 2\mu, \bar{l} = l/(2\mu+1) = 2p^\alpha \pi_{p^\alpha,\mu}/(4,p^\alpha) = 2p^\alpha \pi_{p^\alpha,\mu}$ so the power of b_1 in the Fibonacci orbit is given by

$$z_{l+n} = \frac{2^{n-2} 2p^\alpha \pi_{p^\alpha,\mu}(2p^\alpha \pi_{p^\alpha,\mu}+1)\dots(2p^\alpha \pi_{p^\alpha,\mu}+n-3)}{(n-2)!}.$$

We must have $p^\alpha | z_{l+n}$ (since $|b_1| = p^\alpha$).

Now consider what happens in the $D^{\mu+1}_{2p^\alpha}$ case. Keeping l to be the Fibonacci length of $D^{\mu}_{2p^\alpha}$ but letting $3 \le n \le 2\mu+2$, as above we obtain

$$z_{l+n} = \frac{2^{n-2} 2p^\alpha \pi_{p^\alpha,\mu}(2p^\alpha \pi_{p^\alpha,\mu}+1)\dots(2p^\alpha \pi_{p^\alpha,\mu}+n-3)}{(n-2)!},$$

and since we require $\pi_{p^\alpha,\mu+1} > \pi_{p^\alpha,\mu}$ we have $p^\alpha \nmid z_{l+n}$ for some n, $3 \le n \le 2\mu+2$. For n in the range $3 \le n \le 2\mu$ we have $p^\alpha | z_{l+n}$ (this is just the previous μ case), so p^α does not divide one or both of $z_{l+n}, n = 2\mu+1$ or $2\mu+2$. This means that $(2\mu)!$ contains a power of p, p^γ say. Thus $\pi_{p^\alpha,\mu+1} = p^\gamma \pi_{p^\alpha,\mu}$.

We now examine the powers of b_1 in the Fibonacci orbit of $D^{\mu+1}_{2p}$ and letting $l = LEN(D^{\mu}_{2p})$, $3 \le n \le 2\mu+2$ gives

$$z_{l+n} = \frac{2^{n-2} 2p\pi_{p,\mu}(2p\pi_{p,\mu}+1)\dots(2p\pi_{p,\mu}+n-3)}{(n-2)!}.$$

For $3 \le n \le 2\mu, p | z_{l+n}$ but for either $n = 2\mu+1$ or $2\mu+2$ we introduce a factor of $(2\mu-1)(2\mu)$ in the denominator, as in the $D^{\mu+1}_{2p^\alpha}$ case, which contains p^γ. So $\pi_{p,\mu+1} = p^\gamma \pi_{p,\mu}$.

By the inductive hypothesis we know that $\pi_{p^\alpha,\mu} = \pi_{p,\mu}$. Thus $\pi_{p,\mu+1} = p^\gamma \pi_{p,\mu} = p^\gamma \pi_{p^\alpha,\mu} = \pi_{p^\alpha,\mu+1}$.

Case 3. $\pi_{p,\mu+1} > \pi_{p,\mu}$

Using an analogous argument as the above we see that in this case $\pi_{p^\alpha,\mu+1} = \pi_{p,\mu+1}$.

The second statement of the lemma holds since if we examine the powers of b_1 in the Fibonacci orbit of $D^i_{2(2^\beta)}$ we have, for $l = MinLEN(D^i_{2(2^\beta)}) = 2(2i+1)(2^\beta/(4,2^\beta)), i \geq 2, 3 \leq n \leq 2i$,

$$z_{l+n} = \frac{2^{n-2}2(2^\beta/(4,2^\beta))(2(2^\beta/(4,2^\beta))+1)\dots(2(2^\beta/(4,2^\beta))+n-3)}{(n-2)!}.$$

Now this is always divisible by 2^β. □

Definition: For $i \geq 2$, let $m = 2^{\alpha_0}p_1^{\alpha_1}\dots p_k^{\alpha_k}$ where p_l are distinct odd primes, and let $t_l = \lfloor \log_{p_l}(2i-3) \rfloor$. Define $\Phi_{m,i} = p_1^{t_1}\dots p_k^{t_k}$, and for all $\alpha \geq 1, \Phi_{2^\alpha,i} = 1$.

We note here that $\Phi_{m,i}$ is obviously multiplicative in that for $(m,n) = 1, \Phi_{mn,i} = \Phi_{m,i}\Phi_{n,i}$.

Lemma 3.5: *For all primes p, for* $\alpha \in \mathbb{N}$ *and* $i \geq 2, \Phi_{p^\alpha,i} = \Phi_{p,i}$.

Proof: The definition of $\Phi_{m,i}$ is equivalent to finding for $m = 2^{\alpha_0}p_1^{\alpha_1}\dots p_k^{\alpha_k}$, where p_l are distinct odd primes, the largest natural number t_l such that $(p_l^{t_l}+3)/2 \leq i$ and letting $\Phi_{m,i} = \prod_{p_l|m} p_l^{t_l}$. Since the power of a prime in the prime decomposition of m is not used in computing $\Phi_{m,i}$, the lemma holds. □

Proposition 3.6: *For* $D^i_{2p}, i \geq 2$ *and p a prime,* $\pi_{p,i} = \Phi_{p,i}$.

Proof: Let p be a fixed odd prime. We use induction on i, so $\pi_{p,k} = \Phi_{p,k}$ for $k < i$.

- $i = 2$.

It follows by Lemma 3.4 that $\pi_{p,2} = 1$ and by definition that $\Phi_{p,2} = 1$.

We now prove the result for i. First we note that $\pi_{p,i}$ forms a monotonic increasing sequence by Lemma 3.2. There are two cases to consider:

1. $\pi_{p,i-1} = \pi_{p,i}$;
2. $\pi_{p,i-1} < \pi_{p,i}$.

- Case 1, $\pi_{p,i-1} = \pi_{p,i}$.

There are two possibilities for the value of $\Phi_{p,i}$; either $\Phi_{p,i-1} = \Phi_{p,i}$ or $\Phi_{p,i-1} < \Phi_{p,i}$. We will assume the latter case and reach a contradiction and so obtain $\Phi_{p,i} = \Phi_{p,i-1} = \pi_{p,i-1} = \pi_{p,i}$, the desired result.

Assume that $\Phi_{p,i} > \Phi_{p,i-1}$. If $\Phi_{p,i-1} = p^c$, where $c = \lfloor \log_p(2(i-1)-3) \rfloor$, then $\Phi_{p,i} = p^{c+1}$, where $c+1 = \lfloor \log_p(2i-3) \rfloor$. We know that $\pi_{p,i-1} = \pi_{p,i}$ so, using Proposition 2.5 and Lemma 3.1, the prime p divides

$$z_{l+(2i-1)} = \frac{2^{2i-3}\bar{l}(\bar{l}+1)\dots(\bar{l}+(2i-1)-3)}{((2i-1)-2)!},$$

where $\bar{l} = LEN(D^i_{2p})/(2i+1) = 2p\pi_{p,i}/(4,p)$. But $p^c = \pi_{p,i-1}$, so $\bar{l} = 2p.p^c = 2p^{c+1}$ giving

$$z_{l+(2i-1)} = \frac{2^{2i-3}(2p^{c+1})(2p^{c+1}+1)\dots(2p^{c+1}+2i-4)}{(2i-3)!}$$

$$= 2^{2i-3}\binom{2p^{c+1}}{2i-3}\binom{2p^{c+1}+1}{1}\binom{2p^{c+1}+2}{2}\dots\binom{2p^{c+1}+2i-4}{2i-4}.$$

Note that the bth bracketed term, $2 \le b \le 2i-3$, in the above product may be written as $\left(\frac{2p^{c+1}+p^i m_b}{p^i m_b}\right)$, where $(m_b, p) = 1$ and $0 \le i \le c+1$ (this last inequality follows from the definition of c). Carrying out the obvious simplification we obtain $\left(\frac{2p^{c+1-i}+m_b}{m_b}\right)$. Now notice that $\left(\prod_{b=2}^{2i-3} m_b\right)$ is coprime to p. Recall that $p | z_{l+(2i-1)}$ and, by assumption, $\lfloor \log_p(2i-3) \rfloor = c+1$ or $2i-3 = p^{c+1}$. From this last equation it follows that $p \not| z_{l+(2i-1)}$, a contradiction. Hence $\Phi_{p,i} = p^c = \Phi_{p,i-1} = \pi_{p,i-1} = \pi_{p,i}$.

- Case 2, $\pi_{p,i-1} < \pi_{p,i}$.

Let $l = 2p\pi_{p,i-1}$. So by Lemma 3.1 we have $l = 2p^k$ for some integer k. Using Proposition 2.5 for $3 \le n \le 2i$ gives

$$\begin{aligned} z_{l+n} &= \frac{2^{n-2} 2p^k (2p^k+1) \dots (2p^k+n-3)}{(n-2)!} \\ &= 2^{n-2} \binom{2p^k}{n-2} \binom{2p^k+1}{1} \binom{2p^k+2}{2} \cdots \binom{2p^k+n-3}{n-3}. \end{aligned}$$

Since $\pi_{p,i-1} < \pi_{p,i}$ we must have $p \not| z_{l+2i-1}$ or $p \not| z_{l+2i}$. Using an argument similar to that given above we see that $p^k = (2i-1)-2$ or $p^k = (2i)-2$. Hence $k = \lfloor \log_p(2i-3) \rfloor$. Finally by Lemma 3.1 we have $\pi_{p,i} = p^{k-1}.p = p^k$, and so $p^{k-1} = \pi_{p,i-1} = \Phi_{p,i-1}$ and $p^k = \Phi_{p,i} = \pi_{p,i}$. □

Theorem 3.7: *For $i \ge 2$, $LEN(D^i_{2m}) = \Phi_{m,i} MinLEN(D^i_{2m})$.*

Proof: The theorem follows from Lemmas 3.3, 3.4, 3.5 and Proposition 3.6. □

Corollary 3.8:

$$LEN(D^2_{2m}) = 10m/(4, m)$$

Proof: In this case, for $m = 2^{\alpha_0} p_1^{\alpha_1} \dots p_k^{\alpha_k}$ where p_l are distinct odd primes, we have $t_l = \lfloor \log_{p_l}(1) \rfloor = 0$ giving $\Phi_{m,2} = 1$. So by Theorem 3.7 we have $LEN(D^2_{2m}) = MinLEN(D^2_{2m}) = 10m/(4, m)$. □

Corollary 3.9:

$$LEN(D^3_{2m}) = 14m\Phi_{m,3}/(4, m)$$

where $\Phi_{m,3} = 3$ if $m \equiv 0 \mod 3$, and $\Phi_{m,3} = 1$ otherwise.

Proof: Using the same notation as in the proof of the previous corollary we obtain $t_l = \lfloor \log_{p_l}(3) \rfloor$ which is equal to 1 if $p_l = 3$ and 0 otherwise. □

4. OTHER DIHEDRAL GROUP GENERATORS

We now examine the case of other presentations of D_{2m}^2 when m is odd.

Theorem 4.1: *For m odd, D_{2m}^2 is defined by the presentation*

$$\langle x, y | x^{2m} = 1, (x^m y)^2 = 1, y^{2m} = (xy)^2 \rangle.$$

In this case $LEN_{\{x,y\}}(D_{2m}^2) = 6$.

Proof: That the presentation defines D_{2m}^2 is shown in [4]. We first show that the following relations hold in the presentation for D_{2m}^2, m odd:

$$xy^2xy^2xyxy^2xy = x$$

and

$$yxyxy^2xyxy^2xy^2xyxy^2xy = y.$$

In [4] it is shown that the following relations also hold in the presentation for D_{2m}^2:

$$(i)\ y^{2m} = 1,\ (ii)\ (xy)^2 = 1,\ (iii)\ xy^{m-1}x^{-1} = y^{-m+1},\ (iv)\ y^{-m}xy^m = x^{-1}.$$

Now

$$1 = xy^{m-1}x^{-1}xy^{m+1}x^{-1} \text{ by (i)}$$

and

$$1 = \underline{xy^{m-1}x^{-1}}xy^{m+1}x^{-1} = y^{-m+1}xy^{m+1}x^{-1} \text{ by (iii)}.$$

We now have

$$1 = y\underline{y^{-m}xy^m}yx^{-1} = yx^{-1}yx^{-1} \text{ by (iv)}.$$

Now using (ii) we see that

$$1 = yx^{-1}y\underline{x^{-1}} = yx^{-1}y^2xy = y\underline{x^{-1}}yxyxy^2xy = y^2xy^2xyxy^2xy.$$

Thus $xy^2xy^2xyxy^2xy = x$ in D_{2m}^2, m odd.

The second result follows since

$$yxyxy^2\underline{xyxy^2xy^2xyxy^2xy} = yxyxyyxyx = yyxyx = y.$$

We now recall the well-known result that every finite 2-generator group $G = \langle a, b \rangle$ is a homomorphic image of some Fibonacci group $F(2, n)$, where $n = LEN_{\{a,b\}}(G)$. In the preceding paragraph we have shown that $LEN(D_{2m}^2) \leq 6$ for all m, where D_{2m}^2 is defined by the presentation

$$\langle x, y \mid x^{2m} = 1, (x^m y)^2 = 1, y^{2m} = (xy)^2 \rangle,$$

m odd; the proof is complete on noting that $6 = \min\{n : |F(2, n)| \text{ is infinite }\}$. □

Relating this result to that of Corollary 3.8 we see that the generating set for D_{2m}^2, m odd, used in this section always gives a constant Fibonacci length of 6 while the generating set used in the previous section gives a Fibonacci length of $10m/(4, m)$, which is always greater than 6.

Note: The presentation given in Theorem 4.1 for D_{2m}^2, m odd, is efficient. For odd values of m, D_{2m}^3 has an efficient presentation

$$\langle a, x, z \mid x^{2m} = z^{2m} = (xz^m)^{2m} = a^2 = [x, a] = [a, z^m] = [x, z^{m-1}] = 1, \\ (x^{m-1}z^m)^2 = (az^{m-1})^2 = 1\rangle.$$

In this case our computations, using the GAP computer algebra package [8], imply that the Fibonacci length is equal to $8m$. This remains as a conjecture.

ACKNOWLEDGMENTS

The third author wishes to thank the University for Teacher Education of Iran and the University of St. Andrews for their support while this research was being undertaken.

REFERENCES

[1] Aydin, H. and Smith, G.C. "Finite p-quotients of Some Cyclically Presented Groups." *J. London Math. Soc.*, Vol. *49* (1994): pp. 83-92.

[2] Ayik, H., Campbell, C.M., O'Connor, J.J. and Ruškuc, N. "The Semigroup Efficiency of Direct Powers of Groups." Proc. Internat. Conf. on Semigroups. Edited by P. Smith, E. Giraldes and P. Martins. World Scientific, Singapore, 2000, 19-25.

[3] Campbell, C.M., Doostie, H. and Robertson, E.F. "Fibonacci Lengh of Generating Pairs in Groups." Applications of Fibonacci Numbers. Volume 3. Edited by G.A. Bergum, et. al., Kluwer Academic Press, Dordrecht, 1990, 27-35.

[4] Campbell, C.M., Robertson, E.F. and Williams, P.D. "On the Efficiency of Some Direct Powers of Groups." Groups - Canberra 1989. Edited by L.G. Kovács, Lecture Notes in Math. Volume *1456* Springer, Berlin, 1990, 106-113.

[5] Doostie, H. "Fibonacci-Type Sequences and Classes of Groups." (Ph.D. Thesis). University of St. Andrews, Scotland, (1988).

[6] Doostie, H. and Campbell, C.M. "Fibonacci Length of Automorphism Groups Involving Tribonacci Numbers." *Vietnam J. Math.*, Vol. *28* (2000): pp. 57-65.

[7] Doostie, H. and Golamie, R. "Computing on the Fibonacci Lengths of Finite Groups." *Int. J. Appl. Math.*, Vol. *4* (2000): pp. 149-156.

[8] GAP, *GAP-Groups, Algorithms, and Programming, Version 4.3*, Aachen, St. Andrews, 2002. (http://www-gap.dcs.st-andrews.ac.uk/~gap)

[9] Johnson, D.L. Presentations of Groups. 2nd edition. London Math. Soc. Student Texts. Volume *15*, Cambridge University Press, Cambridge 1997.

[10] Lennox, J.C. and Wiegold, J. "Generators and Killers for Direct and Free Products." *Arch. Math. (Basel)*, Vol. *34* (1980): pp. 296-300.

[11] Sims, C.C. Computation with Finitely Presented Groups. Encyclopedia of Mathematics and its Applications, Volume 48. Cambridge University Press, Cambridge 1994.

[12] Stewart, A.G.R. and Wiegold, J. "Growth Sequences of Finitely Generated Groups II." *Bull. Austral. Math. Soc*, Vol. *40* (1989): pp. 323-329.

[13] Thomas, R.M. "The Fibonacci Groups Revisited." *Groups - St. Andrews 1989*. Edited by C.M. Campbell and E.F. Robertson. Cambridge University Press, Cambridge, 1991, 445-454.

[14] Wall, D.D. "Fibonacci Series modulo m." *Amer. Math. Monthly*, Vol. *67* (1960): pp. 525-532.

[15] Wiegold, J. "The Schur Multiplier: an Elementary Approach." *Groups - St. Andrews 1981*. Edited by C.M. Campbell and E.F. Robertson, Cambridge University Press, Cambridge, 1982, 137-154.

[16] Wilcox, H.J. "Fibonacci Sequences of Period n in Groups." *The Fibonacci Quarterly*, Vol. *24* (1986): pp. 356-361.

AMS Classification Numbers: 20D60

SOME SUMS RELATED TO SUMS OF ORESME NUMBERS

Charles K. Cook

1. INTRODUCTION

In [1] A. F. Horadam presented a history of number attributed to Nicole Oresme, namely the sequence

$$\{O_n\} = \left\{\frac{n}{2^n}\right\} = \left\{\frac{1}{2}, \frac{2}{4}, \frac{3}{8}, \ldots, \frac{n}{2^n}, \ldots\right\}. \tag{1.1}$$

As with all of Horadam's papers, an abundance of properties of these numbers was obtained. Those generalized in this paper are referenced here.

$$O_{n+2} = O_{n+1} - \frac{1}{4}O_n, O_0 = 0, O_1 = \frac{1}{2} \tag{1.2}$$

$$O_{n+2} - \frac{3}{4}O_n + \frac{1}{4}O_{n-1} = 0 \tag{1.3}$$

$$O_{n+2} - \frac{3}{4}O_{n+1} + \frac{1}{16}O_{n-1} = 0 \tag{1.4}$$

$$\sum_{j=0}^{n-1} O_j = 4\left(\frac{1}{2} - O_{n+1}\right) \tag{1.5}$$

This paper is in final form and no version of it will be submitted for publication elsewhere.

F.T. Howard (ed.), Applications of Fibonacci Numbers, Volume 9: Proceedings of the Tenth International Research Conference on Fibonacci Numbers and their Applications, 87–99.

$$\sum_{j+0}^{\infty} O_j = 2 \tag{1.6}$$

$$\sum_{j=0}^{n-1} (-1)^j O_j = \frac{4}{9}\left[-\frac{1}{2} + (-1)^n (O_{n+1} - 2O_n)\right] \tag{1.7}$$

$$\sum_{j=0}^{n-1} O_{2j} = \frac{4}{9}\left[2 + O_{2n-1} - 5O_{2n}\right] \tag{1.8}$$

$$\sum_{j=0}^{n-1} O_{2j+1} = \frac{1}{9}\left[10 + 5O_{2n-1} - 16O_{2n}\right] \tag{1.9}$$

$$O_{n+1}O_{n-1} - O_n^2 = -\left(\frac{1}{4}\right)^n \tag{1.10}$$

and the generating function

$$\sum_{n=0}^{\infty} O_n x^n = \sum_{n=0}^{\infty} \frac{n}{2^n} x^n = \frac{2x}{(2-x)^2}. \tag{1.11}$$

2. ORESME, FIBONACCI, LUCAS AND PELL SUMS

Variations of (1.11) can be obtained from derivatives of the generating function for the sum of the geometric series. The following converge for $|x| < k$.

$$\sum_{n=0}^{\infty} \frac{x^n}{k^n} = \frac{k}{k-x} \tag{2.1}$$

$$\sum_{n=0}^{\infty} \frac{nx^n}{k^n} = \frac{kx}{(k-x)^2} \tag{2.2}$$

$$\sum_{n=0}^{\infty} \frac{n^2x^n}{k^n} = \frac{kx(k+x)}{(k-x)^3} \tag{2.3}$$

$$\sum_{n=0}^{\infty} \frac{n^3 x^n}{k^n} = \frac{kx(k^2+4kx+x^2)}{(k-x)^4} \tag{2.4}$$

$$\sum_{n=0}^{\infty} \frac{n^4 x^n}{k^n} = \frac{kx(k^3+11k^2x+11kx^2+x^3)}{(k-x)^5} \tag{2.5}$$

...

Proof of (2.4):

Let $f_0(x) = \frac{1}{1-x} = \sum x^n$. Then $f_0'(x) = \frac{1}{(1-x)^2}$.

Let $f_1(x) = \frac{x}{(1-x)^2} = \sum nx^n$. Then $f_1'(x) = \frac{1+x}{(1-x)^3}$.

Let $f_2(x) = \frac{x(1+x)}{(1-x)^3} = \sum n^2x^n$. Then $f_2'(x) = \frac{1+4x+x^2}{(1-x)^4}$.

Hence $\sum n^3x^n = \frac{x(1+4x+x^2)}{(1-x)^4}$ so that $\sum n^3\left(\frac{x^n}{k^n}\right) = \frac{kx(k^2+4kx+x^2)}{(k-x)^4}$. □

(2.1) to (2.5) form sequences of sums $S_m(n^m, k, x)$. The special case with $k = 2$ and $x = 1$ yields a sequence of Oresme sums

$$\{S_m(m,2,1)\} = \{s_o, s_1, s_2, s_3, s_4, \dots\} = \left\{\sum_{n=0}^{\infty} \frac{n^m}{2^n}\right\} = \left\{\sum_{n=0}^{\infty} n^{m-1} O_n\right\}$$

$$= \{2, 2, 6, 26, 150, \dots\}. \tag{2.6}$$

By using the generating functions (2.1)-(2.5) analogous sums and sequences of sums can be found for the Fibonacci, Lucas, and Pell numbers.

Thus for the Fibonacci numbers, F_n, since $|x| = |\frac{1\pm\sqrt{5}}{2}| < 2$ the following converge if $k \geq 2$.

$$\sum_{n=0}^{\infty} \frac{F_n}{k^n} = \frac{k}{k^2-k-1} \tag{2.7}$$

$$\sum_{n=0}^{\infty} \frac{nF_n}{k^n} = \frac{k(k^2+1)}{(k^2-k-1)^2} \tag{2.8}$$

$$\sum_{n=0}^{\infty} \frac{n^2F_n}{k^n} = \frac{k(k^4+k^3+6k^2-k+1)}{(k^2-k-1)^3} \tag{2.9}$$

$$\sum_{n=0}^{\infty}\frac{n^3F_n}{k^n}=\frac{k(k^6+4k^5+24k^4+24k^2-4k+1)}{(k^2-k-1)^4} \tag{2.10}$$

$$\sum_{n=0}^{\infty}\frac{n^4F_n}{k^n}=\frac{k(k^8+11k^7+87k^6+48k^5+240k^4-48k^3+87k^2-11k+1)}{(k^2-k-1)^5} \tag{2.11}$$

$$\cdots$$

Proof of (2.10):

$$\sum\frac{n^3F_n}{k^n}=\frac{1}{\sqrt{5}}\sum\left[\left(\frac{\alpha}{k}\right)^n-\left(\frac{\beta}{k}\right)^n\right]=\frac{1}{\sqrt{5}}\left[\frac{k\alpha(k^2+4\alpha k+\alpha^2)}{(k-\alpha)^4}-\frac{k\beta(k^2+4\beta k+\beta^2)}{(k-\beta)^4}\right]$$

$$=\frac{k}{\sqrt{5}(k^2-k-1)^4}\left[\alpha(k^2+4\alpha k+\alpha^2)(k-\beta)^4-\beta(k^2+4\beta k+\beta^2)(k-\alpha)^4\right]$$

The bracketed expression in the numerator becomes

$$(\alpha-\beta)k^6+4(\alpha^2-\beta^2)k^5+[6(\alpha-\beta)+16(\alpha-\beta)+\alpha^3-\beta^3]k^4+$$
$$[\alpha^3-\beta^3+16(\alpha-\beta)+6(\alpha-\beta]k^2+4(\beta^2-\alpha^2)k+\alpha-\beta.$$

Thus the sum simplifies to

$$\sum\frac{n^3F_n}{k^n}=\frac{k(k^6+4k^5+24k^4+24k^2-4k+1)}{(k^2-k-1)^4}. \quad \square$$

(2.7) to (2.11) form sequences $\{F_k^{(m)}(n^m,k,F_n)\}$, $m\geq 0$, $k\geq 2$ from which the following sequences of sums are obtained.

$$F_k^{(0)}=\{fs_2^{(0)},fs_3^{(0)},fs_4^{(0)},\ldots\}$$
$$=\left\{\frac{2}{1},\frac{3}{5},\frac{4}{11},\frac{5}{19},\frac{6}{29},\frac{7}{41},\frac{8}{55},\frac{9}{71},\frac{10}{89},\cdots\right\} \tag{2.12}$$

$$F_k^{(1)}=\{fs_2^{(1)},fs_3^{(1)},fs_4^{(1)},\ldots\}$$
$$=\left\{\frac{10}{1},\frac{30}{5^2},\frac{68}{11^2},\frac{130}{19^2},\frac{222}{29^2},\frac{350}{41^2},\frac{520}{55^2},\frac{738}{71^2},\frac{1010}{89^2},\cdots\right\} \tag{2.13}$$

$$F_k^{(2)} = \{fs_2^{(2)}, fs_3^{(2)}, fs_4^{(2)}, \ldots\}$$

$$= \left\{\frac{94}{1}, \frac{480}{5^3}, \frac{1652}{11^3}, \frac{4480}{19^3}, \frac{10338}{29^3}, \frac{21224}{41^3}, \frac{39880}{55^3}, \frac{69912}{71^3}, \frac{115910}{89^3}, \ldots\right\} \tag{2.14}$$

$$F_k^{(3)} = \{fs_2^{(3)}, fs_3^{(3)}, fs_4^{(3)}, \ldots\}$$

$$= \left\{\frac{1330}{1}, \frac{11550}{5^4}, \frac{58820}{11^4}, \frac{218530}{19^4}, \frac{658230}{29^4}, \frac{1705550}{41^4}, \frac{3944200}{55^4}, \frac{8343090}{71^4}, \frac{16423610}{89^4}, \ldots\right\} \tag{2.15}$$

$$F_k^{(4)} = \{fs_2^{(4)}, fs_3^{(4)}, fs_4^{(4)}, \ldots\}$$

$$= \left\{\frac{25102}{1}, \frac{373800}{5^5}, \frac{2843924}{11^5}, \frac{14527480}{19^5}, \frac{56969826}{29^5}, \frac{185009552}{41^5}, \frac{521513800}{55^5}, \frac{1316481264}{71^5}, \frac{3041605910}{89^5}, \ldots\right\}. \tag{2.16}$$

$$\cdots$$

Similarly for the Lucas numbers, L_n, since again, $|x| = |\frac{1\pm\sqrt{5}}{2}| < 2$ the following converge if $k \geq 2$.

$$\sum_{n=0}^{\infty} \frac{L_n}{k^n} = \frac{k(2k-1)}{k^2-k-1} \tag{2.17}$$

$$\sum_{n=0}^{\infty} \frac{nL_n}{k^n} = \frac{k(k^2+4k-1)}{(k^2-k-1)^2} \tag{2.18}$$

$$\sum_{n=0}^{\infty} \frac{n^2 L_n}{k^n} = \frac{k(k^4+9k^3+9k-1)}{(k^2-k-1)^3} \tag{2.19}$$

$$\sum_{n=0}^{\infty} \frac{n^3 L_n}{k^n} = \frac{k(k^6+20k^5+14k^4+72k^3-14k^2+20k-1)}{(k^2-k-1)^4} \tag{2.20}$$

$$\sum_{n=0}^{\infty}\frac{n^4L_n}{k^n}=\frac{k(k^8+43k^7+89k^6+422k^5+422k^3-89k^2+43k-1)}{(k^2-k-1)^5} \tag{2.21}$$

$$\cdots$$

Proof of (2.20):

$$\sum\frac{n^3L_n}{k^n}=\sum\left[\left(\frac{\alpha}{k}\right)^n+\left(\frac{\beta}{k}\right)^n\right]=\left[\frac{k\alpha(k^2+4\alpha k+\alpha^2)}{(k-\alpha)^4}+\frac{k\beta(k^2+4\beta k+\beta^2)}{(k-\beta)^4}\right]$$

$$=\frac{k}{(k^2-k-1)^4}\left[\alpha(k^2+4\alpha k+\alpha^2)(k-\beta)^4+\beta(k^2+4\beta k+\beta^2)(k-\alpha)^4\right]$$

The bracketed expression in the numerator becomes

$$(\alpha+\beta)k^6+[8+4(\alpha^2+\beta^2)]k^5+[-6(\alpha+\beta)+16(\alpha+\beta)+\alpha^3+\beta^3]k^4+$$
$$[4(\alpha^2+\beta^2)+48+4(\alpha^2+\beta^2)]k^3+[-(\alpha^3+\beta^3)-16(\alpha+\beta)+6(\alpha+\beta)]$$
$$k^2+[4(\beta^2+\alpha^2)+8]k-(\alpha+\beta)].$$

Thus the sum simplifies to

$$\sum\frac{n^3L_n}{k^n}=\frac{k(k^6+20k^5+14k^4+72k^3-14k^2+20k-1)}{(k^2-k-1)^4}. \quad \square$$

(2.17) to (2.21) form sequences $\{L_k^{(m)}(n^m,k,L_n)\}$, $m\geq 0$, $k\geq 2$ from which the following sequences of sums are obtained.

$$L_k^{(0)}=\{1s_2^{(0)},1s_3^{(0)},1s_4^{(0)},\ldots\}$$

$$=\left\{\frac{6}{1},\frac{15}{5},\frac{28}{11},\frac{45}{19},\frac{66}{29},\frac{91}{41},\frac{120}{55},\frac{153}{71},\frac{190}{89},\ldots\right\} \tag{2.22}$$

$$L_k^{(1)}=\{1s_2^{(1)},1s_3^{(1)},1s_4^{(1)},\ldots\}$$

$$=\left\{\frac{22}{1},\frac{60}{5^2},\frac{124}{11^2},\frac{220}{19^2},\frac{354}{29^2},\frac{532}{41^2},\frac{760}{55^2},\frac{1044}{71^2},\frac{1390}{89^2},\ldots\right\} \tag{2.23}$$

$$L_k^{(2)} = \{1s_2^{(2)}, 1s_3^{(2)}, 1s_4^{(2)}, \ldots\}$$

$$= \left\{\frac{210}{1}, \frac{1050}{5^3}, \frac{3468}{11^3}, \frac{8970}{19^3}, \frac{19758}{29^3}, \frac{38850}{41^3}, \frac{70200}{55^3}, \frac{118818}{71^3}, \frac{190890}{89^3}, \ldots\right\} \tag{2.24}$$

$$L_k^{(3)} = \{1s_2^{(3)}, 1s_3^{(3)}, 1s_4^{(3)}, \ldots\}$$

$$= \left\{\frac{2974}{1}, \frac{25800}{5^4}, \frac{130492}{11^4}, \frac{478120}{19^4}, \frac{1412922}{29^4}, \frac{3580864}{41^4}, \frac{8087800}{55^4}, \frac{16702272}{71^4}, \frac{32107990}{89^4}, \ldots\right\} \tag{2.25}$$

$$L_k^{(4)} = \{1s_2^{(4)}, 1s_3^{(4)}, 1s_4^{(4)}, \ldots\}$$

$$= \left\{\frac{56130}{1}, \frac{836250}{5^5}, \frac{6369900}{11^5}, \frac{32550570}{19^5}, \frac{127433550}{29^5}, \frac{412168050}{41^5}, \frac{1154595000}{55^5}, \frac{2891089170}{71^5}, \frac{6616135290}{89^5}, \ldots\right\}. \tag{2.26}$$

$\ldots$

Finally, for the Pell numbers, P_n (See [2]) since $|1 \pm \sqrt{2}| < 3$, the following converge for $k \geq 3$.

$$\sum_{n=0}^{\infty} \frac{P_n}{k^n} = \frac{k}{(k^2 - 2k - 1)} \tag{2.27}$$

$$\sum_{n=0}^{\infty} \frac{nP_n}{k^n} = \frac{k(k^2 + 1)}{(k^2 - 2k - 1)^2} \tag{2.28}$$

$$\sum_{n=0}^{\infty} \frac{n^2 P_n}{k^n} = \frac{k(k^4 + 2k^3 + 6k^2 - 2k + 1)}{(k^2 - 2k - 1)^3} \tag{2.29}$$

$$\sum_{n=0}^{\infty} \frac{n^3 P_n}{k^n} = \frac{k(k^6 + 8k^5 + 27k^4 + 27k^2 - 8k + 1)}{(k^2 - 2k - 1)^4} \tag{2.30}$$

$$\sum_{n=0}^{\infty} \frac{n^4 P_n}{k^n} = \frac{k(k^8 + 22k^7 + 120k^6 + 102k^5 + 270k^4 - 102k^3 + 120k^2 - 22k + 1)}{(k^2 - 2k - 1)^5} \tag{2.31}$$

$\ldots$

Proof of (2.30):

Note here that $\alpha = 1+\sqrt{2}$, and $\beta = 1-\sqrt{2}$, so that $\alpha-\beta = 2\sqrt{2}$, $\alpha^2-\beta^2 = 4\sqrt{2}$ and $\alpha^3-\beta^3 = 10\sqrt{2}$.

$$\sum \frac{n^3 P_n}{k^n} = \frac{1}{2\sqrt{2}} \sum \left[\left(\frac{\alpha}{k}\right)^n - \left(\frac{\beta}{k}\right)^n \right] = \frac{1}{2\sqrt{2}} \left[\frac{k\alpha(k^2+4\alpha k+\alpha^2)}{(k-\alpha)^4} - \frac{k\beta(k^2+4\beta k+\beta^2)}{(k-\beta)^4} \right]$$

$$= \frac{k}{2\sqrt{2}(k^2-k-1)^4} \left[\alpha(k^2+4\alpha k+\alpha^2)(k-\beta)^4 - \beta(k^2+4\beta k+\beta^2)(k-\alpha)^4\right]$$

The bracketed expression in the numerator becomes

$$(\alpha-\beta)k^6 + 4(\alpha^2-\beta^2)k^5 + [6(\alpha-\beta)+16(\alpha-\beta)+\alpha^3-\beta^3]k^4 +$$
$$[\alpha^3-\beta^3+16(\alpha-\beta)+6(\alpha-\beta)]k^2 + 4(\beta^2-\alpha^2)k + \alpha - \beta.$$

Thus the sum simplifies to

$$\sum \frac{n^3 P_n}{k^n} = \frac{k(k^6+8k^5+27k^4+27k^2-8k+1)}{(k^2-k-1)^4}. \quad \square$$

(2.27) to (2.31) form sequences $\{P_k^{(m)}(n^m, k, P_n)\}$, $m \geq 0$, $k \geq 3$ from which the following sequences of sums are obtained.

$$P_k^{(0)} = \{ps_3^{(0)}, ps_4^{(0)}, ps_5^{(0)}, \ldots\}$$

$$= \left\{\frac{3}{2}, \frac{4}{7}, \frac{5}{14}, \frac{6}{23}, \frac{7}{34}, \frac{8}{47}, \frac{9}{62}, \frac{10}{79}, \ldots\right\} \tag{2.32}$$

$$P_k^{(1)} = \{ps_3^{(1)}, ps_4^{(1)}, ps_5^{(1)}, \ldots\}$$

$$= \left\{\frac{30}{2^2}, \frac{68}{7^2}, \frac{130}{14^2}, \frac{222}{23^2}, \frac{350}{34^2}, \frac{520}{47^2}, \frac{738}{62^2}, \frac{1010}{79^2}, \ldots\right\} \tag{2.33}$$

$$P_k^{(2)} = \{ps_3^{(2)}, ps_4^{(2)}, ps_5^{(2)}, \ldots\}$$

$$= \left\{\frac{552}{2^3}, \frac{1892}{7^3}, \frac{5080}{14^3}, \frac{11598}{23^3}, \frac{23576}{34^3}, \frac{43912}{47^3}, \frac{76392}{62^3}, \frac{125810}{79^3}, \ldots\right\} \tag{2.34}$$

$$P_k^{(3)} = \{ps_3^{(3)}, ps_4^{(3)}, ps_5^{(3)}, \dots\}$$

$$= \left\{\frac{15240}{2^4}, \frac{78404}{7^4}, \frac{290680}{14^4}, \frac{868686}{23^4}, \frac{2227400}{34^4}, \frac{5092360}{47^4}, \frac{10647864}{62^4}, \frac{20726210}{79^4}, \dots\right\} \tag{2.35}$$

$$P_k^{(4)} = \{ps_3^{(4)}, ps_4^{(4)}, ps_5^{(4)}, \dots\}$$

$$= \left\{\frac{561216}{2^5}, \frac{4345508}{7^5}, \frac{22310080}{14^5}, \frac{87372942}{23^5}, \frac{282337664}{34^5}, \frac{790203016}{47^5}, \frac{1977971328}{62^5}, \frac{4528097810}{79^5}, \dots\right\}. \tag{2.36}$$

...

3. K-ORESME ORESME NUMBERS

Generalizations can be obtained in more than one way. For example, if $A_n^{(k)} = \frac{n}{k^n}$, then it follows that

$$A_{n+2}^{(k)} = A_{n+1}^{(k)} - \frac{k-1}{k^2} A_n^{(k)} - \frac{k-2}{k^{n+2}} \text{ for } k \geq 2. \tag{3.1}$$

Investigating the solutions of this non-homogeneous equation will not be explored in this paper. Here the k-Oresme numbers are defined analogous to (1.2) as the solutions to the homogeneous equation

$$A_{n+2}^{(k)} = A_{n+1}^{(k)} - \frac{1}{k^2} A_n^{(k)} \text{ where } A_0^{(k)} = 0, \text{ and } A_1^{(k)} = \frac{1}{k} \text{ for } k \geq 3. \tag{3.2}$$

Solutions are found to be

$$A_n^{(k)} = \frac{1}{k^n\sqrt{k^2-4}}\left[\left(\frac{k+\sqrt{k^2-4}}{2}\right)^n - \left(\frac{k-\sqrt{k^2-4}}{2}\right)^n\right] = \frac{\alpha^n - \beta^n}{k^n\sqrt{k^2-4}}. \tag{3.3}$$

The numerators form sequences all beginning with $a_0 = 0$ and $a_1 = 1$. Additional members are shown for $k = 3$ through $k = 13$ in Table 1. Note that the sequence for $k = 3$ yields

Fibonacci numbers, F_{2n} (See 5.1). It will be shown in section 5 that several of these sequences form familiar patterns after a modification of the initial conditions.

k \ n	2	3	4	5	6	7	8	9	...
3	3	8	21	55	144	377	987	2584	...
4	4	15	56	209	780	2911	10864	40545	...
5	5	24	115	551	2640	12649	60605	290376	...
6	6	35	204	1189	6930	40391	235416	1372105	...
7	7	48	329	2255	15456	105937	726103	4976784	...
8	8	63	496	3905	30744	242047	1905632	15003009	...
9	9	80	711	6319	56160	499121	4435929	39424240	...
10	10	99	980	9701	96030	950599	9409960	93149001	...
11	11	120	1309	14279	155760	1699081	18534131	202176360	...
12	12	143	1704	20305	241956	2883167	34356048	409389409	...
13	13	168	2171	28055	362544	4685017	60542677	782369784	...

Table 1 Numerator sequences for k-Oresme numbers

4. SOME PROPERTIES OF K-ORESME NUMBERS

It is a routine, though sometimes laborious, exercise to establish identities analogous to Horadam's given in section 1 as (1.3) - (1.10).

Theorem:

(a) $$A_{n+2}^{(k)} - \left(\frac{k^2-1}{k^2}\right) A_n^{(k)} + \frac{1}{k^2} A_{n-1}^{(k)} = 0 \tag{4.1}$$

(b) $$A_{n+2}^{(k)} - \left(\frac{k^2-1}{k^2}\right) A_{n+1}^{(k)} + \frac{1}{k^4} A_{n-1}^{(k)} = 0 \tag{4.2}$$

(c) $$\sum_{j=0}^{n-1} A_j^{(k)} = k^2 \left(\frac{1}{k} - A_{n+1}^{(k)}\right) \tag{4.3}$$

(d) $$\sum_{j=0}^{\infty} A_j^{(k)} = k \tag{4.4}$$

(e) $$\sum_{j=0}^{n-1} (-1)^j A_j^{(k)} = \frac{k^2}{2k^2+1} \left[-\frac{1}{k} + (-1)^n (A_{n+1}^{(k)} - 2A_n^{(k)})\right] \tag{4.5}$$

(f) $$\sum_{j=0}^{n-1} A_{2j}^{(k)} = \frac{k^2}{2k^2+1}\left[k + A_{2n-1}^{(k)} - (k^2+1)A_{2n}^{(k)}\right] \tag{4.6}$$

(g) $$\sum_{j=0}^{n-1} A_{2j+1}^{(k)} = \frac{k^2}{2k^2+1}\left[\frac{k^2+1}{k} + \frac{k^2+1}{k^2}(A_{2n-1}^{(k)} - k^2 A_{2n}^{(k)}\right] \tag{4.7}$$

(h) $$A_{n+1}^{(k)}A_{n-1}^{(k)} - (A_n^{(k)})^2 = -\left(\frac{1}{k^2}\right)^n . \tag{4.8}$$

<u>**Proof of (e):**</u> Using $\sum_{j=0}^{n-1}(-1)^j x^j = \frac{1+(-1)^{n-1}x^n}{1+x}$ and $A_n^{(k)} = \frac{1}{\sqrt{k^2-4}}\left[\left(\frac{\alpha}{k}\right)^n - \left(\frac{\beta}{k}\right)^n\right]$ and noting that $\alpha+\beta=k$ and $\alpha\cdot\beta=1$, we find that

$$\begin{aligned}
\sum_{j=0}^{n-1}(-1)^j A_j^{(k)} &= \frac{1}{\sqrt{k^2-4}}\left[\frac{1+(-1)^{n-1}\alpha^n}{k^n\left(1+\frac{\alpha}{k}\right)} - \frac{1+(-1)^{n-1}\beta^n}{k^n\left(1+\frac{\beta}{k}\right)}\right] \\
&= \frac{k}{k^n\sqrt{k^2-4}(2k^2+1)}\left[-\sqrt{k^2-4} + (-1)^n\{k(\beta^n-\alpha^n) + (\beta^{n-1}-\alpha^{n-1})\}\right] \\
&= \frac{k}{2k^2+1}\left[-1 + (-1)^n(-kA_n^{(k)} - \frac{1}{k}A_{n-1}^{(k)})\right] \\
&= \frac{k^2}{2k^2+1}\left[-\frac{1}{k} + (-1)^n\left(-A_n^{(k)} - \frac{1}{k^2}A_{n-1}^{(k)}\right)\right] \\
&= \frac{k^2}{2k^2+1}\left[-\frac{1}{k} + (-1)^n\left(A_{n+1}^{(k)} - 2A_n^{(k)}\right)\right]. \qquad \square
\end{aligned}$$

5. SOME VARIATIONS

Returning to the sequences of Table 1, we note that $k=3$ yielded a Fibonacci number pattern. Thus it is seen from (3.3) that

$$A_n^{(3)} = \frac{1}{3^n\sqrt{5}}\left[\left(\frac{3+\sqrt{5}}{2}\right)^n - \left(\frac{3-\sqrt{5}}{2}\right)^n\right] = \frac{F_{2n}}{3^n}. \tag{5.1}$$

If the initial conditions and/or the coefficients of $A_n^{(k)}$ in (3.1) are varied slightly, other familiar sequences are obtained.

These can be represented as generalized Horadam Fibonacci numbers

$$w_n\left(a_0, a_1; 1, \frac{\pm 1}{k^2}\right) \tag{5.2}$$

but the emphasis here is to display the difference equation solutions as variations of k-Oresme numbers. For example, it is routine but in some cases tedious to show that the following hold.

Theorem:

$$\text{(a)}\ \frac{F_{2n-1}}{3^n} \text{ satisfies } A_{n+2}^{(3)} = A_{n+1}^{(3)} - \frac{1}{9}A_n^{(3)};\ A_0^{(3)} = \frac{1}{1},\ A_1^{(3)} = \frac{1}{3} \tag{5.3}$$

$$\text{(b)}\ \frac{L_{3n-4}}{4^n} \text{ satisfies } A_{n+2}^{(4)} = A_{n+1}^{(4)} - \frac{1}{16}A_n^{(4)};\ A_0^{(4)} = \frac{7}{1},\ A_1^{(4)} = \frac{-1}{4} \tag{5.4}$$

$$\text{(c)}\ \frac{P_{2n-1}}{6^n} \text{ satisfies } A_{n+2}^{(6)} = A_{n+1}^{(6)} - \frac{1}{36}A_n^{(6)};\ A_0^{(6)} = \frac{1}{1},\ A_1^{(6)} = \frac{1}{6} \tag{5.5}$$

$$\text{(d)}\ \frac{F_{4n-1}}{7^{n+1}} \text{ satisfies } A_{n+2}^{(7)} = A_{n+1}^{(7)} - \frac{1}{49}A_n^{(7)};\ A_0^{(7)} = \frac{1}{7},\ A_1^{(7)} = \frac{2}{49} \tag{5.6}$$

$$\text{(e)}\ \frac{F_{5n-2}}{11^{n+1}} \text{ satisfies } A_{n+2}^{(11)} = A_{n+1}^{(11)} + \frac{1}{121}A_n^{(11)};\ A_0^{(11)} = \frac{-1}{11},\ A_1^{(11)} = \frac{2}{121}. \tag{5.7}$$

Proof of (c):

Using the recurrence relation for Pell numbers (See [2]) where needed, we have

$$A_{n+1}^{(6)} - \frac{1}{6^2}A_n^{(6)} = \frac{P_{2n+1}}{6^{n+1}} - \frac{P_{2n-1}}{6^{n+2}} = \frac{6P_{2n+1} - P_{2n-1}}{6^{n+2}} = \frac{5P_{2n+1} + 2P_{2n}}{6^{n+2}}$$

$$= \frac{P_{2n+1} + 2P_{2n+2}}{6^{n+2}} = \frac{P_{2n+3}}{6^{n+2}} = A_{n+2}^{(6)}. \quad \square$$

6. CONCLUSION

The interested reader may discover further properties analogous to those presented by Horadam in [1], and obtain sums involving higher powers of n, as well as sums involving other recurring sequences. Varying the initial conditions for the k-Oresme numbers difference equation may also yield some interesting sequences.

Finally note that additional historical and bibliographical information on Nicole Oresme and his work can be found on the ORESME Reading Group Web Page

⟨http://www.nku.edu/%7Ecurtin/oresme.html⟩. Or just search under Oresme for other interesting sites.

ACKNOWLEDGMENTS

The author wishes to thank Fred Howard and the anonymous referee for valuable comments and suggestions for improving the quality of this paper.

Support for this project was provided in part by The University of South Carolina Sumter and in part by The Sumter Partnership of the use Educational Foundation.

REFERENCES

[1] Horadam, A.F. "Oresme Numbers." *The Fibonacci Quarterly*, Vol. *12.3* (1974): pp. 267-271.

[2] Bicknell-Johnson, M. "A Primer on the Pell Sequence and Related Sequences." *The Fibonacci Quarterly*, Vol. *13.4* (1975): pp. 345-349.

AMS Classification Numbers: 11B37, 11B39, 11B60

SOME THOUGHTS ON ROOK POLYNOMIALS ON SQUARE CHESSBOARDS

Dan Fielder

1. INTRODUCTION

A rook polynomial is a polynomial whose x^k coefficient is the number of ways k rooks can be placed on the squares of an arbitrarily shaped chessboard so that no rooks share the same rows or columns. The k rooks are called *non-taking*. Rook polynomials pattern combinatorial situations, especially those involving restricted permutations. The conventional square board used in the game, chess, is but one configuration.

When all possible board shapes are considered, a wide variety of rook polynomials result. They can be obtained through interesting algebraic, semi-algebraic, or algorithmic operations on the (row,column) matrix designation numbers of the individual squares of the board [1], [3], [7]. Since here we use square boards, our task is simplified because square boards have easily derived, unique rook polynomials.

ROOK POLYNOMIALS OF SQUARE BOARDS

Consider, for example, finding the x^2 coefficient of the rook polynomial of a 4×4 chessboard. The value is the number of ways 2 non-taking rooks can be placed on the board, one of which is shown below.

This paper is in final form and no version of it will be submitted for publication elsewhere.

F.T. Howard (ed.), Applications of Fibonacci Numbers, Volume 9: Proceedings of the Tenth International Research Conference on Fibonacci Numbers and their Applications, 101–108.

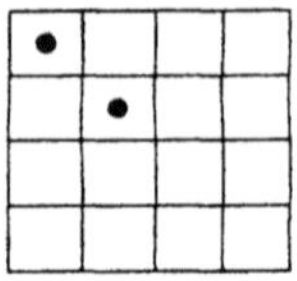

In one row, a rook can be in any of 4 positions and in a second row can be in any of 3 positions. There are $P(4,2)$ ways to place 2 non-taking rooks on any set of two rows. But, there are $C(4,2)$ sets of two rows, so the desired coefficient is $C(4,2)P(4,2) = 72$. Similar reasoning for the general coefficient of x^k in the rook polynomial for an $r \times r$ board yields $C(r,k)P(r,k)$, and a general expression for the rook polynomial becomes $\sum_{k=0}^{r} C(r,k)P(r,k)x^k$. The rook polynomial of our example is

$$1 + 16x + 72x^2 + 96x^3 + 24x^4 \tag{1}$$

THE SQUARE BOARD ROOK POLYNOMIAL TRIANGLE (SBRP)

Note that the numerical coefficients of (1) form the 4th row of Pascal's triangle with each kth entry multiplied by $P(4,k)$. It is interesting to explore some of the extensions to a SBRP triangle, an example of which appears below. The rows start with $r = 0$ and continue through $r = 5$. Positions start at $k = 0$ on the left and continue to the right through $k = r$ for that row. The general term is $C(rk)P(rk) = k![C(r,k)]^2 = \frac{1}{k!}[P(r,k)]^2$ as compared with $C(r,k)$ for Pascal's triangle. An interesting, if minor, feature of the rth row is that the 0th coefficient is always 1, the 1st is r^2, and the rth is $r!$.

$$\begin{array}{ccccccccccc}
 & & & & & 1 & & & & & \\
 & & & & 1 & & 1 & & & & \\
 & & & 1 & & 4 & & 2 & & & \\
 & & 1 & & 9 & & 18 & & 6 & & \\
 & 1 & & 16 & & 72 & & 96 & & 24 & \\
1 & & 25 & & 200 & & 600 & & 600 & & 120
\end{array} \tag{2}$$

ROW GENERATION

Pascal's triangle seems to express its universality by being central to several classes of integer triangles [4], [5]. The SBRP triangle is a member of one such class in that each entry of a next row is a function of the sum two entries immediately above it in the preceding row (as in Pascal's triangle).

Multiplication of the sum of two adjacent r^{th} row entries by the factor, $f_1(r,k)$, yields the immediately below $(r+1)^{th}$ row entry and factor as

$$[C(r,k)P(r,k) + C(r,k+1)P(r,k+1)]f_1(r,k) = C(r+1,k+1)P(r+1,k+1) \tag{3}$$

$$f_1(r,k) = \frac{(r+1)^2}{(k+1)+(r-k)^2} \tag{4}$$

For $r = 4, k = 2$, (3) and (4) check out as $(72+96)\frac{25}{7} = 600$ on triangle (2). We can similarly determine the $(r-1)^{th}$ row entry of (2) from the sum of the two immediately below r^{th} row entries. The $f_2(r,k)$ becomes

$$f_2(r,k) = \frac{(r-k)^2(k+1)}{r^2[(k+1)+(r-k)^2]}. \tag{5}$$

(Although rarely done in Pascal triangles, the previous $(r-1)^{th}$ row entry from the sum of the two immediately below r^{th} row entries is easily found to be $f_3(r,k) = \frac{(r-k)(k+1)}{(r+1)r}$.)

PROPERTIES OF RIGHT-DESCENDING DIAGONALS

Integers of the right-descending diagonals of (2) appear in Sloane and Plouffe's [8] encyclopedia of sequences as absolute value coefficients of Laguerre polynomials. To be consistent with conventional terminology, we will express the sequence terms in index n starting with $n = 1$, by first finding the diagonal terms as functions of the row number, s, of the left terminus and the row number, r, of a sequence term. It appears that the first term of a Laguerre sequence is 1 and the second term is always a square. For agreement with our computations and terminology, the Laguerre sequence which has $(s+1)^2$ as its second term is the sequence associated with our parameter s. For example, the diagonal of (2) which reads 1, 9, 72, $\cdots$ has $s = 2$ since $(2+1)^2 = 9$. For the right-descending diagonals, r increases by one for the next term. Recall also that we have to take into account that we start indexing r with 0, not 1. By noting this, we can derive equivalent forms of a general diagonal term as

$$C(r,r-s)P(r,r-s) = [C(r,r-s)]^2(r-s)! = \frac{(r)(r-1)\cdots(r-(s-1))}{(s!)^2}r! \tag{6}$$

The expresssions of (6) have no value for $r < s$, i.e., until $r = s$. However, if r is replaced by $r = n+s-1$, we have the sequence values in the more useful and conventional index n starting with 1 for $n = 1$. With the substitution for r, (6) has the closed form

$$\frac{((n+s-1)!)^2}{(n-1)!(s!)^2} \tag{7}$$

The 0th right-descending diagonal of (2) is 1, 1, 2, 6, 24, $\cdots$, $(n-1)!$, $\cdots$. Sloane and Plouffe [8] list this sequence as M1675 but use $n!$ as the general term because they start the sequence with the value 1 for $n = 0$. The 1st $(s = 1)$ right-descending diagonal of (2) is 1, 4, 18, 96,

600, $\cdots, n \cdot n!, \cdots$ and appears as M3545 in [8]. The 2nd ($s = 2$) right-descending diagonal of (2) is 1, 9, 72, 600, 5400, $\cdots \frac{((n+1)!)^2}{(n-1)!(2!)^2}$. Reference [8] lists this sequence as M4649. At this point, reference [8] includes a closed generating function in the form of a ratio of two polynomials in x for M4649. M4649 is listed as an *exponential* generating function. This and higher s-valued right-descending diagonals are identified as Laguerre polynomial coefficient sequences. However, only M4649 includes a closed generating function. We welcome this chance to contribute missing exponential generating functions in succeeding sections and to develop a general generating function applicable to all higher Laguerre polynomials found in, and suggested by, [8].

GENERATING FUNCTIONS FOR COEFFICIENTS OF LAGUERRE POLYNOMIALS

A quotient of polynomials serves as either an *ordinary* closed generating function or an *exponential* closed generating function depending on how the quotient is expressed. In the ordinary function, the general term is $a_k x^k$ while in the exponential function the same general term is $k!a_k\left(\frac{x^k}{k!}\right) = b_k\left(\frac{x^k}{k!}\right)$. While it is easy to get a_k's by direct division, it is not possible to get b_k's directly this way. (Both Liu [7, pp. 33-34] and Brualdi [2, pp. 237-243] have interesting discussions on this apparent impasse.) A fruitful approach to the inverse problem of finding a closed function from either ordinary or exponential coefficients first involves conversion (if needed) to ordinary coefficients. Finding the closed form generating function is then often a matter of judgment, experience, and exceptionally good luck. Reference [8] devotes an interesting chapter to various aspects of this problem.

The simplest solution attempt is indicated when we have a linear sequence of constant coefficients, and we are reasonably sure that a closed form exists as a ratio of two polynomials. If this is indeed the case, the sequence is index-invariant and a set of successive forward finite differences from the sequence leads to a closed form ratio of polynomials. We use the systematic z-Transform [6] approach which served so well in a similar situation in [4].

As stated earlier, the generating function listed in [8] is an exponential generating function. It seems reasonable to assume that generating functions from higher order right-descending diagonals are also exponential. To avoid difficulties with exponential generating functions, we divide the coefficients of known exponential generating functions by $k!$ for the appropriate index k and treat them as coefficients of ordinary generating functions for which we might find a closed form ratio of polynomials.

To review how the z-Transform approach [4], [6] works and to check our calculations, we examine Sloane and Plouffe's M4649. For completeness, we will then obtain the two generating functions missing from [8] and develop a general expression for the closed generating function for all right-descending diagonals of (2).

When working with z-Transforms [6], it is necessary to include a_0 and any other lowest-indexed initial conditions. Since here $a_0 = 0$, computations are simplified. We start with a triangular table of forward differences of the coefficients based on a finite number of coefficients. The size of the table needed also depends on the order of recursion, which in this case is 4. The first values of M4649 have been divided by the appropriate $n!$ to appear as a_n. The $\Delta^k a_n$'s are k^{th} forward differences where we let $a_n = \Delta^0 a_n$ to complete the table.

$$\begin{array}{cccccc}
n & a_n & \Delta^1 a_n & \Delta^2 a_n & \Delta^3 a_n & \Delta^4 a_n \\
0 & 0 & & & & \\
 & & 1 & & & \\
1 & 1 & & 2.5 & & \\
 & & 3.5 & & 1.5 & \\
2 & 4.5 & & 4 & & 0 \\
 & & 7.5 & & 1.5 & \\
3 & 12 & & 5.5 & & 0 \\
 & & 13 & & 1.5 & \\
4 & 25 & & 7 & & 0 \\
 & & 20 & & 1.5 & \\
5 & 45 & & 8.5 & &
\end{array} \tag{8}$$

In (8), $\Delta^4 a_n = 0, \Delta^3 a_{n+1} - \Delta^3 a_n = 0$. Since $\Delta^2 a_{n+2} - \Delta^2 a_{n+1} = \Delta^3 a_{n+1}$ and $\Delta^2 a_{n+1} - \Delta^2 a_n = \Delta^3 a_n, \Delta^2 a_{n+2} - 2\Delta^2 a_{n+1} + \Delta^2 a_n = 0$. We reduce to $\Delta^1 a_{n+3} - 3\Delta^1 a_{n+2} + 3\Delta^1 a_{n+1} + \Delta^1 a_n = 0$.

Continuing with computing the zero differences leads to the homogeneous difference equation

$$a_{n+4} - 4a_{n+3} + 6a_{n+2} - 4a_{n+1} + a_n = 0 \tag{9}$$

The z-Transform [4], [6] of (9) yields

$$\{z^4 Z(a_n) - z^4 a_0 - z^3 a_1 - 2^2 a_2 - z a_3\} - 4\{z^3 Z(a_n) - z^3 a_0 - z^2 a_1 - z a_2\} +$$

$$6\{z^2 Z(a_n) - z^2 a_0 - z a_1\} - 4\{z Z(a_n) - z a_0\} + Z(a_n) = 0 \tag{10}$$

After factoring out $Z(a_n)$, rearranging terms, and substituting for a_0 through a_3 from (8), we have the formula for $Z(a_n)$ and the numerical closed form ordinary generating function in z as

$$Z(a_n) = \frac{z^4 a_0 + z^3(a_1 - 4a_0) + z^2(a_2 - 4a_1 + 6a_0) + z(a_3 - 4a_2 + 6a_1 - 4a_0)}{(z-1)^4}$$

$$= \frac{z^3 + \frac{1}{2}z^2}{(z-1)^4} \tag{11}$$

If the right side of (11) were expanded, the sequence terms of the ordinary generating function would appear as coefficients of powers of $\frac{1}{z}$. By replacing z by $\frac{1}{x}$ in (11), we obtain as (12) the ordinary generating funciton for our a's in x and at the same time the exponential generating function for the terms of the second right-descending diagonal. Recall this verifies M4649 of Sloane and Plouffe [8].

$$\frac{x + \frac{1}{2}x^2}{(1-x)^4} \tag{12}$$

The "4-ness" and relationship to coefficients of the 4th row of Pascal's triangle in (8)-(11), and the fact that the recursion order is $4 = 2s$ for the $s = 2$ diagonal is not a coincidence. All of the right descending diagonals have the same recursion, $2s$. We observe the orderly appearance and arrangement of binomial coefficients from Pascal's triangle and use the inverse z-Transform to get the common generating function in z and x.

$$\frac{z^5 + 2z^4 + \frac{z^3}{3}}{(z-1)^6} \to \frac{x + 2x^2 + \frac{x^3}{3}}{(1-x)^6} \tag{14}$$

By applying the same procedures, we obtain generating functions in z and x for the right-descending diagonals for $s = 4$, 5, and 6.

$$s = 4 : \frac{z^7 + \frac{9z^6}{2} + 3z^5 + \frac{z^4}{4}}{(z-1)^8} \to \frac{x + \frac{9x^2}{2} + 3x^3 + \frac{x^4}{4}}{(1-x)^8} \tag{15}$$

$$s = 5 : \frac{z^9 + 8z^8 + 12z^7 + 4z^6 + \frac{z^5}{5}}{(z-1)^{10}} \to \frac{x + 8x^2 + 12x^3 + 4x^4 + \frac{x^5}{5}}{(1-x)^{10}} \tag{16}$$

$$s = 6 : \frac{z^{11} + \frac{25z^{10}}{2} + \frac{100z^9}{9} + 25x^8 + 5z^7 + \frac{z^6}{6}}{(z-1)^{12}} \to \frac{x + \frac{25x^2}{2} + \frac{100x^3}{3} + 25x^4 + 5x^5 + \frac{x^6}{6}}{(1-x)^{12}} \tag{17}$$

GENERAL GENERATING FUNCTION IN S

Consider the triangle formed by and implied by successive rows of the numerator terms of the right sides of generating functions (14)-(17).

$$\begin{array}{ccccccccccc}
 & & & & & 1x & & & & & \\
 & & & & 1x & & \frac{1}{2}x^2 & & & & \\
 & & & 1x & & 2x^2 & & \frac{1}{3}x^3 & & & \\
 & & 1x & & \frac{9}{2}x^2 & & 3x^3 & & \frac{1}{4}x^4 & & \\
 & 1x & & 8x^2 & & 12x^3 & & 4x^4 & & \frac{1}{5}x^5 & \\
1x & & \frac{25}{2}x^2 & & \frac{100}{3}x^3 & & 25x^4 & & 5x^5 & & \frac{1}{6}x^6
\end{array} \tag{18}$$

A significant fact about (18) is its triangular form whose rows are the successive numerators of the closed (either exponential or ordinary) generating functions like those which we exemplified in (13)-(17). Moreover, should we multiply a rook polynomial by x and divide each coefficient by the factorial of the power of its associated x in the new expression, the result would also be the entries of (18). Another interesting feature of (18) triangle in the development of the extension of (11).

We purposefully chose $s = 2$ as our derivation parameter for demonstrating the z-Transform approach because the derived generating function could then be compared with the only function listed in [8].

The case for $s = 0$ is unique in that the generating function, $-\log(1-x)$, is logarithmic. The terms of its series expansion may be observed as the 0th diagonal of (18), while the adjusted coefficients of a series "exponential" expansion appear as the 0th diagonal of (2).

For $s = 1$, the generating function is $\frac{x}{(1-x)^2}$. The terms of the ordinary series expansion can be seen from the 1st diagonal of (18), while the coefficients of the exponential series expansion appear as the 1st diagonal of (2).

SOME MISSING CLOSED GENERATING FUNCTIONS

The 3rd ($s = 3$) right-descending diagonal of (2) can be extended to yield series coefficients, 1, 16, 200, 2400, 29400 , $\cdots$, $\frac{((n+2)!)^2}{(n-1)!(3!)^2}, \cdots$. This sequence is listed in [8] as M5019 with, of course, the Laguerre polynomial connection. The ordinary generating function which shares the same closed form with the exponential generating function of M5019 yields 1, 8, $\frac{100}{3}$, 100, 245, $\frac{1568}{3}$, 1008, 1800, $\cdots$, $\frac{((n+2)!)^2}{(n-1)!(3!)^2 n!}, \cdots$. Because $s = 3$, and the recursion order is 6, the difference triangle requires only $a_0 = 0$, and a_1 through a_5 from the above series to find the closed form ordinary generating function. We learned earlier [4] that we can bypass much of the development and substitute the sequence values in

$$\frac{\begin{array}{c} z^6 a_0 + z^5(a_1 - 6a_0) + z^4(a_2 - 6a_1 + 15a_0) + z^3(a_3 - 6a_2 + 15a_1 - 20a_0) + \\ z^2(a_4 - 6a_3 + 15a_2 - 20a_1 + 15a_0) + z(a_5 - 6a_4 + 15a_3 - 20a_2 + 15a_1 - 6a_0) \end{array}}{(z-1)^6} \tag{13}$$

is that it has a terms of its right-descending diagonals, the terms of series ordinary generating functions. In terms of s, (7) indicates that the general series exponential generating function for right-descending diagonals and corresponding series ordinary generating functions in n, starting with $n = 1$ are, respectively,

$$\sum_{n=1}^{\infty} \frac{((n+s-1)!)^2}{(n-1)!(s!)^2} x^n, \quad \sum_{n=1}^{\infty} \frac{((n+s-1)!)^2}{(n-1)!(s!)^2} x^n \tag{19}$$

For $s = 3$, for example, the generating function numerator of (14) occupies the $s = 2$ row in (18). This difference of one is s for row occupation in general. However, the denominator term is $(1-x)^4$ or, in general, $(1-x)^{2s}$. Hence, for a given s, the general generating function for either of the parts of (19), depending on exponential or ordinary interpretation, becomes the desired general generating function.

$$\frac{\sum_{k=0}^{s-1} \frac{((s-1)!)^2}{((s-k-1)!)^2(k!)^2(k+1)} x^{(k+1)}}{(1-x)^{2s}} \tag{20}$$

SUMMARY

We have shown elementary, but interesting, comparisons of Square Board Rook Polynomial (SBRP) triangles with Pascal triangles. We have investigated and extended the sequence properties of the right-descending diagonals of SBRP triangles. In so doing, we have added to and generalized known Laguerre sequence listings.

ACKNOWLEDGMENTS

We thank the anonymous referee for his careful and constructive review. Incorporation of his suggestions has very much improved the paper. We are happy!

REFERENCES

[1] Anderson, P. G. Private correspondence, October 2000.

[2] Brualdi, R.A. Introductory Combinatorics. (Second edition). Englewood Cliffs, NJ: Prentice-Hall. 1992.

[3] Fielder, D.C. "A Generator of Rook Polynomials" currently under review by *The Mathematica Journal*.

[4] Fielder, D.C. and Alford, C.O. "An Investigation of Sequences Derived From Hoggatt Sums and Hoggatt Triangles." Third International Conference on Fibonacci Numbers and Their Applications. Pisa, Italy, July 25-29, 1988. Also in Application of Fibonacci Numbers, Volume **3**. Edited by G.E. Bergum et. al. Dordrecht, Holland; Kluwer, 1990.

[5] Fielder, D.C. and Alford, C.O. "Pascal's Triangle: Top Gun or Just One of the Gang?." Fourth International Conference on Fibonacci Numbers and Their Applications. Wake Forest, NC, July 30-August 3, 1990. Also in Application of Fibonacci Numbers, Volume **4**. Edited by G.E. Bergum et. al. Dordrecht, Holland; Kluwer, 1991.

[6] Jury, E.I. Theory and Application of the z-Transform Method. New York: John Wiley and Sons, 1964.

[7] Liu, C.L. Introduction to Combinatorial Mathematics. New York: McGraw-Hill, 1968.

[8] Sloane, N.J.A. and Plouffe, S. The Encyclopedia of Integer Sequences. New York: Academic Press, 1995.

AMS Classification Numbers: 05A15, 11Y55, 12E10

PYTHAGOREAN QUADRILATERALS

Robert Hochberg and Glenn Hurlbert

1. INTRODUCTION

The Pythagorean Theorem says that if a and b are the leg lengths of a right triangle with hypotenuse c, then $a^2 + b^2 = c^2$. The infinitely many integral solutions are well classified, and numerous generalizations have been thoroughly studied. Lagrange [6] proved that every positive integer was the sum of 4 squares of integers. Waring [9] conjectured and Hilbert [5] proved that for every positive integer n there was a constant $c(n)$ such that every positive integers was the sum of $c(n)$ non-negative nth powers. Fermat, Euler, Gauss and Jacobi studied the number of solutions to $a^2 + b^2 = n$ for fixed n (see eg. [3]). Lucas [7] challenged his readers to find all solutions n and c to $\sum_{i=1}^{n} i^2 = c^2$, (for a nice solution see [1]) and Pell's equation $a^2 - kb^2 = \pm 1$ for fixed k occupied the attention of many mathematicians. And finally there is the problem posed by Fermat of representing nth powers of integers as the sum of two smaller nth powers for $n > 2$, which was recently solved by Wiles [10].

The above generalizations are all algebraic in nature. Here we offer a *geometric* generalization of Pythagorean triples. It too has a nice associated Diophantine equation, (see equation (5)) and its solution leads to some interesting number theoretic questions. We ask: What 4-tuples of integers (A, B, C, D) are there such that a convex quadrilateral with these side lengths can be inscribed in a circle with diameter D? Since all triples of integers (A, B, C) which satisfy the property that a triangle with side lengths A, B and C can be inscribed in a circle of diameter C are precisely the Pythagorean triples (such a triangle is necessarily right), we call our solutions *Pythagorean 4-tuples*. We look for *primitive* Pythagorean 4-tuples, i.e., those for which $\gcd(A, B, C, D) = 1$, for example, $(1, 1, 1, 2)$, the top half of a regular hexagon of side length 1. The reader might enjoy finding a Pythagorean 4-tuple on his or her own before proceeding. Many similar problems involving geometric figures with integer lengths can be found in [4], questions D18-D22, including the notorious Perfect Cubiod Problem.

This paper is in final form and no version of it will be submitted for publication elsewhere.

F.T. Howard (ed.), Applications of Fibonacci Numbers, Volume 9: Proceedings of the Tenth International Research Conference on Fibonacci Numbers and their Applications, 109–119.

The formulas which generate all primitive Pythagorean triples ($A = u^2 - v^2, B = 2uv, C = u^2 + v^2$) require only that the generators u and v be relatively prime and of different parity. We are able to find equally simple formulas (Theorem 3) only in the case that some two sides are equal. We include in Section 3 our derivation in this special case, with our primary emphasis on the recursive nature of the family of all such primitive solutions. In contrast to the recursion involved with Pell's equation, our solutions split into several recursive families.

In Section 4 we prove two theorems about impossible diameters of Pythagorean 4-tuples whose sides are distinct. Section 5 is devoted to questions and conjectures regarding future directions of research.

2. PRELIMINARIES

We begin by labelling the diameter D, its opposite side A, the remaining sides B and C, and the diagonals M and N as in Figure 1.

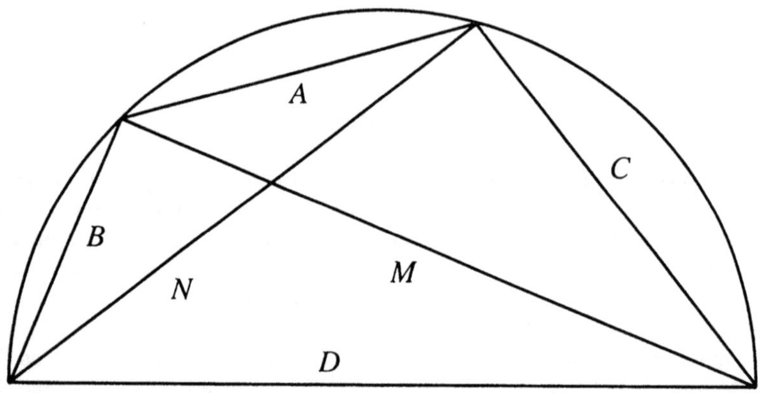

Figure 1.

The smallest distinct-side Pythagorean 4-tuple is (2,7,11,14). How does one verify that this is indeed a Pythagorean 4-tuple? Since the triangles with sides B, M, D and C, N, D are both right, we find that $M = 7\sqrt{3}$ and $N = 5\sqrt{3}$. Now we use

Proposition 1 (Ptolemy's Theorem): *The sum of the products of opposite sides of a cyclic quadrilateral is equal to the product of the diagonals.*

With the variables as shown in Figure 1, we have

$$AD + BC = MN. \tag{1}$$

The converse of Ptolemy's theorem (see [8]) is not true in general (eg. $(A, C, D, B, M, N) = (3, 1, 1, 1, \sqrt{2}, 2\sqrt{2})$), but is true whenever (2) holds below - equation (2) inscribes the right triangles (B, M, D) and (C, N, D) and (1) insures that $A > 0$ so that M and N are the diagonals. Thus, we can now check our example to see that (2,7,11,14) is indeed a Pythagorean 4-tuple.

Glancing at the right triangles in Figure 1, we see that

$$M^2 + B^2 = D^2 = N^2 + C^2, \tag{2}$$

so that M and N are both square roots of integers. Equation (1) tells us that the product MN is an integer, and thus we can write $M = S\sqrt{k}$ and $N = T\sqrt{k}$ for some integers S, T, and square-free k, which we call the *surd* of our Pythagorean 4-tuple. Substituting into (1) we get

$$AD + BC = STk. \tag{3}$$

A more useful equation is obtained by substituting for M and N as follows:

$$AD + BC = \sqrt{(D^2 - B^2)(D^2 - C^2)}, \tag{4}$$

which becomes, as a polynomial in D,

$$D^3 - (A^2 + B^2 + C^2)D - 2ABC = 0. \tag{5}$$

This is our generalization of the Pythagorean equation $C^2 - A^2 - B^2 = 0$ for right triangles. Geometrically, it is clear that given positive real numbers A, B, and C, there is a unique positive D (not necessarily integral or rational) so that its corresponding quadrilateral inscribes in a circle of diameter D. Algebraically, (5) describes the same information: the derivative (in D) of the lefthand side is strictly positive so its real root is unique. It is of course positive.

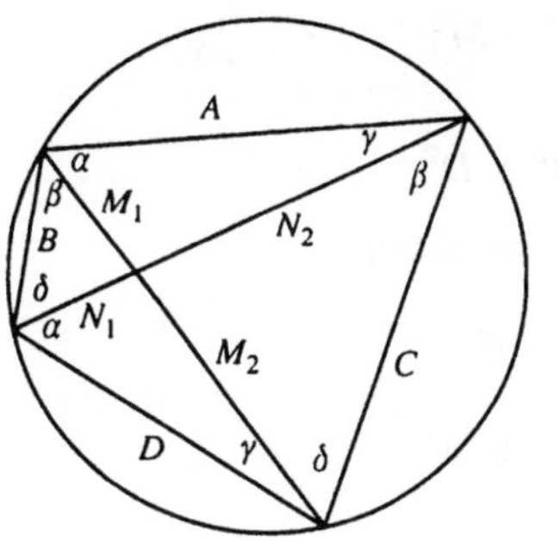

Figure 2a.

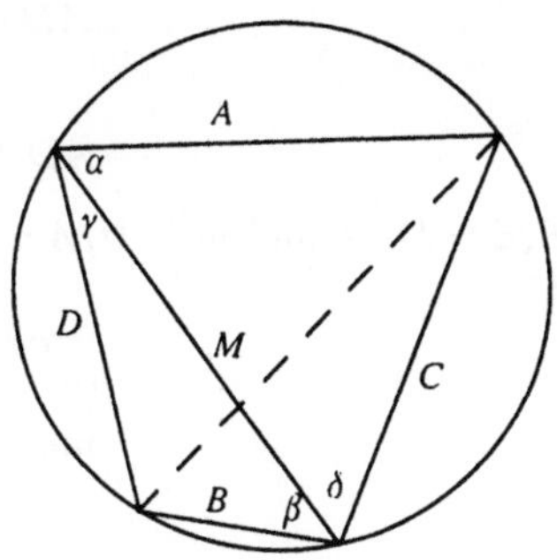

Figure 2b.

We end this section with a nice quick proof of Ptolemy's Theorem.

Proof of Proposition 1: Let the diagonals cut one another into lengths M_1, M_2, N_1, and N_2 so that $M = M_1 + M_2$ and $N = N_1 + N_2$ (see Figure 2a). We caluclate the area H of the quadrilateral in 2 ways. First we sum the areas of the four triangles, getting

$$\begin{aligned} H = &\frac{1}{2}M_1N_1\sin(\pi - \beta - \gamma) + \frac{1}{2}M_1N_2\sin(\pi - \beta - \delta) \\ &+ \frac{1}{2}M_2N_1\sin(\pi - \beta - \delta) + \frac{1}{2}M_2N_2\sin(\pi - \beta - \gamma) \\ = &\frac{1}{2}(M_1N_1 + M_1N_2 + M_2N_1 + M_2N_2)\sin(\alpha + \gamma) \\ = &\frac{1}{2}MN\sin(\alpha + \gamma). \end{aligned}$$

Second, if we flip the tirangle with sides MBD (see Figure 2b), we calculate this same area to be

$$H = \frac{1}{2}AD\sin(\alpha+\gamma) + \frac{1}{2}BC\sin(\beta+\delta)$$

$$= \frac{1}{2}(AD+BC)\sin(\alpha+\gamma).$$

Hence, $AD + BC = MN$. □

3. EQUAL SIDES

Of course, (1,1,1,2) is the unique primitive Pythagorean 4-tuple with $A = B = C$, so let us consider the situation in which exactly 2 sides are equal. If the equal sides are not B and C, but rather, say, A and B, then we may flip the triangle with sides A, B and N along the side N (as we did in the proof of Ptolemy's theorem) to make the equal sides opposite one another. So without loss of generality, we assume $A \neq B = C$. But $B = C$ implies $M = N$, so (1) becomes

$$AD + B^2 = M^2,$$

and then

$$AD + 2B^2 = D^2.$$

With $a = A/B, d = D/B$, and $m = D/A = d/a$, we obtain

$$ad + 2 = d^2,$$

$$ma^2 + 2 = m^2a^2,$$

and

$$a^2 = 2/(m^2 - m) = 1/\binom{m}{2}. \tag{6}$$

Thus we conclude

Lemma 2: *There exists a Pythagorean 4-tuple with two equal sides if and only if $\binom{m}{2}$ is the square of a rational number for some rational $m > 1$.* □

Such solutions, with $m = p/q$, become $(q, q\sqrt{\binom{m}{2}}, q\sqrt{\binom{m}{2}}, p)$. With $m = 32/7$, for example, we have the solution (7,20,20,32).

As in the case of Pythagorean triples, we have simple formulas to generate all Pythagorean 4-tuples with 2 equal sides. Lemma 2 yields

Theorem 3: *All primitive Pythagorean 4-tuples with equal sides are of the form $(|2s^2 - r^2|, sr, sr, max\{2s^2, r^2\})$, where s and r are any relatively prime natural numbers with r odd.*

Proof: Write $m = p/q$ in lowest terms. Then $\binom{m}{2} = p(p-q)/2q^2$ so that

$$\sqrt{\binom{m}{2}} = \frac{1}{q}\sqrt{\frac{p(p-q)}{2}}. \tag{7}$$

Now, $gcd(p,q) = 1$, so $gcd(p, p-q) = 1$ and $\sqrt{p(p-q)/2} \in \mathbf{Z}$ if and only if both $p/2$ and $(p-q)$ are squares, or both p and $(p-q)/2$ are squares, depending upon whether p is even or odd (q is necessarily odd since $p(p-q)$ must be even in (7)). If p is even then let $p/2 = s^2$ and $(p-q) = r^2$, and if p is odd then let $(p-q)/2 = s^2$ and $p = r^2$. In either case, r is odd and (7) becomes

$$\sqrt{\binom{m}{2}} = sr/q, \tag{8}$$

so $a = q/sr$ and $d = p/sr$. With p even we obtain $q = 2s^2 - r^2$, and with p odd we have $q = r^2 - 2s^2$. □

The problem becomes interesting if we fix an odd q and consider how many solutions (*q-solutions*) there are of the form $A = q, B = C$, and $D = p$. We already know that q is odd, but since $r^2 \equiv 1 \bmod 8$ and $2s^2 \equiv 0$ or $2 \bmod 8$, we have that $q \equiv \pm 1 \bmod 8$. In fact, since $q = |2s^2 - r^2|$, $(r/s)^2 \equiv 2 \bmod q'$ for each prime q' dividing q. Thus 2 is a quadratic residue of q', so by Gauss' Lemma, $q' \equiv \pm 1 \bmod 8$. Let us call an integer q *admissible* if each of its prime divisors is congruent to $\pm 1 \bmod 8$. Then it so happens that we can recursively generate infinitely many primitive q-solutions (in fact, all of them) whenever q is admissible.

Before plunging into the recursive algorithm, let us first describe its inverse. The idea is to show how all q-solutions are derived from "smaller" q-solutions until a smallest *seed* is found, in somewhat the same fashion as solutions to Pell's equation are constructed. To this end, suppose we have the q-solution $(q, rs, rs, p) = (q, r_n s_n, r_n s_n, p_n)$.

First assume that p_n is even. Then $p_n = 2s_n^2$ and $q = 2s_n^2 - r_n^2$ for some $r_n, s_n > 0$. If $r_n \le s_n$ then we will stop, so let us suppose that $s_n < r_n$. We let $s_{n-1} = r_n - s_n$ and $r_{n-1} = \sqrt{2s_{n-1}^2 + q}$, so that $q = r_{n-1}^2 - 2s_{n-1}^2$. Then $(s_{n-1} + s_n)^2 = r_n^2 = 2s_n^2 - q$, and the quadratic formula shows that $s_n = s_{n-1} \pm r_{n-1}$. But since $r_{n-1} > s_{n-1}$ we actually have $s_n = s_{n-1} + r_{n-1}$. Finally, one checks that if $(q, r_n s_n, r_n s_n, p_n)$ is a Pythagorean 4-tuple,then so is $(q, r_{n-1}s_{n-1}, r_{n-1}s_{n-1}, p_{n-1})$, where $p_{n-1} = r_{n-1}^2$. Call this algorithm EVEN.

Next assume that p_n is odd. Then $p_n = r_n^2$ and $q = r_n^2 - 2s_n^2$ for some $r_n, s_n > 0$, and here we suppose not only that $s_n < r_n$, but $s_n > 2r_n/3$ as well. Similarly, let $s_{n-1} = r_n - s_n$ and $r_{n-1} = \sqrt{2s_{n-1}^2 - q}$, so that $q = 2s_{n-1}^2 - r_{n-1}^2$. Then $(s_{n-1} + s_n)^2 = 2s_n^2 + q$ and $s_n = s_{n-1} \pm r_{n-1}$. But the extra condition $r_n > 3s_n/2$ (which maintains the condition $r_{n-1} > s_{n-1}$) implies that $r_{n-1} > s_{n-1}$, so $s_n = s_{n-1} + r_{n-1}$. And now with $p_{n-1} = 2s_{n-1}^2$

we know that $(q, r_{n-1}s_{n-1}, r_{n-1}s_{n-1}, p_{n-1})$ is Pythagorean if $(q, r_n s_n, r_n s_n, p_n)$ is. Call this algorithm ODD.

Finally we describe algorithm ODD*. Its only deviance from ODD is the condition $r_n \leq 3s_n/2$, and now $s_n = s_{n-1} \pm r_{n-1}$ since $r_{n-1} \leq s_{n-1}$.

The inverse of our recursive algorithm is the combination of EVEN, ODD, and ODD*, since exactly one of the initial conditions will be satisfied. The key observations to make are that EVEN will always be followed by ODD or ODD*, ODD will always be followed by EVEN, and ODD* is the terminating algorithm. Also, we always have $s_{n-1} < s_n$ in EVEN and ODD, so no q-solutions are repeated and thus the algorithm does halt. Because of the $\pm$ in ODD*, the inverse algorithm can be pictured as in Figure 3.

The only exception to Figure 3 occurs when equality holds in ODD*. In this case $r_1 = 3$ and $s_1 = 2$ and ODD* produces the trivial hexagonal solution. Here we have only one infinite string of q-solutions, rather than the two pictured.

The values s_0 will be called a *seed* of q, and r_0 its *root*. From the above discussion one sees that all roots are even. One can also seed that s_0 is a seed of q if and only if it satisfies

$$\begin{array}{ll} (i) & gcd(s_0, q) = 1, \\ (ii) & s_0 \leq \sqrt{q}, \text{ and} \\ (iii) & 2s_0^2 - q = r_0^2 \text{ for some integer } r_0. \end{array} \tag{9}$$

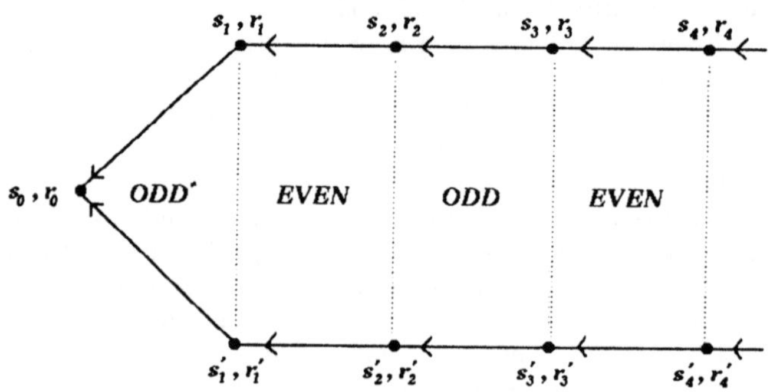

Figure 3.

So to find all primitive q solutions, we find all seeds of q and for each seed s_0 let $r_0 = \sqrt{2s_0^2 - q}$,

$$s_1 = s_0 - r_0, \quad s_1' = s_0 + r_0,$$

and

$$r_1 = s_1 + s_0, \quad r_1' = s_1' + s_0.$$

Then let

$$s_n = s_{n-1} + r_{n-1}, \quad s_n' = s_{n-1}' + r_{n-1}',$$

and

$$r_n = s_{n-1} + s_n, \quad r_n' = s_{n-1}' + s_n'.$$

We then obtain solutions $A_n = q, B_n = C_n = s_n r_n$, and $D_n = r_n^2$ for odd n, $2s_n^2$ for even n (likewise for A'_n, B'_n, C'_n, D'_n,). Table 4 gives an example with $q = 7$, the only seed of which is $s_0 = 2$.

n	s_n	r_n	B_n	D_n	s'_n	r'_n	B'_n	D'_n
0	2	1	2	8				
1	1	3	3	9	3	5	15	25
2	4	5	20	32	8	11	88	128
3	9	13	117	169	19	27	513	729
4	22	31	682	968	46	65	2990	4132
5	53	75	3975	5625	111	157	17427	24649
6	128	181	23168	32768	268	379	101572	143648
⋮	⋮		⋮		⋮		⋮	

Table 4.

q	seeds
7^5	92
17^4	205
$7^2 \cdot 17^2$	89, 101
$17 \cdot 31 \cdot 79$	151, 173, 177, 191
$7^2 \cdot 17^2 \cdot 23$	404, 426, 486, 558
$17 \cdot 23 \cdot 47 \cdot 103$	976, 984, 1004, 1026, 1130, 1166, 1194, 1246

Table 5.

As $n \to \infty$, $D_n \to \infty$, so that geometrically our quadrilaterals tend to approximate an isosceles right triangle (remember $A = q$ is fixed.) Thus, it should be the case that $D_n/B_n \to \sqrt{2}$, and likewise for D'_n/B'_n. Algebraically, we have $D_n/B_n = r_n/s_n = \sqrt{2 \pm p/s_n^2} \to 2$.

One natural question that arises is, just how many seeds does a fixed admissible q have? The keen observer of Table 5 might guess the correct answer.

Theorem 4: *Let $\omega = \omega(q)$ be the number of distinct prime divisors of q. Then the number of seeds for fixed admissible q is $2^{\omega-1}$.*

As an obvious corollary, we have

Corollary 5: *(A, B, B, D) is a Pythagorean 4-tuple for some B and D if and only if A is admissible.* □

We will prove Theorem 4 using the following two claims.

Claim 6: *For prime q there is at least one seed.*

Claim 7: *For prime q there is at most one seed.*

Proof of Claim 6: Choose some $\delta \in \{0, 1, \ldots, q-1\}$ so that $\delta^2 \equiv 2 \bmod q$, and for $i = 0, 1, \ldots, \lfloor\sqrt{q}\rfloor$ define $a_i \in \{0, 1, \ldots, q-1\}$ by $a_i = i\delta \bmod q$ (notice that $\delta > \lfloor\sqrt{q}\rfloor$). Of the $\lceil\sqrt{q}\rceil$ distinct a_i, some two have difference less than $\sqrt{q}$ mod q, say a_i and a_j, with $i < j$. But then $0 = a_0$ and a_{j-i} are also less than $\sqrt{q}$ apart, so we may assume that $i = 0$. Moreover, by considering δ versus $-\delta$, we may also ensure that $0 < a_j < \sqrt{q}$.

Now since $2j^2 \equiv (j\delta)^2 \equiv a_j^2 \mod q$, we know that $2j^2 - a_j^2 = mq$ for some nonzero integer m (since 2 is not square). But $2j^2 - a_j^2 > 0 - q$ implies that $m \geq 0$, while $2j^2 - a_j^2 < 2q$ implies that $m < 2$. Hence $m = 1$ and, by (9), j is a seed. ◇

Proof of Claim 7: Suppose we have two seeds $t_1 > t_2$ and let $\tau_i^2 = 2t_i^2 - q$. Then $2(t_1^2 - t_2^2) = (\tau_1^2 - \tau_2^2)$. If we let $f = (t_1 + t_2), g = (t_1 - t_2), h = (\tau_1 + \tau_2)$,

	$h/2$	l
f	u	v
g	x	y

Figure 6. $fg = uvxy = hl/2$.

and $l = (\tau_1 - \tau_2)$, then this equation becomes $2fg = hl$. Since

$$q = 2t_1^2 - \tau_1^2 = 2\left(\frac{f+g}{2}\right)^2 - \left(\frac{h+l}{2}\right)^2$$

we have

$$4q = 2(f+g)^2 - (h+l)^2.$$

Let us make some substitutions (see Figure 6). We let $f = uv$ and $g = xy$ so that $h = 2ux$ and $l = vy$ (h is even since τ_1 and τ_2 are both odd). Then

$$4q = (2u^2 - y^2)(v^2 - 2x^2).$$

Notice that $uv = f > g = xy$ and $2ux = h > l = vy$, so that $(uv)(2ux) > (xy)(vy)$ and $2u^2 > y^2$. Thus also $v^2 > 2x^2$. Now suppose on the contrary that $q|(v^2 - 2x^2)$ (the case that $q|(2u^2 - y^2)$ would be handled similary). Then $(v^2 - 2x^2) \geq q$ so $v^2 > q$ and $v > \sqrt{q}$. But $uv = f = t_1 + t_2 \leq 2\sqrt{q}$ so that $u \leq 2\sqrt{q}/v < 2$. Hence $u = 1$, $(2u^2 - y^2) = (2 - y^2) > 0$ and $y = 1$. Thus, $(2u^2 - y^2) = 1$ and $(v^2 - 2x^2) = 4q$. But $v = f \leq 2\sqrt{q}$, so $v^2 - 2x^2 < v^2 \leq 4q$, a contradiction. ◇

Proof of Theorem 4 The number of seeds found in the proof of Claim 6 depended on the number of square roots δ of 2. For admissible prime q (and admissible prime powers), the number of square roots is two, δ and $-\delta$, one of which yields the seed. For non-prime admissible q, each distinct prime power factor q_j has two square roots of 2 mod q_j, so the Chinese Remainder Theorem guarantees exactly $2^{\omega(q)}$ square roots of 2 mod q occuring, of course, in pairs $\pm\delta_j$. Each pair yields a distinct seed, and thus there are exactly $2^{\omega-1}$ seeds. □

4. DISTINCT SIDES

It is perhaps surprising to note that a diameter of a Pythagorean 4-tuple cannot be an odd prime. Suppose D is some odd prime, p, and rewrite (5) as $p(p^2 - A^2 - B^2 - C^2) = 2ABC$

to see that p divides the quantity $2ABC$. But $p = D > 2, A, B, C$, a contradiction. This is summarized below.

Theorem 8: *The only Pythagorean 4-tuple with prime diameter is the trivial* $(1,1,1,2)$. □

Assume that $B \neq C$ in a Pythagorean 4-tuple of surd k. Then we have two distinct integral solutions (x, y) to the equation $D^2 = x^2 + ky^2$, these being (B, S) and $C, T)$ (from equation 2), and A is then given by Ptolemy's theorem, $A = (kST - BC)/D$, and must be integral. When D is prime, the above theorem says this is all impossible. Why? It turns out that asking for two distinct solutions to $D^2 = x^2 + ky^2$ is too much when D is prime, as the following theoerm shows.

Theorem 9: *For prime p and square-free k, the diophantine equation $p^2 = x^2 + ky^2$ has at most one solution (x, y).*

Proof: Assume $p > 2$, and suppose

$$u^2 + kv^2 = p^2 \tag{10}$$

$$w^2 + kz^2 = p^2.$$

We will show that $u = w$. Moving u^2 and w^2 to the other sides and multiplying, we find

$$k^2v^2z^2 = (p^2 - u^2)(p^2 - w^2), \tag{11}$$

which can be arranged as

$$(kvz - uw)(kvz + uw) = p^2(p^2 - u^2 - w^2). \tag{12}$$

Thus p^2 must divide the left side of (12). If p divided each factor, then p would divide the difference, $2uw$. But glancing at (10), we see that $p > u, w$, whence p can't divide $2uw$, and so p^2 must divide one of the factors. Taking the square root of both sides of (11) one sees that $kvz < p^2$, so p^2 certainly cannot divide $(kvz - uw)$ and hence must divide $(kvz + uw)$. But $u, w < p$ and $kvz < p^2$ imply $(kvz + uw) < 2p^2$. Hence $(kvz + uw) = p^2$, and substitution into (12) yields

$$(kvz - uw)p^2 = p^2(p^2 - u^2 - w^2),$$

$$kvz - uw = p^2 - u^2 - w^2,$$

and

$$p^2 = u^2 - uw + w^2 + kvz = (u - w)^2 + uw + kvz.$$

But since we know that $p^2 = uw + kvz$, we have $p^2 = (u - w)^2 + p^2$ and thus $u = w$. □

We can say more about the diameter of a Pythagorean 4-tuple:

Theorem 10: *A diameter of a distinct-side Pythagorean 4-tuple cannot be the square of a prime.*

Proof: We will show this by proving that $D = p^2$, for p an odd prime, implies that some two sides must be the same. If p is even, then $D = 4$, and A, B, C must be 1,2,3, which doesn't satisfy equation (5). So take p to be odd. From (5) we have

$$p^2(p^4 - A^2 - B^2 - C^2) = 2ABC. \tag{13}$$

Thus $p^2|2ABC$, while $p\nmid 2$ and $p^2\nmid A, B, C$. Hence p must divide two of the sides, say, A and B. If p also divided C, then we could divide all sides by p obtaining an integral solution to (5) with prime diameter, contradicting Theorem 8. So $p\nmid C$.

Let $A = p\alpha$ and $B = p\beta$ so that (13) becomes

$$p^2(p^4 - p^2\alpha^2 - p^2\beta^2 - C^2) = 2p^2\alpha\beta C. \tag{14}$$

Dividing by p^2, moving C^2, and factoring again, we obtain

$$p^2(p^2 - \alpha^2 - \beta^2) = C(2\alpha\beta + C), \tag{15}$$

so $p\nmid C$ implies $p^2|(2\alpha\beta + C)$. But $A, B < p^2$ implies that $\alpha, \beta < p$, which implies that $2\alpha\beta < 2p^2$. Furthermore, since $C < D = p^2$, we have $2\alpha\beta + C < 3p^2$. Thus, either

$$I: \quad 2\alpha\beta + C = p^2, \tag{16}$$

or

$$II: \quad 2\alpha\beta + C = 2p^2. \tag{17}$$

Case I: We substitute (16) into (15) and divide by p^2 to obtain $p^2 = C + \alpha^2 + \beta^2$. Solving (16) for C and substituting, we get $p^2 = p^2 - 2\alpha\beta + \alpha^2 + \beta^2 = p^2 + (\alpha - \beta)^2$, implying $\alpha = \beta$ and $A = B$.

Case II: We make similar substitutions. Combining (17) and (15) we get $p^2 = 2C + \alpha^2 + \beta^2 = 2(2p^2 - 2\alpha\beta) + \alpha^2 + \beta^2 = 4p^2 + (\alpha - \beta)^2 - 2\alpha\beta$, which implies $2\alpha\beta = 3p^2 + (\alpha - \beta)^2 > 3p^2$. However this contradicts $2\alpha\beta = 2p^2 - C < 2p^2$, from (17). Therefore this case is impossible and Theorem 10 is proven. □

We note that Theorem 10 cannot be improved to the cube of a prime, since (18,161,294,343) is Pythagorean.

5. REMARKS

Aside from asking for a complete classification of primitive Pythagorean 4-tuples, several other questions arise from this work, the most obvious being whether one can find Pythagorean 5-tuples, and so on.

Conjecture 11: *For every $n \geq 3$ there are infinitely many primitive Pythagorean n-tuples.*

We say that a Pythagorean n-gon (and its corresponding n-tuple of side lengths) has k *parameters* if its set of side lengths takes on exactly k values. Primitive Pythagorean triples have three parameters and are completely characterized. Primitive Pythagorean 4-tuples with two sides equal have three parameters and are completely characterized in Section 3.

Problem 12: *Characterize primitive Pythagorean n-tuples having three parameters.*

The interesting thing about the 5-tuple (169,520,561,425,1105) is that the three chords closest to the diameter geometrically are integral, meaning that it contains as a subfigure the Pythagorean 4-tuple (169,943,425,1105) of surd 1 (its diagonals are 1020 and 1092). Likewise, this 4-tuple contains two Pythagorean triples. Such 5-tuples (in general, n-tuples) are called *nested.*

Question 13: *Are there infinitely many n for which there eixsts a nested Pythagorean n-tuple?*

Notice that this question doesn't require primitivity. The following related question does, however.

Question 14: *Is there some n for which there exist infinitely many nested primitive Pythagorean n-tuples?*

It is also possible to prove that all 5 chords interior to a Pythagorean pentagon are rational. This raises the following natural problem.

Problem 15: *Classify all primitve Pythagorean 4-tuples of surd 1 (i.e., nested primitive 4-tuples).*

Fässler [2] considers questions about Pythagorean triples such as how many there are with a common hypothenuse, perimeter, area, etc. One can ask the same questions regarding 4-tuples. In particular, can one characterize those integers P which are perimeters of primitive Pythagorean 4-tuples? And for such P, how many distinct 4-tuples are there? Also, can one classify all integers r which are the root of some admissible q? More questions abound.

ACKNOWLEDGMENT

The authors wish to thank Lucille and David Oakstreet for some of the inspiration which led to this work.

REFERENCES

[1] Anglin, W.S. "The Square Pyramid Puzzle." *Amer. Math. Monthly*, Vol. *97* (1990): pp. 120-124.

[2] Fässler, A. "Multiple Pythagorean Number Triples" *Amer. Math. Monthly*, Vol. *98* (1991): pp. 505-517.

[3] Grosswald, Emil Representations of Integers as Sums of Squares. Springer-Verlag, New York, 1985.

[4] Guy, R. Unsolved Problems in Number Theory. (2nd ed.), Springer-Verlag, New York, 1994.

[5] Hilbert, D., in *Math. Annelen*, Vol. *67* (1909): pp. 61-75.

[6] Lagrange, Nouv. Mem. Acad. Roy. Sc. de Berlin. Année 1770, Berlin, 1772 pp. 123-133.

[7] Lucas, Edouard, "Question 1180." *Nouvelles Annales de Mathématiques, ser. 2*, Vol. *14* (1875): p. 336.

[8] Coxeter, H.S.M. and Greitzer, S.L. Geometry Revisited. Random House, New York, 1967.

[9] Waring, E. Meditationes Algebraicae. Cambridge, 1770, pp. 204-205.

[10] Wiles, Andrew "Modular Elliptic Curves and Fermat's Last Theorem." *Annals of Math*, Vol. *142* (1995): pp. 443-551.

AMS Classification Numbers: 11B37, 11D25, 51M04

A GENERAL LACUNARY RECURRENCE FORMULA

F. T. Howard

1. INTRODUCTION

The Bernoulli numbers B_n may be defined by means of the generating function

$$\frac{x}{e^x - 1} = \sum_{n=0}^{\infty} B_n \frac{x^n}{n!}. \tag{1.1}$$

An example of a "lacunary" recurrence for these numbers is

$$\sum_{j=0}^{n} \binom{6n+3}{6j} B_{6j} = 2n + 1. \tag{1.2}$$

This recurrence has lacunae, or gaps, or length 6. That is, to compute B_{6n}, it is not necessary to know the values of B_j for all $j < 6n$; we need only know the values of B_{6j} for $j = 0, 1, \cdots, n-1$.

The purpose of this paper is to prove a general lacunary recurrence, for arbitrary gaps, that is applicable to the Bernoulli numbers, the Genocchi numbers, the Eulerian numbers, the Fibonacci numbers, and many other special sequences. The writer believes that the method used in this paper is new and that the formulas are especially easy to use. It is interesting that some of the formulas with gaps of 5 involve the Lucas numbers.

This paper is in final form and no version of it will be submitted for publication elsewhere.

F.T. Howard (ed.), *Applications of Fibonacci Numbers, Volume 9: Proceedings of the Tenth International Research Conference on Fibonacci Numbers and their Applications*, 121–135.

The problem of finding lacunary recurrences for the Bernoulli numbers has a long history, with the motivation being to find quick ways of computing the numbers. Using different methods, Ramanujan [4], [11], Lehmer [10], Riordan [12, pp. 138–140], Chellali [2], Yalavigi [13], and Berndt [1] have all worked out formulas. References to the nineteenth century work of van den Berg and Haussner, and other historical information, can be found in [10].

2. A GENERAL FORMULA

Let $F(x)$ be a function, not identically 0, that can be represented by a power series with a positive radius of convergence:

$$F(x) = \sum_{n=0}^{\infty} f_n \frac{x^n}{n!}. \tag{2.1}$$

If $n < 0$, we define $f_n = 0$. Define the numbers a_n by means of the generating function

$$\frac{hx^t}{F(x)} = \sum_{n=0}^{\infty} a_n \frac{x^n}{n!} \tag{2.2}$$

where t is an arbitrary nonnegative integer and h is an arbitrary rational number. The following lemma [5] is essential, and for completeness we include the proof.

Lemma 2.1: Let m be a positive integer, and let $\theta = e^{2\pi i/m}$, a primitive m^{th} root of unity. Let

$$F(x)F(\theta x)\cdots F(\theta^{m-1}x) = \sum_{n=0}^{\infty} b_n \frac{x^n}{n!}.$$

Then $b_n = 0$ unless m divides n. That is,

$$F(x)F(\theta x)\cdots F(\theta^{m-1}x) = \sum_{n=0}^{\infty} b_{mn} \frac{x^{mn}}{(mn)!}. \tag{2.3}$$

Proof: Let $H(x) = F(x)F(\theta x)\cdots F(\theta^{m-1}x)$. Clearly $H(x) = H(\theta x)$, so $b_n = \theta^n b_n$ for $n = 0, 1, 2, \cdots$. Since $\theta^n = 1$ only when m divides n, we see that the lemma is valid. This completes the proof. □

For our main result, Theorem 2.1 below, we also need the numbers c_n, defined by

$$F(\theta x)F(\theta^2 x)\cdots F(\theta^{m-1}x) = \sum_{n=0}^{\infty} c_n \frac{x^n}{n!}, \tag{2.4}$$

with $c_n = 0$ for $n < 0$. Also, throughout the paper we use the notation

$$(x)_t = x(x-1)(x-2)\cdots(x-t+1). \tag{2.5}$$

Theorem 2.1: Let m be a positive integer and let θ be a primitive m^{th} root of unity. Let a_n, b_n, c_n be defined by (2.2), (2.3) and (2.4), respectively. Then for $0 \le r < m$,

$$\sum_{j=0}^{n} \binom{mn+r}{mj+r} b_{m(n-j)} a_{mj+r} = h(mn+r)_t c_{mn+r-t}. \tag{2.6}$$

Proof: In (2.3) divide both sides by $F(x)$ to obtain

$$F(\theta x)\cdots F(\theta^{m-1}x) = \frac{1}{F(x)} \sum_{n=0}^{\infty} b_{mn} \frac{x^{mn}}{(mn)!}. \tag{2.7}$$

Now multiply both sides of (2.6) by hx^t, and use (2.4), to get

$$hx^t \sum_{n=0}^{\infty} c_n \frac{x^n}{n!} = \frac{hx^t}{F(x)} \sum_{n=0}^{\infty} b_{mn} \frac{x^{mn}}{(mn)!}. \tag{2.8}$$

Substitute (2.2) into (2.8), and compare coefficients of $x^{mn+r}/(mn+r)!$ to complete the proof.
□

We now look at two simple special cases. If $m = 1$, then $\theta = 1$, $b_n = f_n$, $c_0 = 1$ and $c_n = 0$ if $n \neq 0$. Theorem 2.1 gives the recurrence

$$\sum_{j=0}^{n} \binom{n}{j} f_{n-j} a_j = h(n)_t c_{n-t}. \tag{2.9}$$

For example, for the Bernoulli numbers, $f_0 = 0$ and $f_n = 1$ for $n > 0$. Thus (2.9) gives us $B_0 = 1$, and for $n > 1$:

$$\sum_{j=0}^{n-1} \binom{n}{j} B_j = 0.$$

If $m = 2$, then $\theta = -1$, $b_{2n} = \sum_{j=0}^{2n} \binom{2n}{j} (-1)^j f_j f_{2n-j}$, and $c_n = (-1)^n f_n$. Thus for $r = 0$ or $r = 1$,

$$\sum_{j=0}^{n} \binom{2n+r}{2j+r} b_{2(n-j)} a_{2j+r} = h(2n+r)_t (-1)^{r-t} f_{2n+r-t}. \tag{2.10}$$

For the Bernoulli numbers, $b_0 = 0$ and $b_{2n} = -2$, so (2.10) gives us: for $n \geq 1$

$$\sum_{j=0}^{n-1} \binom{2n}{2j} B_{2j} = n.$$

3. BERNOULLI, GENOCCHI AND EULERIAN NUMBERS

In this section, and section 4, we show how Theorem 2.1 can be applied to the Bernoulli, Genocchi and Eulerian numbers for $m = 3$, 4 and 5. More generally, the results of these sections are for numbers a_n defined by the following type of generating function: Let h, v and q be nonzero rational numbers, let w be an arbitrary rational number, and let t be a nonnegative integer. Let

$$F(x) = ve^{qx} + w, \tag{3.1}$$

in definition (2.2). Note that if $v + w \neq 0$, then $a_n = 0$ for $n < t$ and $a_t = t!h/(v + w)$. If $v + w = 0$, then $a_t = 0$ for $n < (t-1)$ and $a_{t-1} = (t-1)!h/(qv)$. It follows from (2.9), with $m = 1$, that

$$(v + w)a_n + v\sum_{j=0}^{n-1} \binom{n}{j} a_j q^{n-j} = \begin{cases} 0 \text{ if } n \neq t \\ h(n!) \text{ if } n = t. \end{cases}$$

When $h = t = v = q = 1$ and $w = -1$, we have $a_n = B_n$, the Bernoulli number defined by (1.1). When $h = 2$, $t = v = q = 1$ and $w = 1$, then $a_n = G_n$, the Genocchi number:

$$\frac{2x}{e^x + 1} = \sum_{n=0}^{\infty} G_n \frac{x^n}{n!}.$$

When $t = 0$, $h = u - 1$, $v = u$, $q = 1 - u$ and $w = -1$, then $a_n = A_n(u)$, the Eulerian number:

$$\frac{u - 1}{ue^{(1-u)x} - 1} = \sum_{n=0}^{\infty} A_n(u) \frac{x^n}{n!}.$$

A good reference for all of these special numbers is [3, pp. 48-50].

Using the notation of Section 2, let $m = 3$ and let θ be a primitive third root of unity. Let $F(x)$ be defined by (3.1), and note that $1 + \theta + \theta^2 = 0$. Then

$$\sum_{n=0}^{\infty} b_n \frac{x^n}{n!} = F(x)F(\theta x)F(\theta^2 x) = (v^3 + w^3) + v^2 w \sum_{j=0}^{2} e^{-\theta^j qx} + vw^2 \sum_{j=0}^{2} e^{\theta^j qx}.$$

Thus $b_0 = (v+w)^3$, and for $n > 0$

$$\begin{aligned} b_{3n} &= v^2w(-q)^{3n}(1+\theta^{3n}+\theta^{6n}) + vw^2q^{3n}(1+\theta^{3n}+\theta^{6n}) \\ &= 3vwq^{3n}[(-1)^n v + w]. \end{aligned} \tag{3.2}$$

We now compute c_n. We have

$$\sum_{n=0}^{\infty} c_n \frac{x^n}{n!} = F(\theta x)F(\theta^2 x) = v^2 e^{-qx} + vw(e^{\theta qx} + e^{\theta^2 qx}) + w^2.$$

Thus $c_0 = (v+w)^2$, and for $n > 0$

$$c_n = v^2(-q)^n + vwq^n(\theta^n + \theta^{2n}) = \begin{cases} v^2(-q)^n + 2vwq^n, & \text{if } n \equiv 0 \pmod 3 \\ v^2(-q)^n - vwq^n, & \text{if } n \equiv 1 \text{ or } 2 \pmod 3 \end{cases}. \tag{3.3}$$

By Theorem 2.1, we have for $r = 0,\ 1$, and 2:

$$\sum_{j=0}^{n} \binom{3n+r}{3j+r} b_{3(n-j)} a_{3j+r} = h(3n+r)_t c_{3n+r-t}, \tag{3.4}$$

with $b_{3(n-j)}$ and c_{3n+r-t} given by (3.2) and (3.3), respectively. For the Bernoulli numbers, the case $r = 0$ gives formula (1.2), and we get similar formulas for $r = 1$ and $r = 2$. We get gaps of 6, instead of 3, because $B_n = 0$ if n is odd, $n > 1$. That is, in (3.4) we can assume $3j + r$ is even for the Bernoulli numbers.

We proceed in the same way when $m = 4$ and $\theta = i$, a primitive fourth root of unity. We have

$$\begin{aligned} \sum_{n=0}^{\infty} b_n \frac{x^n}{n!} &= F(x)F(\theta x)F(\theta^2 x)F(\theta^3 x) \\ &= (v^4+w^4) + v^3 w \sum_{j=0}^{3} e^{-\theta^j qx} + v^2w^2 \sum_{0 \le j < s \le 3} e^{(\theta^j + \theta^s)qx} + vw^3 \sum_{j=0}^{3} e^{\theta^j qx}. \end{aligned} \tag{3.5}$$

Now since

$$1 + i^2 = i + i^3 = 0,$$

$$i^2 + i^3 = -(i+1),\ i + i^2 = i - 1,\ 1 + i^3 = 1 - i,$$

$$(1+i)^4 = (1-i)^4 = -4,$$

from (3.5) we have for $n > 0$

$$\begin{aligned} b_{4n} &= v^3w[4q^{4n}] + v^2w^2[4(-4)^n q^{4n}] + vw^3[4q^{4n}] \\ &= 4vwq^{4n}[vw(-4)^n + w^2 + v^2], \\ \text{and } b_0 &= (v+w)^4. \end{aligned} \tag{3.6}$$

To compute c_n, we examine

$$\sum_{n=0}^{\infty} c_n \frac{x^n}{n!} = F(\theta x)F(\theta^2 x)F(\theta^3 x)$$

$$= w^3 + v^3 e^{(\theta+\theta^2+\theta^3)qx} + v^2 w \sum_{0<j<s\leq 3} e^{(\theta^j+\theta^s)qx} + vw^2 \sum_{j=1}^{3} e^{\theta^j qx}.$$

Thus we have $c_0 = (v+w)^3$, and for $n > 0$:

$$c_n = q^n\{v^3(-1)^n + v^2w[(i-1)^n + (-i-1)^n] + vw^2[(-1)^n + i^n + (-i)^n]\}.$$

This gives

$$c_{4n} = q^{4n}v[v^2 + 2vw(-4)^n + 3w^2], \tag{3.7}$$
$$c_{4n+1} = q^{4n+1}v[-v^2 - 2vw(-4)^n - w^2], \tag{3.8}$$
$$c_{4n+2} = q^{4n+2}v[v^2 - w^2], \tag{3.9}$$
$$c_{4n+3} = q^{4n+3}v[-v^2 + 4v^2w(-4)^n - w^2]. \tag{3.10}$$

Thus when $m = 4$, for $r = 0,\ 1,\ 2,\ 3$ we have:

$$\sum_{j=0}^{n} \binom{4n+r}{4j+r} b_{4(n-j)} a_{4j+r} = h(4n+r)_t c_{4n+r-t},$$

where $b_{4(n-j)}$ and c_{4n+r-t} are given by (3.6) and (3.7) – (3.10), respectively. When $r = 0$, for example, we get the following formula for the Bernoulli numbers:

$$\sum_{j=0}^{n-1} \binom{4n}{4j} [4(-4)^{n-j} - 8] B_{4j} = 4n[(-4)^n - 2],$$

and there are similar formulas for $r = 1,\ 2$ and 3.

4. GAPS OF LENGTH 5

For the numbers a_n defined by (3.1), the lacunary formulas with gaps of length 5 involve Lucas numbers. This happens because of the following relationships: Let $\theta = e^{2\pi i/5}$, a primitive fifth root of unity. Then

$$\theta + \theta^4 = \frac{-1+\sqrt{5}}{2}, \quad \theta^2 + \theta^3 = \frac{-1-\sqrt{5}}{2}.$$

so

$$L_n = (-1)^n[(\theta+\theta^4)^n + (\theta^2+\theta^3)^n], \tag{4.1}$$

where L_n is the nth Lucas number. Since θ is a fifth root of unity, the following equations are obvious:

$$\left.\begin{array}{ll} \theta^3+\theta^4 = \theta(\theta^2+\theta^3), & \theta^2+1 = \theta(\theta+\theta^4) \\ \theta^4+1 = \theta^2(\theta^2+\theta^3), & \theta^3+\theta = \theta^2(\theta+\theta^4) \\ 1+\theta = \theta^3(\theta^2+\theta^3), & \theta^4+\theta^2 = \theta^3(\theta+\theta^4) \\ \theta+\theta^2 = \theta^4(\theta^2+\theta^3), & 1+\theta^3 = \theta^4(\theta+\theta^4) \end{array}\right\} \tag{4.2}$$

We also note that

$$\theta^5 = 1 \text{ and } 1+\theta+\theta^2+\theta^3+\theta^4 = 0. \tag{4.3}$$

Using the notation of section 3, with $m = 5$, we have

$$\sum_{n=0}^{\infty} b_n \frac{x^n}{n!} = F(x)F(\theta x)F(\theta^2 x)F(\theta^3 x)F(\theta^4 x)$$

$$= (v^5+w^5) + v^4 w \sum_{j=0}^{4} e^{-\theta^j qx} + v^3 w^2 \sum_{0\le j<s\le 4} e^{-(\theta^j+\theta^s)qx}$$

$$+ v^2 w^3 \sum_{0\le j<s\le 4} e^{(\theta^j+\theta^s)qx} + vw^4 \sum_{j=0}^{4} e^{\theta^j qx}. \tag{4.4}$$

By (4.1), (4.2), (4.3) and (4.4), we have $b_0 = (v+w)^5$ and for $n > 0$:

$$\begin{aligned} b_{5n} &= v^4 w q^{5n}[5(-1)^n] + v^3 w^2 q^{5n}[5(-1)^n\{(\theta^2+\theta^3)^{5n} + (\theta+\theta^4)^{5n}\}] \\ &\quad + v^2 w^3 q^{5n}[5\{(\theta^2+\theta^3)^{5n} + (\theta+\theta^4)^{5n}\}] + 5vw^4 q^{5n} \\ &= 5vwq^{5n}[(-1)^n v^3 + v^2 w L_{5n} + vw^2(-1)^n L_{5n} + w^3] \end{aligned} \tag{4.5}$$

To compute c_n, we will use (4.1)–(4.3) and the fact that

$$(\theta+\theta^4)(\theta^2+\theta^3) = -1. \tag{4.6}$$

We have

$$\sum_{n=0}^{\infty} c_n \frac{x^n}{n!} = F(\theta x)F(\theta^2 x)F(\theta^3 x)F(\theta^4 x)$$

$$= w^4 + v^4 e^{-qx} + v^3 w \sum_{s=1}^{4} e^{-(1+\theta^s)qx} + v^2 w^2 \sum_{1 \le j < s \le 4} e^{(\theta^j + \theta^s)qx}$$

$$+ v w^3 \sum_{j=1}^{4} e^{\theta^j qx}. \tag{4.7}$$

Now we observe that, by (4.2),

$$\sum_{s=1}^{4} e^{-(1+\theta^s)qx} = \sum_{n=0}^{\infty} y_n \frac{x^n}{n!},$$

where

$$y_n = (-q)^n[(\theta^{2n} + \theta^{3n})(\theta^2 + \theta^3)^n + (\theta^n + \theta^{4n})(\theta + \theta^4)^n]$$

$$= \begin{cases} 2q^n L_n & \text{if } n \equiv 0 \pmod 5, \\ -q^n L_{n+1} & \text{if } n \equiv 1 \text{ or } 4 \pmod 5, \\ q^n L_{n-1} & \text{if } n \equiv 2 \text{ or } 3 \pmod 5. \end{cases} \tag{4.8}$$

Also by (4.2),

$$\sum_{1 \le j < s \le 4} e^{(\theta^j + \theta^s)qx} = \sum_{n=0}^{\infty} p_n \frac{x^n}{n!},$$

where

$$p_n = q^n[(-1)^n L_n + (\theta^n + \theta^{4n})(\theta^2 + \theta^3)^n + (\theta^{2n} + \theta^{3n})(\theta + \theta^4)^n]$$

$$= \begin{cases} 3(-q)^n L_n & \text{if } n \equiv 0 \pmod 5, \\ (-q)^n L_{n+1} & \text{if } n \equiv 1 \text{ or } 4 \pmod 5, \\ -(-q)^n L_{n-1} & \text{if } n \equiv 2 \text{ or } 3 \pmod 5. \end{cases} \tag{4.9}$$

By (4.7), (4.8) and (4.9), $c_0 = (v+w)^4$ and

$$c_n = \begin{cases} q^n[(-1)^n v^4 + 4vw^3 + 2v^3wL_n + 3(-1)^n v^2w^2L_n], & \text{if } n \equiv 0 \pmod 5, \\ q^n[(-1)^n v^4 - vw^3 - v^3wL_{n+1} + (-1)^n v^2w^2L_{n+1}], & \text{if } n \equiv 1 \text{ or } 4 \pmod 5, \\ q^n[(-1)^n v^4 - vw^3 + v^3wL_{n-1} + (-1)^{n+1} v^2w^2L_{n-1}], & \text{if } n \equiv 2 \text{ or } 3 \pmod 5. \end{cases} \tag{4.10}$$

Thus for $m = 5$ and $r = 0,1, 2, 3$, or 4:

$$\sum_{j=0}^{n} \binom{5n+r}{5j+r} b_{5(n-j)} a_{5j+r} = h(5n+r)_t c_{5n+r-t},$$

where $b_{5(n-j)}$ and c_{5n+r-t} are given by (4.5) and (4.10), respectively. For example, for the Bernoulli numbers with $r = 0$, we have

$$5\sum_{j=0}^{n} \binom{10n+5}{10j+5} (1 + L_{10(n-j)+5}) B_{10j} = (10n+5)(1 + L_{10n+5}).$$

5. THE RECIPROCAL OF $e^x - 1 - x - \cdots - \frac{x^{k-1}}{(k-1)!}$

Let $k \geq 0$, and define

$$F_k(x) = e^x - 1 - x - \cdots - \frac{x^{k-1}}{(k-1)!}. \tag{5.1}$$

In this section we show how Theorem 2.1 can be applied to the numbers $B_{k,n}$ defined by

$$\frac{x^k/k!}{F_k(x)} = \sum_{n=0}^{\infty} B_{k,n} \frac{x^n}{n!}. \tag{5.2}$$

We first observe that $B_{0,n} = (-1)^n$, and $B_{1,n} = B_n$, the Bernoulli number. The numbers $B_{2,n}$ have been examined in some detail [7], [9], and $B_{k,n}$, for general k, has also been studied [8]. To avoid confusion with the Eulerian numbers, in the present paper we have changed the notation of [7]–[9] from A_n to $B_{2,n}$ and from $A_{k,n}$ to $B_{k,n}$.

By (2.9) we have the simple recurrence: $B_{k,0} = 1$ and for $n > k$

$$\sum_{j=0}^{n-k} \binom{n}{j} B_{k,j} = 0.$$

By (2.10) we have for $r = 0$, $r = 1$, and $n \geq k$:

$$\sum_{j=0}^{n-k} \binom{2n+r}{2j+r} b_{2(n-j)} B_{k,2j+r} = \binom{2n+r}{k} (-1)^{r-k},$$

where

$$b_{2(n-j)} = \sum_{s=k}^{2n-2j-k} (-1)^s \binom{2n-2j}{s} = 2\sum_{s=0}^{k-1} (-1)^{s+1} \binom{2n-2j}{s}.$$

Let $m = 3$, let θ be a primitive third root of unity and let $F_k(x)$ be defined by (5.1). Define $b_{k,n}$, $c_{k,n}$, $G_k(x)$ and $H_k(x)$ in the following way.

$$G_k(x) = F_k(x)F_k(\theta x)F_k(\theta^2 x) = \sum_{n=0}^{\infty} b_{k,n} \frac{x^n}{n!}, \tag{5.3}$$

$$H_k(x) = F_k(\theta x)F_k(\theta^2 x) = \sum_{n=0}^{\infty} c_{k,n} \frac{x^n}{n!}. \tag{5.4}$$

Thus

$$\begin{cases} b_{0,0} = 1, \text{ and } b_{0,n} = 0 \text{ for } n > 0, \\ c_{0,n} = (-1)^n. \end{cases} \tag{5.5}$$

Our goal here is to use an inductive method to find formulas for $b_{k,n}$ and $c_{k,n}$ for general k. Since

$$G_{k+1}(x) = \prod_{j=0}^{2} \left(F_k(\theta^j x) - \frac{(\theta^j x)^k}{k!} \right),$$

we have

$$G_{k+1}(x) =$$

$$G_k(x) - \frac{x^k}{k!}\sum_{j=0}^{2}\left[\theta^{jk}H_k(\theta^j x)\right] + \frac{x^{2k}}{(k!)(k!)}\sum_{j=0}^{2}\left[\theta^{(3-j)k}F_k(\theta^j x)\right] - \frac{x^{3k}}{(k!)(k!)(k!)}. \quad (5.6)$$

Now

$$x^k\sum_{j=0}^{2}\left[\theta^{jk}H_k(\theta^j x)\right] = \sum_{n=2k}^{\infty}\left[1+\theta^{n+k}+\theta^{2n+2k}\right]c_{k,n}\frac{x^{n+k}}{n!}$$

$$= \sum_{n=3k}^{\infty}\left[1+\theta^{n}+\theta^{2n}\right](n)_k c_{k,n-k}\frac{x^n}{n!}, \quad (5.7)$$

and

$$x^{2k}\sum_{j=0}^{2}\left[\theta^{(3-j)k}F_k(\theta^j x)\right] = \sum_{n=k}^{\infty}\left[1+\theta^{n+2k}+\theta^{2n+k}\right]\frac{x^{n+2k}}{n!}$$

$$= \sum_{n=3k}^{\infty}\left[1+\theta^{n}+\theta^{2n}\right](n)_{2k}\frac{x^n}{n!}. \quad (5.8)$$

By (5.6), (5.7) and (5.8) we have, for $n > k$:

$$b_{k+1,3n} = b_{k,3n} - 3\binom{3n}{k}c_{k,3n-k} + 3\binom{3n}{2k}\binom{2k}{k}. \quad (5.9)$$

If $n = k$, then $(3k)!/(k!)^3$ must be subtracted from (5.9).
Next we find a recurrence for $c_{k,n}$. Since

$$H_{k+1}(x) = \prod_{j=1}^{2}\left(F_k(\theta^j x) - \frac{(\theta^j x)^k}{k!}\right),$$

it is clear that

$$H_{k+1}(x) = H_k(x) - \frac{x^k}{k!}\sum_{j=1}^{2}\left[\theta^{(3-j)k}F_k(\theta^j x)\right] + \frac{x^{2k}}{(k!)(k!)}.$$

Thus for $n > k$:

$$c_{k+1,n} = c_{k,n} - \binom{n}{k}(\theta^{n+k} + \theta^{2n-k})$$

$$= \begin{cases} c_{k,n} - 2\binom{n}{k} \text{ if } n+k \equiv 0 \pmod 3, \\ c_{k,n} + \binom{n}{k} \text{ if } n+k \equiv 1 \text{ or } 2 \pmod 3. \end{cases} \tag{5.10}$$

If $n = 2k$, then $\binom{2k}{k}$ must be added to (5.10).

Thus we can say: For $r = 0, 1, 2$

$$\sum_{j=0}^{n}\binom{3n+r}{3j+r} b_{k,(n-j)} B_{k,3j+r} = \binom{3n+r}{k} c_{k,3n+r-k}, \tag{5.11}$$

where $b_{k,(n-j)}$ and $c_{k,3n+r-k}$ are given recursively by (5.9) and (5.10). For example, using (5.5) as a starting point, we have

$$b_{1,3n} = -3(-1)^n + 3 = \begin{cases} 0 \text{ if } n \text{ is even,} \\ 6 \text{ if } n \text{ is odd,} \end{cases}$$

and

$$c_{1,n} = \begin{cases} (-1)^n - 2 \text{ if } n \equiv 0 \pmod 3, \\ (-1)^n + 1 \text{ if } n \equiv 1 \text{ or } 2 \pmod 3, \end{cases}$$

and these values agree with (3.2) and (3.3). For $k = 2$, we obtain

$$b_{2,3n} = 3(3n-1)\left[3n - 1 + (-1)^n\right], \tag{5.12}$$

$$c_{2,n} = \begin{cases} (-1)^n - 2 + n \text{ if } n \equiv 0 \pmod 3, \\ (-1)^n + 1 + n \text{ if } n \equiv 1 \pmod 3, \\ (-1)^n + 1 - 2n \text{ if } n \equiv 2 \pmod 3. \end{cases} \tag{5.13}$$

For example, let $k = 2, r = 0$ and define $g(n) = (3n-1)[3n-1+(-1)^n]$. By (5.11), (5.12) and (5.13), we have for $n > 0$:

$$\sum_{j=0}^{n} \binom{3n+6}{3j} g(n+2-j) B_{2,3j} = \frac{1}{2}(n+2)g(n+2). \tag{5.14}$$

To illustrate how (5.14) can be used to compute $B_{2,3n}$, we first note that $g(2) = 30$, $g(3) = 56$ and $g(4) = 132$. Then by (5.14), for $n = 0$ we get $B_{2,0} = 1$. For $n = 1$, we get

$$\binom{9}{3} g(2) B_{2,3} + g(3) B_{2,0} = \frac{3}{2} g(3),$$

so

$$B_{2,3} = \frac{1}{90}.$$

For $n = 2$, we have

$$\binom{12}{6} g(2) B_{2,6} + \binom{12}{3} g(3) B_{2,3} + g(4) B_{2,0} = 2g(4),$$

which gives

$$B_{2,6} = \frac{-1}{5670}.$$

The recursive method used in this section can also be used for $m = 4$ and $m = 5$, but the formulas become complicated and cumbersome. For example, here is the formula (without proof) for $b_{2,5n}$.
For n odd, $n > 1$: $b_{2,5n} = 5(5n-1)(5n-2)(25n^2 - 20n + 1 + L_{5n} - 5nL_{5n-3})$;
For n even, $n > 0$: $b_{2,5n} = 25n(5n-1)(5n-3)(5n-3+L_{5n-3})$.
Also, $c_{2,5n} = c_{1,5n} + 5n[(1+2(-1)^{n-1})L_{5n} + (5n-1)(-1)^n L_{5n-1} + (25n^2 - 25n + 7)]$ where the value of $c_{1,5n}$ can be computed from (4.10). There are similar formulas for $c_{2,5n+r}$, $(r = 1, 2, 3, 4)$.

6. FINAL COMMENTS

The generating function (2.1) can be generalized by defining polynomials $a_n(z)$ by means of

$$\frac{hx^t e^{xz}}{F(x)} = \sum_{n=0}^{\infty} a_n(z) \frac{x^n}{n!}. \tag{6.1}$$

Thus

$$a_n(z) = \sum_{j=0}^{n} \binom{n}{j} a_j z^{n-j}. \tag{6.2}$$

One well known example is the Bernoulli polynomial $B_n(z)$, defined by

$$\frac{xe^{xz}}{e^x - 1} = \sum_{n=0}^{\infty} B_n(z) \frac{x^n}{n!}.$$

Another example is the polynomial $B_{2,n}(z)$, studied in [7] and [9]:

$$\frac{(x^2/2)e^{xz}}{e^x - 1 - x} = \sum_{n=0}^{\infty} B_{2,n}(z) \frac{x^n}{n!}.$$

Define $c_n(z)$ by

$$c_n(z) = \sum_{j=0}^{n} \binom{n}{j} c_j z^{n-j}. \tag{6.3}$$

It is easy to see that for the polynomials defined by (6.1) and (6.3), we can extend Theorem 2.1 in the following way.

Theorem 6.1: Let m be a positive integer and let θ be a primitive m^{th} root of unity. Let $a_n, b_n, c_n, a_n(z), c_n(z)$ be defined by (2.2), (2.3), (2.4) (6.1), (6.3), respectively. Then for $0 \le r < n$,

$$\sum_{j=0}^{n} \binom{mn+r}{mj+r} b_{m(n-j)} a_{mj+r}(z) = h(mn+r)_t c_{mn+r-t}(z).$$

We also observe that the Genocchi polynomial $G_n(z)$ defined by

$$\frac{2xe^{xz}}{e^x + 1} = \sum_{n=0}^{\infty} G_n(z) \frac{x^n}{n!}$$

is closely related to the Euler number E_n [3, p. 48]:

$$\frac{2e^x}{e^{2x} + 1} = \sum_{n=0}^{\infty} E_n \frac{x^n}{n!}.$$

It is clear that we have the relationship

$$(n+1)E_n = 2^n G_{n+1}\left(\frac{1}{2}\right),$$

and Theorem 6.1 can be used to find a lacunary recurrence for the Euler numbers.

In a letter to the writer, A. Granville made the following observation. If $F(x)$ in the present paper is replaced by $G(x, e^x)$, where $G(x, y)$ is a polynomial in two variables with integer coefficients, then b_n and c_n are both linear combinations of elements of linear recurrence sequences of order dividing $\phi(n)$. We reserve this topic for a later paper.

Finally, we note that the general method of this paper was used by the writer [5], [6] to find lacunary recurrences for the Tribonacci numbers and generalized Fibonacci numbers. To the writer's knowledge, Theorem 2.1 and Theorem 6.1 are new, and the lacunary recurrences for the Eulerian numbers and the numbers $B_{k,n}$ have not appeared in the literature.

REFERENCES

[1] Berndt, B.C. Ramanujan's Notebooks, Part IV. New York: Springer-Verlag, 1994.

[2] Chellali, M. "Accélération de calcul de nombres de Bernoulli." *J. Number Theory*, Vol. *28* (1988): pp. 347-362.

[3] Comtet, L. Advanced Combinatorics. Dordrecht: Reidel, 1974.

[4] Howard, F.T. "Formulas of Ramanujan Involving Lucas Numbers, Pell Numbers, and Bernoulli Numbers." Applications of Fibonacci Numbers. Volume 6. Edited by G.E. Bergum, A.N. Philippou and A.F. Horadam, Kluwer Academic Publishers, Dordrecht, The Netherlands, 1991.

[5] Howard, F.T. "Generalizations of a Fibonacci Identity." Applications of Fibonacci Numbers. Volume 8. Edited by Fredric T. Howard, Kluwer Academic Publishers, Dordrecht, The Netherlands, 1999.

[6] Howard, F.T. "A Tribonacci Identity." *The Fibonacci Quarterly* Vol. *39* (2001): pp. 352-357.

[7] Howard, F.T. "A Sequence of Numbers Related to the Exponential Function." *Duke Mathematical Journal*, Vol. *34* (1967): pp. 599-616.

[8] Howard, F.T. "Some Sequences of Rational Numbers Related to the Exponential Function." *Duke Mathematical Journal*, Vol. *34* (1967): pp. 701-716.

[9] Howard, F.T. "Numbers Generated by the Reciprocal of $e^x - x - 1$." *Mathematics of Computation*, Vol. *31* (1977): pp. 581-598.

[10] Lehmer, D.H. "Lacunary Recurrence Formulas for the Numbers of Bernoulli and Euler." *Annals of Math.*, Vol. *36* (1935): pp. 637-649.

[11] Ramanujan, S. "Some Properties of Bernoulli's Numbers." *J. Indian Math. Society*, Vol. *3* (1911): pp. 219-234.

[12] Riordan, J. Combinatorial Identities. New York: Wiley, 1968.

[13] Yalavigi, C.C. "Bernoulli and Lucas Numbers." *The Mathematics Education*, Vol. *4* (1971): pp. 99-102.

AMS Classification Numbers: 11B68, 11B39

ORDERING WORDS AND SETS OF NUMBERS: THE FIBONACCI CASE

Clark Kimberling

1. INTRODUCTION

Let S be the set of numbers defined by these rules: $0 \in S$ and $1 \in S$, and if $x \in S - \{0\}$, then $2x \in S$ and $4x+1 \in S$. The elements of S form a sequence $s = (0, 1, 2, 4, 5, 8, 9, 10, 16, \ldots)$ that has remarkable connections with the sequence of Fibonacci numbers. For example, the positions of the even numbers in s are the positions of the 0's in the infinite Fibonacci word, which begins with 0100010100. This ordering also matches that of the set of binary words under a certain order relation. The purpose of this paper is to explore various sequences having this ordering.

2. ORDERING BINARY WORDS

Consider the set of non-empty words defined on the alphabet (0,1). The set of all such words can be ordered in groups, represented as rows in Table 1:

row 1	0							
row 2	0^2	1						
row 3	0^3	01	10					
row 4	0^4	0^21	010	10^2	1^2			
row 5	0^5	0^31	0^210	010^2	01^2	10^3	101	1^20.

Table 1. Binary words ordered

This paper is in final form and no version of it will be submitted for publication elsewhere.

F.T. Howard (ed.), Applications of Fibonacci Numbers, Volume 9: Proceedings of the Tenth International Research Conference on Fibonacci Numbers and their Applications, 137–144.

To generate row n for $n \geq 3$, prefix each word in row $n-1$ with 0, prefix each word in row $n-2$ with 1, and order the resulting words by the relation $\prec$ defined as follows:

$$x_1 x_2 \ldots x_h \prec y_1 y_2 \ldots y_k$$

if one of these conditions holds:

i. $x_1 < y_1$;

ii. $x_i = y_i$ for $i = 1, 2, \ldots, h'$ and $x_{h'+1} < y_{h'+1}$ for some $h' < k$.

In Table 1, replace each word by its final letter, obtaining these rows:

$$0; \quad 0,1; \quad 0,1,0; \quad 0,1,0,0,1; \quad 0,1,0,0,1,0,1,0.$$

Here, for $n \geq 3$, row n consists of row $n-1$ followed by row $n-2$, so that as $n \to \infty$, an infinite word beginning with 010010100100101001010 is defined. This is the well known *infinite Fibonacci word* $\ell_1 \ell_2 \ell_3 \ldots$ (*A*003849 in Sloane [2]) given by $\ell_i = 0$ or 1 according as i is of the form $\lfloor n\tau \rfloor$ or $\lfloor n\tau \rfloor + n$, where $\tau = (1+\sqrt{5})/2$ is the golden mean.

We return now to Table 1. The method of generating of row n from rows $n-1$ and $n-2$ yields the following for $n \geq 3$: row n consists of F_{n+1} words, of which F_n begin with 0, F_n end with 0, F_{n-1} begin with 1, and F_{n-1} end with 1. Moreover, if $x(n,a,b)$ denotes the number of words $\ell_1 \ell_2 \ldots \ell_k$ in row n of Table 1 that satisfy $\ell_1 = a$ and $\ell_k = b$, then the recurrences

$$x(n,0,0) = x(n-1,0,0) + x(n-1,1,0), \quad x(n,0,1) = x(n-1,0,1) + x(n-1,1,1);$$
$$x(n,1,0) = x(n-2,0,0) + x(n-2,1,0), \quad x(n,1,1) = x(n-2,1,0) + x(n-2,1,1)$$

with initial values in Table 1 yield, for $n \geq 3$:

$$x(n,0,0) = F_{n-1}, \quad x(n,0,1) = F_{n-2};$$
$$x(n,1,0) = F_{n-2}, \quad x(n,1,1) = F_{n-3}.$$

It can be shown that the number of 0's in row n is given by the sequence (1,2,5,10,20,38,71,...), the self-convolution (A001629) of the Fibonacci sequence (A000045).

2. ORDERING SEQUENCES OF NUMBERS

We turn now to the set S defined in the Introduction. The elements of S can be grouped as in Table 1, with an additional row 0:

row 0	0	1						
row 1	2							
row 2	4	5						
row 3	8	9	10					
row 4	16	17	18	20	21			
row 5	32	33	34	36	37	40	41	42

Table 2. The set S ordered

As in Table 1, note that row n, for $n \geq 3$, consists of F_{n+1} terms, of which F_n are $2x$ for x in row $n-1$ and F_{n-1} are $4x+1$ for x in row $n-2$. The sequence obtained by arranging the numbers in S in increasing order appears in several guises (see Sloane [2], sequence A003714); for example, these are integers whose binary representations have no adjacent 1's. A connection between them and the infinite Fibonacci word has been noted in the Introduction. In order to state and prove the connection in more general form, let $S := S(a,b,c,d,e)$ be the set of numbers defined by these rules:

i. $0 \in S$ and $e \in S$,

ii. if $x \in S - \{0\}$, then $ax+b \in S$ and $cx+d \in S$,

where a,b,c,d,e are numbers for which the sets

$$E := \{ax+b : x \in S\} \cup \{0\} \text{ and } F := \{cx+d : x \in S\} \cup \{e\}$$

are disjoint. The set S can be partitioned into rows, as indicated in Table 3:

row 0	0	e	
row 1	$ae+b$		
row 2	$a^2e+(a+1)b$	$ce+d$	
row 3	$a^3e+(a^2+a+1)b$	$ace+bc+d$	$ace+b+ad$

Table 3. Numbers in $S(a,b,c,d,e)$

For $n \geq 3$ of Table 3 extended, row n consists of the numbers $ax+b$ for x in row $n-1$ and the numbers $cx+d$ for x in row $n-2$. Within row n, the ordering of terms is here defined to be the ordering given by Table 2; that is, the terms are ordered so that for $(a,b,c,d,e) = (2,0,4,1,1)$, Table 3 is identical to Table 2. Under this ordering, there are choices of (a,b,c,d,e) for which each row of Table 3 is increasing. In that case, we shall call $S(a,b,c,d,e)$ a *strict Fibonacci set.*

3. STRICT FIBONACCI SETS

In order to characterize strict Fibonacci sets, it will be helpful to begin with some notation. First, we rewrite Table 3 with a^2 substituted for c:

row 0	0	e	
row 1	$ae+b$		
row 2	$a^2e+(1+a)b$	a^2e+d	
row 3	$a^3e+(1+a+a^2)b$	a^3e+a^2b+d	$a^3e+b+ad$

Table 4. Numbers in $S(a,b,a^2,d,e)$

Each entry in row n of Table 4, extended, has the form $a^ne+p(a)b+q(a)d$, where $p(a)$ and $q(a)$ are polynomials in a in which every coefficient is 0 or 1. A second manner in which row

n arises from rows $n-1$ and $n-2$ is given by Lemma 1. We write $R_n :=$ the finite sequence of numbers in row n of Table 4, extended, and

$$f(x) := ax + b \qquad g(x) := a^2x + d$$
$$h_n(x) := x + (a^n - a^{n-1})e + a^{n-1}b \quad k_n(x) := x + (a^n - a^{n-2})e + a^{n-2}d$$

Lemma 1: For $n \geq 2$, if $a^{n-1}e+p(a)b+q(a)d \in R_{n-1}$, then $a^n e+(p(a)+a^{n-1})b+q(a)d \in R_n$, and if $a^{n-2}e + p(a)b + q(a)d \in R_{n-2}$, then $a^n e + p(a)b + (q(a) + a^{n-2})d \in R_n$.

Proof: The assertions clearly hold for $n = 2$ and $n = 3$. Assume for arbitrary $n \geq 4$ that they hold for $m = 2, 3, \ldots, n$.

Suppose $s = a^{n-1}e + p(a)b + q(a)d \in R_{n-1}$, and write $s' = a^n e + (p(a) + a^{n-1})b + q(a)d$. If $q(0) \neq 1$, then $s = f(s_1)$ for some s_1 in R_{n-2}, so that

$$s_1 = a^{n-2}e + [(p(a) - 1)/a]b + [q(a)/a]d,$$

and by induction hypothesis, $h_{n-1}(s_1) \in R_{n-1}$, so that $s' = f(h_{n-1}(s_1)) \in R_n$, by the definition of R_n. However, if $q(0) = 1$, then $s = g(s_1)$ for some s_1 in R_{n-3}, and $h_{n-2}(s_1) \in R_{n-2}$, so that $s' = g(h_{n-2}(s_1)) \in R_n$.

Next, suppose $s = a^{n-2}e + p(a)b + q(a)d \in R_{n-2}$, and write

$$s' = a^n e + p(a)b + (q(a) + a^{n-2})d.$$

If $q(0) \neq 1$, then $s = f(s_1)$ for some s_1 in R_{n-3}, so that $k_{n-1}(s_1) \in R_{n-1}$ and $s' = f(k_{n-1}(s_1)) \in R_n$; if $q(0) = 1$, then $s = g(s_1)$ for some s_1 in R_{n-4}, and $k_{n-2}(s_1) \in R_{n-2}$, so that $s' = g(k_{n-2}(s_1)) \in R_n$. □

Theorem 1: Suppose a, b, d, e are real numbers satisfying these conditions:

$$a \geq \tau, \tag{1}$$
$$(a-1)e + b > 0, \tag{2}$$
$$(a+1)b < d, \tag{3}$$
$$d \leq (a+1)[ab + (a-1)^2 e]. \tag{4}$$

Then $S(a, b, a^2, d, e)$ is a strict Fibonacci set.

Proof: With reference to rows 1,2, and 3 of Table 4, and as an initial step in an induction argument, inequalities (1)-(4) imply that

$$e < ae + b < a^2e + (a+1)b < a^2e + d < a^3e + (a^2 + a + 1)b.$$

Assume for arbitrary $n \geq 4$ that each of $R_3, R_4, \ldots, R_{n-1}$ is increasing. To prove R_n increasing, we apply Lemma 1 and the fact that the functions h_n and k_n preserve order: the first F_n numbers are in increasing order, as they are the numbers $h_n(x)$ as x runs through R_{n-1}; likewise, the remaining F_{n-1} numbers in R_n are in increasing order, as they are the numbers $k_n(x)$ as x runs through R_{n-2}. We shall prove that the greatest, $G(n)$, of the first F_n

numbers is less than the least, $L(n)$, of the remaining F_{n-1} numbers. If n is odd, one easily finds

$$G(n) = a^n e + a^{n-1}b + (a^{n-3} + a^{n-5} + \cdots + 1)d,$$
$$L(n) = a^n e + (a^{n-3} + a^{n-4} + \cdots + 1)b + a^{n-2}d,$$

so that the required inequality is equivalent to

$$(a^{n-1} - a^{n-3} - a^{n-4} - \cdots - 1)b < (a^{n-2} - a^{n-3} - a^{n-5} - \cdots - 1)d. \tag{5}$$

To prove (5), we note first that

$$a^n - a^{n-1} - a^{n-2} + 1 = a^{n-2}(a^2 - a - 1) + 1 > 0, \tag{6}$$

by (1), since τ is the greatest root of the polynomial $x^2 - x - 1$. Then by (3),

$$(a-1)(a^n - a^{n-1} - a^{n-2} + 1)b < (a^n - a^{n-1} - a^{n-2} + 1)d.$$

Dividing both sides by $a^2 - 1$ and using the identity $(a^k - 1)/(a-1) = a^{k-1} + a^{k-2} + \cdots + 1$ leads directly to (5). If n is even, then

$$G(n) = a^n e + (a^{n-1} + 1)b + (a^{n-3} + a^{n-5} + \cdots + a)d,$$
$$L(n) = a^n e + (a^{n-3} + a^{n-4} + \cdots + a + 1)b + a^{n-2}d,$$

and the inequality $G(n) < L(n)$ is equivalent to

$$(a^{n-2} - a^{n-4} - a^{n-5} - \cdots - 1)b < (a^{n-3} - a^{n-4} - a^{n-6} - \cdots - 1)d,$$

for which the proof given for (5) applies.

Next we shall prove that $\max(R(n)) < \min(R(n+1))$ for $n \geq 0$. The desired inequalities clearly hold for $n = 0$ and $n = 1$. If n is even and ≥ 2, then the desired inequality is

$$a^n e + (a^{n-2} + a^{n-4} + \cdots + 1)d < a^{n+1}e + (a^n + a^{n-1} + \cdots + 1)b. \tag{7}$$

Inequality (3) gives $-d/a^n < -(a+1)b/a^n$; adding this to both sides of inequality (4) and simplifying gives

$$0 < \frac{a^{n+1} - 1}{a - 1}b - \frac{a^n - 1}{a^2 - 1}d + a^n(a-1)e,$$

which leads directly to (7). If n is odd and ≥ 3, then the desired inequality is

$$a^n e + b + (a^{n-2} + a^{n-4} + \cdots + a)d < a^{n+1}e + (a^n + a^{n-1} + \cdots + 1)b, \tag{8}$$

which is proved in much the same way that (7) was proved for even n.

In Table 4, the entry in row 1 is in the set E, corresponding to the last letter, 0, of the entry in row 1 of Table 1. The entries in row 2 of Table 4 are in the sets E and F, respectively, corresponding to the respective last letters 0 and 1 of the words in row 2 of Table 1. Assume for arbitrary $n \geq 3$ that the numbers in each of $R_0, R_1, \ldots, R_{n-1}$ are in E or F according as the matching words in Table 1, extended end in 0 or 1. The latter ordering for rows n is given by the juxtaposition $W_{n-1}W_{n-2}$, where W_k denotes the word consisting of the initial F_{k+1} letters of the infinite Fibonacci word.

Now consider R_n. By induction hypothesis, its first F_n terms match the letters of W_{n-1}, and its remaining F_{n-1} terms match the letters of W_{n-2}. By induction, the match holds for R_n for all $n \geq 1$. Therefore, beginning with $ae + b$, each term of s is in E or F according as the matching final letter in Table 1 is 0 or 1. This 01-sequence of final letters comprises the subword w of the infinite Fibonacci word $01w$. Since $0 \in E$ and $e \in F$, row 0 of Table 4 completes the match between s and the infinite Fibonacci word. In other words, $S(a, b, a^2, d, e)$ is a strict Fibonacci set. □

The value τ in inequality (1) is best possible. To see this, note that if $0 < a < \tau$, then $a^2 - a - 1 < 0$ in (7), so that for all m beyond some value, the expression to the left of "=" in (6) is negative, and consequently, the required inequality (5) fails to hold.

Example 1: $a = \tau, b = 0, c = \tau^2, d = 1, e = 1$. Since $\tau^2 = \tau + 1$, the numbers in S are of the form $u + v\tau$, where u and v are nonnegative integers.

row 1	0, 1												
row 2	1, 1	2, 1											
row 3	1, 2	2, 2	1, 3										
row 4	2, 3	3, 3	2, 4	3, 4	4, 4								
row 5	3, 5	4, 5	3, 6	4, 6	5, 6	4, 7	5, 7	4, 8					
row 6	5, 8	6, 8	5, 9	6, 9	7, 9	6, 10	7, 10	6, 11	7, 11	8, 11	7, 12	8, 12	9, 12

Table 5. $S(\tau, 0, \tau^2, 1, 1)$ **by rows;** $u + v\tau$ **is written as** u, v

Writing row k of Table 5, extended, as $p_{k,i}, q_{k,i}$ for $i = 1, 2, \ldots, F_{k+1}$, we have the following recurrences:

$$p_{k,i} = \begin{cases} p_{k-1,i} + F_{k-3} & \text{if } 1 \leq i \leq F_k \\ p_{k-2,i-F_k} + F_{k-1} & \text{if } F_k + 1 \leq i \leq i \leq F_{k+1} \end{cases}$$

and

$$q_{k,i} = \begin{cases} q_{k-1,i} + F_{k-2} & \text{if } 1 \leq i \leq F_k \\ q_{k-2,i-F_k} + F_k & \text{if } F_k + 1 \leq i \leq F_{k+1}, \end{cases}$$

with initial values as in Table 5. Solutions are given by

$$p_{k,i} = F_{k-1} + \lfloor i/\tau \rfloor - \lfloor i/\tau^2 \rfloor \text{ and } q_{k,i} = F_k + \lfloor i/\tau^2 \rfloor.$$

These formulas can be used to prove that all Wythoff pairs occur in Table 5, extended, and that the columns in which they occur are numbered 1,6,11,14,19,24,27, ... The difference sequence for these column numbers beginning with $11 - 6, 14 - 11, 19 - 14$ is the infinite Fibonacci word on the alphabet (5,3).

Examples 2-6: More strict Fibonacci sets, as sequences.

a	b	a^2	d	e	initial terms	index in [2]
2	0	4	1	1	0,1,2,4,5,8,9,10,16,17,18,20,21,32	A003714
2	−1	4	0	2	0,2,3,5,8,9,12,15,17,20,23,29,32,33	A060138
3	−1	9	0	1	0,1,2,5,9,14,18,26,41,45,53,77,81,122	A060139
3	0	9	1	1	0,1,3,9,10,27,28,30,81,82,84,90,91,243	A060140
3	0	9	2	1	0,1,3,9,11,27,29,33,81,83,87,99,101,243	A060141

Table 6. $S(a, b, a^2, d, e)$ for selected integers a, b, d, e

4. FIBONACCI SETS, NOT NECESSARILY STRICT

Recall that S is a strict Fibonacci set if, when its elements are arranged in increasing order, the positions of elements of E are identical to the positions of 0 in the infinite Fibonacci word. Theorem 1 recognizes a large class of strict Fibonacci sets $S(a, b, c, d, e)$ for which $c = a^2$. The question, whether the condition $c = a^2$ is necessary, leads to Theorem 2 below. Let us say that $S(a, b, c, d, e)$ has *Fibonacci form* if row n of Table 3, extended, has least element $L_n = a^n e + b(a^n - 1)/(a - 1)$ and greatest element

$$G_n - \begin{cases} aec^{(n-1)/2} + b + ad(c^{(n-1)/2} - 1)/(c - 1), & n \text{ odd} \\ c^{n/2}e + d(c^{n/2} - 1)/(c - 1), & n \text{ even,} \end{cases}$$

and $G_n < L_{n+1}$, for $n \geq 1$. Since a strict Fibonacci set has Fibonacci form, we know by Theorem 1 that there are sets $S(a, b, c, d, e)$ of Fibonacci form for which $c = a^2$. Next, we shall see that this condition, if accompanied by inequalities (1)-(3), is necessary.

Theorem 2: Suppose

$$a \geq \tau \tag{1}$$

$$(a - 1)e + b > 0, \tag{2}$$

$$(a + 1)b < d. \tag{3}$$

If $S = S(a, b, c, d, e)$ has Fibonacci form, then $c = a^2$.

Proof: First, suppose $a^2 < c$ and n is even. Dividing the terms of the desired inequality

$$L_n < G_n < L_{n+1} \tag{9}$$

by $c^{n/2}$ we obtain $0 \leq e + d/(c - 1) \leq 0$, or equivalently, $(c - 1)e = -d$. However, inequalities (2) and (3) imply $(c - 1)e > (a^2 - 1)e > -b(a + 1) > -d$.

Next, suppose $a^2 < c$ and n is odd. Divide in (9) by $c^{n/2}$ and take the limit to obtain $(c - 1)(ae + b) = -d$. However, $ae + b > e$ by (2), so that $(c - 1)(ae + b) > (a^2 - 1)e > -d$. The contradictions imply $c \leq a^2$.

If $c < a^2$, we divide in (9) by a^n and obtain $(a - 1)e + b = 0$, contrary to (2). Consequently, $c = a^2$. □

Example 7: $S = S(2,1,4,0,1)$; $s = (0,1,3,4,7,9,12,15,16,19,\ldots)$. Substituting into Table 3 gives a table in which R_n is decreasing for $n \geq 1$. The sequence formed by juxtaposing these R_n, for $n \geq 0$, namely (0,1,3,7,4,15,12,9,31,28,25,19,16, ...), is a strict Fibonacci set.

Example 8: $S = S(2,-1,4,0,1)$; s is essentially sequence A048297 in [2]. In this case, inequality (2) fails to hold. However, the set $S(2,-1,4,0,4)$ equals $S - \{1\}$ and is a strict Fibonacci set. In general, if for given a, b, c, d, e inequalities (1) and (3)hold but (2) and (4) do not, then, as is easily verified, there exists x_0 such that for all $x > x_0$, the numbers $a, b, c, d, e+x$ satisfy (1)-(4), so that $S(a, b, c, d, e+x)$ is a strict Fibonacci set.

Example 9: $S = S(2,0,4,-1,1)$. Here, inequality (3) fails to hold. Substituting into Table 3, we obtain R_n, for $n \geq 2$, matching $w_{n-2}w_{n-1}$, rather than $w_{n-1}w_{n-2}$, where, as in the proof of Theorem 1, w_k denotes the word consisting of the initial F_{k+1} letters of the infinite Fibonacci word. It is proved in [1] that the positions *after the initial* 1 of the even numbers in S are identical to the positions of 0 in the infinite Fibonacci word. The interested reader may wish to prove that is S' is the set of positive integers whose binary representation has no adjacent 0's, then $S = \{x+1 : x \in S'\}$.

REFERENCES

[1] Kimberling, Clark. "A Self-Generating Set and the Golden Mean." *Journal of Integer Sequences*, Vol. *3* (2000): http://www.research.att.com/~njas/sequences/JIS/VOL3/goldentext.html.

[2] Sloane, N.J.A. On-line Encyclopedia of Integer Sequences. Published electronically at http://www.research.att.com/~njas/sequences.

AMS Classification Numbers: 11B39, 11B37, 11B83

SOME BASIC PROPERTIES OF A TRIBONACCI LINE-SEQUENCE

Jack Y. Lee

In this paper we investigate some basic properties of a general Tribonacci sequence, namely, a third order homogeneous anharmonic recurrence sequence, by way of the line-sequential formalism developed previously for Fibonacci sequence, and report some new results. Some results are related or are reducible to known results as special cases. For consistency, we use the same nomenclatures, formats and conventions adopted in our previous works, see [3]-[5]. For publications in this area, see [1], [2], [9]-[14] and the references contained therein.

1. THE COEFFICIENT FORMULA

The recurrence relation of a general Tribonacci line-sequence is given by

$$tu_n + su_{n+1} + ru_{n+2} = u_{n+3}, \tag{1.1}$$

where t, s and r are non-zero constant coefficients, and u_n denotes the nth element in the line sequence $\{u_n\}_{n=-\infty}^{\infty}$.

For simplicity, let $P(n)_{xyzw} := u_{n+x}u_{n+y} - u_{n+z}u_{n+w}$, following the same procedure used in obtaining (3) and (4) in [7], we find the expressions of the recurrence coefficients in terms of the elements, assuming non-vanishing denominators, as follows:

$$r = (P_{0211}P_{1524} - P_{1322}P_{0413})/(P_{0211}P_{1423} - P_{1322}P_{0312}), \tag{1.2a}$$

$$s = (P_{1423}P_{0413} - P_{1524}P_{0312})/(P_{0211}P_{1423} - P_{1322}P_{0312}), \tag{1.2b}$$

$$t = (P_{1423}P_{2433} - P_{1322}P_{2534})/(P_{1322}P_{1322} - P_{0211}P_{2433}), \tag{1.2c}$$

This paper is in final form and no version of it will be submitted for publication elsewhere.

F.T. Howard (ed.), Applications of Fibonacci Numbers, Volume 9: Proceedings of the Tenth International Research Conference on Fibonacci Numbers and their Applications, 145–157.

Where, since the coefficients remain the same for any arbitrary integer n, that is, translationally invariant, we have dropped the translational parameter n. On the other hand, these expressions are not unique, for example, we may also have $r = (P_{1322}P_{1423} - P_{0211}P_{2534})/(P_{1322}P_{1322} - P_{0211}P_{2433})$, and so forth. To summarize, any six consecutive terms in a Tribonacci sequence determine the set of coefficients. Once the coefficients are found, the entire sequence is determined, up to a translational choice of the generator.

It is clear from (1.1) that any line-sequence is determined by any three consecutive terms in the sequence. In the following, we shall also see that any third-order line-sequence can be expressed as a linear combination of three basis line-sequences.

2. THE TRIBONACCI SPACE

The set of basis vectors which forms the basis array of the system (ref. Section 5 in [5], also Section 2 in [1]) is given by:

$$\begin{aligned} U_1 &= U_{[1,0,0]} \\ &= \{u_{1,n}\}_{n=-\infty}^{\infty} : \cdots, (s^2 - rt)/t^2, -s/t, [1,0,0],\ t,\ tr,\ \ t(r^2+s),\ \cdots \end{aligned} \tag{2.1a}$$

$$\begin{aligned} U_2 &= U_{[0,1,0]} \\ &= \{u_{2,n}\}_{n=-\infty}^{\infty} : \cdots, (rs+t)/t^2, -r/t, [0,1,0],\ s,\ sr+t,\ \ s(r^2+s)+rt,\ \cdots \end{aligned} \tag{2.1b}$$

$$\begin{aligned} U_3 &= U_{[0,0,1]} \\ &= \{u_{3,n}\}_{n=-\infty}^{\infty} : \cdots, -s/t^2,\ 1/t,\ [0,0,1],\ r,\ r^2+s,\ r(r^2+s)+rs+t,\ \cdots \end{aligned} \tag{2.1c}$$

where $[u_{i,0},\ u_{i,1},\ u_{i,2}]$ denotes the generator of the ith line-sequence, and $u_{i,n} \in \{u_{i,n}\}_{n=-\infty}^{\infty}$, $i = 1, 2$ and 3, denotes an element in the ith row and nth column in the basis array. Note that these vectors correspond to sequence $T_n^{(6)}$, $T_n^{(5)}$ and $T_n^{(4)}$, respectively, in [14], whereas $T_n^{(1)}$, $T_n^{(2)}$ and $T_n^{(3)}$ therein are linear combinations of these.

With the usual definition for addition and multiplication by scalar (ref. Sections 2.1 and 2.2 in [3]), the set, (2.1a), (2.1b) and (2.1c), thus forms the basis of a 3-dimensional Tribonacci space.

3. TRANSLATIONAL RELATIONS

From the basis array, we observe that

$$tu_{3,n} = u_{1,n+1}, \quad u_{1,n} + su_{3,n} = u_{2,n+1}, \quad u_{2,n} + ru_{3,n} = u_{3,n+1}. \tag{3.1a}$$

Let T_i denote a translation operator, where i is an integer (ref. Section 3 in [3], also see Section 2 in [5]), which operates on the nth element of a line-sequence producing the ith higher order element: $T_i u_n = u_{n+i}$. For $i = 1$, we write $T = T_1$ for simplicity. Thus (3.1a) can be written as

$$tu_{3,n} = Tu_{1,n}, \quad u_{1,n} + su_{3,n} = Tu_{2,n}, \quad u_{2,n} + ru_{3,n} = Tu_{3,n}. \tag{3.1b}$$

In terms of line-sequences, these relations correspond, respectively, to the following relations:

$$TU_1 = tU_3, \tag{3.2a}$$

$$TU_2 = U_1 + sU_3, \tag{3.2b}$$

$$TU_3 = U_2 + rU_3, \tag{3.2c}$$

Note that (3.2a) is equivalent to relation (5.1) in [14]. Note also that relation (3.2c) is equivalent to

$$U_2 = (T - r)U_3, \tag{3.3}$$

which is equivalent to relation (4.1) in [14] (with a sign correction). In the homogeneous case, $r = 1$, it is then equivalent to the relation between sequence $\{R_n\}$ and sequence $\{P_n\}$, see (2) and (4) in [13].

Since an arbitrary line-sequence in the Tribonacci space is a linear combination of its basis line-sequences, relations (3.2a), (3.2b) and (3.2c) enables it to be expressed in terms of a single basis line-sequence by means of a combination of translation and dilation operations (ref. (1.9a) in [6]). For example, we have

$$U_{[i,j,k]} = iU_1 + jU_2 + kU_3$$

$$= (i + j^{-1}T^2 - (jr - k)t^{-1}T)U_1 \tag{3.4a}$$

$$= ((itT^{-1} + k)(T - r)^{-1} + j)U_2 \tag{3.4b}$$

$$= (itT^{-1} + jT - (jr - k))U_3 \tag{3.4c}$$

In particular, we have $U_{[0,1,1]} = (T - (r - 1))U_3$ and $U_{[1,1,2]} = (tT^{-1} + T - (r - 2))U_3$, both of which have, in different occasions, been referred to as "Tribonacci sequence"; see, for example [9], [10], and [11]. However, since in the harmonic case, $r = s = t = 1$, both $U_{[0,1,1]}$ and $U_{[1,1,2]}$ are congruent (ref. Theorem 3 in [3]), to the basis U_3, that is,

$$(a)\ U_{[0,1,1]} = TU_3 \quad (b)\ U_{[1,1,2]} = TU_{[0,1,1]} = T^2U_3, \tag{3.5}$$

we shall instead refer to the basis U_3 as the "Tribonacci line-sequence". This nomenclature is consistent with the convention of referring to $U_{0,1}$ as Fibonacci line-sequence. Also note that the sequence $\{u_n\}$ in [11] is a one-step translation of U_3, namely,

$$TU_3 = TU_{[0,0,1]} = U_{[0,1,r]}, \tag{3.6}$$

in the anharmonic case, $r \neq 1$.

4. THE T- AND C-MATRICES OF TRANSLATION

Because a line-sequence, and hence the basis array, is customarily load out horizontally, yet a matrix operation is usually applied onto a column vector, we need to distinguish the mechanism between the horizontal and the vertical translation operations.

From the basis array, we define the translation matrix (ref. [1])

$$T_n := \begin{pmatrix} u_{1,n} & u_{1,n+1} & u_{1,n+2} \\ u_{2,n} & u_{2,n+1} & u_{2,n+2} \\ u_{3,n} & u_{3,n+1} & u_{3,n+2} \end{pmatrix}. \tag{4.1a}$$

Applying relations (3.1), it takes the form

$$T_n = \begin{pmatrix} tu_{3,n-1} & tu_{3,n} & tu_{3,n+1} \\ tu_{3,n-2} + su_{3,n-1} & tu_{3,n-1} + su_{3,n} & tu_{3,n} + su_{3,n+1} \\ u_{3,n} & u_{3,n+1} & u_{3,n+2} \end{pmatrix}, \tag{4.1b}$$

which corresponds to (4) in [11]. We also define the following vectors:

$$U_n := \begin{pmatrix} u_{1,n} \\ u_{2,n} \\ u_{3,n}, \end{pmatrix} \quad U_{i,n} := \begin{pmatrix} u_{i,n} \\ u_{i,n+1} \\ u_{i,n+2} \end{pmatrix}. \tag{4.2}$$

From the basis array, we find

$$T_0 = \begin{pmatrix} 1 & 0 & 0 \\ 0 & 1 & 0 \\ 0 & 0 & 1 \end{pmatrix} \quad \text{and } T := T_1 = \begin{pmatrix} 0 & 0 & t \\ 1 & 0 & s \\ 0 & 1 & r \end{pmatrix}. \tag{4.3}$$

So the identity operator T_0 is the "generator" matrix of the basis array (ref. Section 2, [1]). Applying relation (1.1), it is easy to verify that the operation of T onto the nth column matrix, U_n, produces the $(n+1)^{th}$ column matrix, U_{n+1}, in the basis array,

$$TU_n = U_{n+1}. \tag{4.4}$$

So we identify the operator T to be the (horizontal) translation matrix of the basis array. It follows that

$$T^m U_n = U_{m+n}, \tag{4.5}$$

which corresponds to (3) in [11].

On the other hand, we have in our previous work defined a C-matrix, see Section 5 in [5]. The C-matrix operates on the nth *column* vector $U_{i,n}$ producing the *column* vector in the next higher order, $U_{i,n+1}$, *in the same* line-sequence. Namely, we have

$$C\,U_{i,n} = U_{i,n+1} \tag{4.6a}$$

Or explicitly,

$$\begin{pmatrix} 0 & 1 & 0 \\ 0 & 0 & 1 \\ t & s & r \end{pmatrix} \begin{pmatrix} u_{i,n} \\ u_{i,n+1} \\ u_{i,n+2} \end{pmatrix} = \begin{pmatrix} u_{i,n+1} \\ u_{i,n+2} \\ u_{i,n+3} \end{pmatrix}. \tag{4.6b}$$

Clearly the C-matrix relates to the T-matrix through a transpose operation denoted by the symbol "†", $C = T^{\dagger}$. In contradistinction to referring to the function of the T operator as "translation (horizontally along the basis array)", we refer to the function of the C operator as "escalation (vertically along a given line-sequence)".

Obviously, we have

$$C^m U_{i,n} = U_{i,n+m}, \tag{4.7}$$

It is easy to verify that the following "Generalized Pleasant Equation" holds:

$$tI + sT + rT^2 - T^3 = 0,$$

which is the Tribonacci case of (4) in [8]. Similarly for C, we have

$$tI + sC + rC^2 - C^3 = 0.$$

5. THE CHARACTERISTIC EQUATION

The characteristic equation, see [2], is given by

$$x^3 - rx^2 - sx - t = 0. \tag{5.1}$$

Let α, β and γ be the roots. Applying Cardan's formula, we obtain: $\alpha = A^{1/3}+B^{1/3}+r/3$, $\beta = \omega A^{1/3}+\omega^2 B^{1/3}+r/3$ and $\gamma = \omega^2 A^{1/3}+\omega B^{1/3}+r/3$, where $A = -q/2+(q^2/4+p^3/27)^{1/2}$, $B = -q/2-(q^2/4+p^3/27)^{1/2}$, and $p = -(r^2+3s)/3$ and $q = -(2r^3+9rs-27t)/27$; and where ω denotes the complex cubic root of 1, $\omega = (-1+i\sqrt{3})/2$. For generality, we shall assume in our current work that the three roots are all non-zero and different.

6. THE GEOMETRIC ARRAY

With the three roots of the characteristic equation (5.1), we form the following three geometric line-sequences, denoted by U_i, $i = \alpha$, β and γ, which together constitute what we shall call the geometric array (ref. (2.6) and (2.7) in [4]):

$$U_\alpha := \ldots,\ [1,\ \alpha,\ \alpha^2],\ \alpha^3,\ \alpha^4,\ \ldots \tag{6.1a}$$

$$U_\beta := \dots, \; [1, \; \beta, \; \beta^2], \; \beta^3, \; \beta^4, \; \dots \tag{6.1b}$$

$$U_\gamma := \dots, \; [1, \; \gamma, \; \gamma^2], \; \gamma^3, \; \gamma^4, \; \dots \tag{6.1c}$$

where the subscript "α" in U_α stands for the generator $[1, \; \alpha, \; \alpha^2]$, and so on, and as before the specification of recurrence coefficients $(t, \; s, \; r)$ is dropped.

As in the 2-dimensional case, see (4.5) in [4], we denoted by M_n the nth order matrix of the geometric array, and G_n the nth column vector in the geometric array:

$$M_n := \begin{pmatrix} \alpha^n & \alpha^{n+1} & \alpha^{n+2} \\ \beta^n & \beta^{n+1} & \beta^{n+2} \\ \gamma^n & \gamma^{n+1} & \gamma^{n+2} \end{pmatrix} \quad \text{and } G_n := \begin{pmatrix} \alpha^n \\ \beta^n \\ \gamma^n \end{pmatrix}. \tag{6.2}$$

So we have

$$M := M_0 = \begin{pmatrix} 1 & \alpha & \alpha^2 \\ 1 & \beta & \beta^2 \\ 1 & \gamma & \gamma^2 \end{pmatrix} \quad \text{and } G_0 = \begin{pmatrix} 1 \\ 1 \\ 1 \end{pmatrix}. \tag{6.3}$$

For later use, we also define the matrices

$$D := \begin{pmatrix} \alpha & 0 & 0 \\ 0 & \beta & 0 \\ 0 & 0 & \gamma \end{pmatrix} \quad \text{and } J_n := \begin{pmatrix} J_n \\ J_{n+1} \\ J_{n+2} \end{pmatrix}, \tag{6.4}$$

where $j_n := \alpha^n + \beta^n + \gamma^n$.

The successive application of the diagonal matrix D onto the M_n-matrix keeps shifting the M_n-matrix to the right along the geometric array and hence represents the translation operation of the geometric array. That is, we have

$$D^m M_n = M_{m+n}, \text{ and also } D^m G_n = G_{m+n}. \tag{6.5}$$

7. THE BINET'S FORMULAS

Since there are three basis line-sequences, they correspond to each a Binet's formula, each being a linear combination of the three geometric line-sequences. Let

$$U_{0,0,1} = aU_\alpha + bU_\beta + cU_\gamma, \tag{7.1}$$

where a, b, and c are the unknown coefficients to be determined. Equating the generating elements on both sides of the equation, we obtain a set of three simultaneous equations in a, b, and c, which can be expressed in the form of a matrix equation in terms of the transposed M-matrix:

$$\begin{pmatrix} 1 & 1 & 1 \\ \alpha & \beta & \gamma \\ \alpha^2 & \beta^2 & \gamma^2 \end{pmatrix} \begin{pmatrix} a \\ b \\ c \end{pmatrix} = \begin{pmatrix} 0 \\ 0 \\ 1 \end{pmatrix}. \tag{7.2}$$

Solving (7.2) for a, b, and c in terms of α, β, and γ, and substituting back into (7.1), we obtain the Binet's formula for the basis line-sequence U_3:

$$U_3 = -[(\beta-\gamma)U_\alpha + (\gamma-\alpha)U_\beta + (\alpha-\beta)U_\gamma]/\Delta, \tag{7.3a}$$

where $\Delta = (\alpha-\beta)(\beta-\gamma)(\gamma-\alpha) \neq 0$.

So, in terms of the elements, we have

$$u_{3,n} = -[(\beta-\gamma)\alpha^n + (\gamma-\alpha)\beta^n + (\alpha-\beta)\gamma^n]/\Delta. \tag{7.3b}$$

This is (3.2) in [14]. In the homogeneous case, $r = s = t = 1$, it reduces to the Binet's formula in [10] (with a second degree translation, see (3.5b)).

Similarly, we obtain the Binet's formula for the line sequence U_2:

$$U_2 = [(\beta^2-\gamma^2)U_\alpha + (\gamma^2-\alpha^2)U_\beta + (\alpha^2-\beta^2)U_\gamma]/\Delta. \tag{7.4a}$$

In terms of the elements, we therefore have

$$u_{2,n} = [(\beta^2-\gamma^2)\alpha^n + (\gamma^2-\alpha^2)\beta^n + (\alpha^2-\beta^2)\gamma^n]/\Delta. \tag{7.4b}$$

Further more, applying (3.2a) to (7.3a), we find Binet's formula for U_1:

$$U_1 = -[\beta\gamma(\beta-\gamma)U_\alpha + \gamma\alpha(\gamma-\alpha)U_\beta + \alpha\beta(\alpha-\beta)U_\gamma]/\Delta. \tag{7.5a}$$

In terms of the elements, we then have

$$u_{1,n} = -[\beta\gamma(\beta-\gamma)\alpha^n + \gamma\alpha(\gamma-\alpha)\beta^n + \alpha\beta(\alpha-\beta)\gamma^n]/\Delta. \tag{7.5b}$$

A simpler way to obtain these results is to apply M-matrix transformation. Define the following column matrices:

$$U := \begin{pmatrix} U_1 \\ U_2 \\ U_3 \end{pmatrix} \quad \text{and } G := \begin{pmatrix} U_\alpha \\ U_\beta \\ U_\gamma \end{pmatrix}, \tag{7.6}$$

where the matrix elements are the whole line-sequences. It can be easily verified that the geometric array relates to the basis array through the M-matrix transformation,

$$G = MU, \tag{7.7a}$$

or in terms of the elements,

$$G_n = MU_n, \tag{7.7b}$$

so we obtain Binet's formulas for the basis line-sequences (ref. (4.9) in [4]) through the inverse transformation,

$$U = M^{-1}G, \text{ or } U_n = M^{-1}G_n. \tag{7.8}$$

The Binet's formula for an arbitrary Tribonacci line-sequence can then be expressed as a linear combination of U_1, U_2 and U_3.

8. THE CONJUGATE TRIPLE

We need the following definition.

Definition 1: Parallel to the 2-dimensional case, see Section 2.3 in [3], we call two line-sequences $U_{a,b,c}$ and $U_{d,e,f}$ orthogonal if their inner product vanishes, namely, $(U_{a,b,c}, U_{d,e,f}) := ad + be + cf = 0$; and the "length" of a line-sequence is the square root of its inner product with itself. A line-sequence is said to be "normal" if its length is 1.

It is seen that the sum of (6.1a), (6.1b) and (6.1c), denoted in our notation by U_J, where the subscript J stands for the generator, $J := [3, r, r^2 + 2s]$, is equivalent to the line-sequence denoted by $\{J_n\}$ in [11], also see (1.3) in [2]. It is clear from Definition 1, that the line-sequence U_K, where $K := [-r, 3, 0]$, is orthogonal to U_J.

Applying cross multiplication of vectors, a third line-sequence U_L, where $L := [-3(r^2 + 2s), -r(r^2 + 2s), r^2 + 9]$, is found which is orthogonal to both U_J and U_K. Thus the triple, U_J, U_K and U_L, also forms a set of (orthogonal but not normal) basis, which we shall call the conjugate (Tribonacci) triple. The conjugate (Tribonacci) array is then given by:

$$U_J := \ldots, \ [3, r, r^2 + 2s], \ 3t + 3rs + r^3, \ 4rt + 4r^2s + 2s^2 + r^4, \ldots \tag{8.1a}$$

$$U_K := \ldots, \ [-r, 3, 0], \ -rt + 3s, \ 3t - r^2t + 3rs, \ \ldots \tag{8.1b}$$

$$U_L := \ldots, \ [-3(r^2 + 2s), \ -r(r^2 + 2s), \ r^2 + 9], \ -(3t + rs - r)r^2 - 2s(3t + rs) + 9r, \ldots \tag{8.1c}$$

The conjugate array matrix, denoted by H, and the vector of the conjugate line-sequences, denoted by W, are, respectively,

$$H := \begin{pmatrix} 3 & r & r^2 + 2s \\ -r & 3 & 0 \\ -3(r^2 + 2s) & -r(r^2 + 2s) & r^2 + 9 \end{pmatrix}, \ W := \begin{pmatrix} U_J \\ U_K \\ U_L \end{pmatrix} \text{ and } W_n := \begin{pmatrix} u_{J,n} \\ u_{K,n} \\ u_{L,n} \end{pmatrix}. \tag{8.2a}$$

We also define the following generator vectors, namely:

$$U_J := \begin{pmatrix} 3 \\ r \\ r^2 + 2s \end{pmatrix}, \ U_K := \begin{pmatrix} -r \\ 3 \\ 0 \end{pmatrix}, \ U_L := \begin{pmatrix} -3(r^2 + 2s) \\ -r(r^2 + 2s) \\ r^2 + 9 \end{pmatrix}. \tag{8.2b}$$

Then, the conversion between the two sets of basis, parallel to (7.7a), is given by $HU = W$. Applying (7.8), the set of Binet's formulas for the conjugate triple is then given by

$$W = HM^{-1}G, \text{ or } W_n = HM^{-1}G_n. \tag{8.3b}$$

The explicit forms of these formulas are left for the interested readers to work out.

9. THE GENERATING FUNCTION

The general formula for the generating function (ref. [9]) is found to be

$$(u_0 + (u_1 - ru_0)x + (u_2 - ru_1 - su_0)x^2)/(1 - rx - sx^2 - tx^3) = \sum_{n=0} u_n x^n, \tag{9.1}$$

where the coefficients in the numerator will be referred to as the coefficients of generating function. The generating function of Tribonacci triple, denoted by R_i for the basis line-sequence U_i, $i = 1, 2$, and 3, are respectively:

$$R_1 = (1 - rx - sx^2)/(1 - rx - sx^2 - tx^3) = \sum_{n=0} u_{1,n} x^n, \tag{9.2a}$$

$$R_2 = (x - rx^2)/(1 - rx - sx^2 - tx^3) = \sum_{n=0} u_{2,n} x^n, \tag{9.2b}$$

$$R_3 = x^2/(1 - rx - sx^2 - tx^3) = \sum_{n=0} u_{3,n} x^n. \tag{9.2c}$$

These functions can be regarded as the three basis components of the general generating function (9.1). From their coefficients we form a coefficient matrix, denoted by R, of the array as follows:

$$R := \begin{pmatrix} 1 & -r & -s \\ 0 & 1 & -r \\ 0 & 0 & 1 \end{pmatrix}. \tag{9.3}$$

Since the generator of any line-sequence can be expressed as a linear combination of the generators of the basis line-sequences, the matrix R then maps the generator of a line-sequence to the coefficient vector of its generating function. For example R_J, the coefficient vector of U_J, is given by $R_J = R^{\dagger} U_J = \begin{pmatrix} 3 \\ -2r \\ -s \end{pmatrix}$, then the generating funciton R_J is given by

$$R_J = (3 - 2rx - sx^2)/(1 - rx - sx^2 - tx^3) = \sum_{n=0} u_{J,n} x^n. \tag{9.4a}$$

Similarly we have

$$R_K = (-r + (r^2+3)x + r(s-3)x^2)/(1 - rx - sx^2 - tx^3) = \sum_{n=0} u_{K,n}x^n. \tag{9.4b}$$

$$R_L = (-3(r^2+2s) + 2r(r^2+2s)x + ((r+2s)(r^2+3s) + r^2 + 9)x^2)/(1 - rx - sx^2 - tx^3)$$
$$= \sum_{n=0} u_{L,n}x^n. \tag{9.4c}$$

10. SUMMARY OF RELATIONS

Following is a summary of some simple relations which either have been developed above or can be easily verified.

$$\begin{array}{lll} (a) T = C^\dagger; & (b) T_m T_n = T_{m+n}; & (c) T_m U_n = U_{m+n}; \\ (d) M_m T_n = M_{m+n}; & (e) D^m M_n = M_{m+n}; & (f) D^m G_n = G_{m+n}; \\ (g) C^m U_{i,n} = U_{i,m+n}; & (h) MU_n = G_n; & (i) M_m^\dagger G_0 = M^\dagger G_m = J_m; \\ (j) G_0^\dagger G_n = j_n; & (k) G_0^\dagger M_n = J_n^\dagger; & (l) R^\dagger U_J = R_J. \end{array} \tag{10}$$

11. SOME APPLICATIONS

Following are some examples of applying the relations obtained above.

Example 1: From (10c), we have the translational operation onto a basis array vector

$$U_{m+n+k} = T_m U_{n+k}. \tag{11.1a}$$

From the definitions of T_m in (4.1a) and U_{n+k} in (4.2a), we obtain, in terms of the elements, the following relation:

$$u_{i,m+n+k} = u_{i,m}u_{1,n+k} + u_{i,m+1}u_{2,n+k} + u_{i,m+2}u_{3,n+k}. \tag{11.1b}$$

where $i = 1$, 2 or 3.

Example 2: Putting $k = 0$ in (11.1b), we obtain

$$u_{i,m+n} = u_{i,m}u_{1,n} + u_{i,m+1}u_{2,n} + u_{i,m+2}u_{3,n}, \tag{11.2a}$$

Applying (3.1), we obtain

$$u_{i,m+n} = tu_{i,m}u_{3,n-1} + u_{i,m+1}(tu_{3,n-2} + su_{3,n-1}) + u_{i,m+2}u_{3,n}, \tag{11.2b}$$

which relates to (7) in [11]. Putting $i = 3$, and $m = n - 2$ in (11.2b), we obtain

$$u_{3,2n-2} = (u_{3,n})^2 + s(u_{3,n-1})^2 + 2tu_{3,n-1}u_{3,n-2}. \tag{11.2c}$$

This relates to (15) in [11].

Example 3: Noting that (11.1a) in general states a cubic relation, which, in terms of the elements, reads:

$$\begin{aligned} u_{i,m+n+k} = u_{i,m} &+ (u_{1,k}u_{1,n} + u_{1,k+1}u_{2,n} + u_{1,k+2}u_{3,n}) \\ &+ u_{i,m+1}(u_{2,k}u_{1,n} + u_{2,k+1}u_{2,n} + u_{2,k+2}u_{3,n}) \\ &+ u_{i,m+2}(u_{3,k}u_{1,n} + u_{3,k+1}u_{2,n} + u_{3,k+2}u_{3,n}) \end{aligned} \tag{11.3}$$

This cubic relation is similar, but not equivalent, to relation (19) in [11].

Example 4: Parallel to the translational operation onto a basis array vector as exemplified by (11.1a), we may also have a translational operation onto a line-sequential vector, primarily a basis line-sequential vector. From (10g), we have

$$U_{i,m+n+k} = C^m(C^nU_{i,k}), \; i = 1, \; 2 \text{ or } 3. \tag{11.4a}$$

In terms of the elements, using (10a), (4.1b) and (4.2b), we find

$$\begin{aligned} u_{i,m+n+k} = \; &tu_{3,m-1}(tu_{3,n-1}u_{i,k} + (su_{3,n-1} + tu_{3,n-2})u_{i,k+1} + u_{3,n}u_{i,k+2}) \\ &+ (su_{3,m-1} + tu_{3,m-2})(tu_{3,n}u_{i,k} + (su_{3,n} + tu_{3,n-1})u_{i,k+1} + u_{3,n+1}u_{i,k+2}) \\ &+ u_{3,m}(tu_{3,n+1}u_{i,k} + (su_{3,n+1} + tu_{3,n})u_{i,k+1} + u_{3,n+2}u_{i,k+2}). \end{aligned} \tag{11.4b}$$

Letting $i = 3$ and changing the subscripts m, k, n to $n - 1$, we find

$$\begin{aligned} u_{3,3n-3} = \; &tu_{3,n-2}(tu_{3,n-2}u_{3,n-1} + (su_{3,n-2} + tu_{3,n-3})u_{3,n} + u_{3,n-1}u_{3,n+1}) \\ &+ (su_{3,n-2} + tu_{3,n-3})(tu_{3,n-1}u_{3,n-1} + (su_{3,n-1} + tu_{3,n-2})u_{3,n} + u_{3,n}u_{3,n+1}) \\ &+ u_{3,n-1}(tu_{3,n}u_{3,n-1} + (su_{3,n} + tu_{3,n-1})u_{3,n} + u_{3,n+1}u_{3,n+1}). \end{aligned} \tag{11.5}$$

This identity relates to (19) in [11].

Example 5:

$$J_m^\dagger U_n = G_0^\dagger G_{m+n}; \tag{11.6a}$$

Or equivalently,

$$j_m u_{1,n} + j_{m+1}u_{2,n} + j_{m+2}u_{3,n} = j_{m+n}. \tag{11.6b}$$

Proof: From (10e) an (10i), recognizing $J_m^\dagger = G_0^\dagger D^m M$, and from (10h), recognizing $U_n = M^{-1}G_n$, then, by applying (10f), we have

$$J_m^\dagger U_n = (G_0^\dagger D^m M)(M^{-1}G_n) = G_0^\dagger(D^m G_n) = G_0^\dagger G_{m+n}.$$

From definition (4.2a) and (6.3b), the left hand side of (11.6a) gives

$$J_m^\dagger U_n = j_m u_{1,n} + j_{m+1} u_{2,n} + j_{m+2} u_{3,n};$$

and from (10j), the right hand side of (11.6a) gives $G_0^\dagger G_{m+n} = j_{m+n}$. Substituting these into each side of (11.6a), we obtain (11.6b). □

Example 6: From $m = 0$, relation (11.6b) reduces to

$$3u_{1,n} + ru_{2,n} + (r^2 + 2s)u_{3,n} = j_n; \tag{11.7a}$$

which gives the nth element of the line-sequence U_J. Hence we have the identity

$$3u_{1,n} + ru_{2,n} + (r^2 + 2s)u_{3,n} = \alpha^n + \beta^n + \gamma^n. \tag{11.7b}$$

Example 7: Replacing n by $m + n$ in (11.6a) and (11.6b), we obtain

$$J_m^\dagger U_{m+n} = j_{2m+n}, \tag{11.8}$$

which relates to Howard's relation (1.5) in [2]. And so forth.

12. FURTHER STUDY

A number of points arising from this investigation worth mentioning. The conjugate triple given in Section 8 above is not unique, there exists other sets of conjugate triples. For example, the set of generators $[3, r, r^2 + 2s]$, $[\frac{2}{3}s, r, -1]$ and $[-r - 2rs - r^3, \frac{2}{3}r^2 s + \frac{4}{3}s^2 + 3, 3r - \frac{2}{3}rs]$ also forms a conjugate triple. The meaning of this multiplicity is not clear as of this writing. Also, in our work on the second order case, we see that parity property plays an important role, see for example (1.3a) to (1.4d) in [6]. Why then in the third order case, this important property is apparently missing? These and other related questions require further investigation.

The author wishes to express his gratitude to the anonymous referee for his/her painstaking corrections of the manuscirpt and for the numerous suggestions resulting in the much improved version of this report.

REFERENCES

[1] Ando, S. "On a System of Sequences Defined by a Recurrence Relation." *The Fibonacci Quarterly*, Vol. *33.3* (1995): pp. 279-282.

[2] Howard, F.T. "A Tribonacci Identity." *The Fibonacci Quarterly*, Vol. *39.4* (2001): pp. 352-357.

[3] Lee, Jack Y. "Some Basic Properties of the Fibonacci Line-sequence", Applications of Fibonacci Numbers, Volume **4**. Ed. G.E. Bergum, A.N. Philippou and A.F. Horadam. Dordrecht: Kluwer, 1990, pp. 203-214.

[4] Lee, Jack Y. "The Golden-Fibonacci Equivalence." *The Fibonacci Quarterly*, Vol. *30.3* (1992): pp. 216-220.

[5] Lee, Jack Y. "Some Basic Translational Properties of the General Fiboancci Sequence", Applications of Fibonacci Numbers, Volume **6**. Ed. G.E. Bergum, A.N. Philippou and A.F. Horadam. Dordrecht: Kluwer, 1994, pp. 339-347.

[6] Lee, Jack Y. "Some General Formulas Associated with the Second-Order Homogeneous Polynomial Line-Sequences." *The Fibonacci Quarterly*, Vol. *39.5* (2001): pp. 419-429.

[7] Lee, Jack Y. "On the Product of Line-Sequences." *The Fibonacci Quarterly*, Vol. *40.5* (2002): pp. 438-440.

[8] Liu, Bolian. "A Matrix Method to Solve Linear Recurrences with Constant Coefficients." *The Fibonacci Quarterly*, Vol. *30.1* (1992): pp. 2-8.

[9] Scott, A., Delaney, T. and Hoggatt, V.E. Jr. "The Tribonacci Sequence." *The Fibonacci Quarterly*, Vol. *15.3* (1977): pp. 193-201.

[10] Spickerman, W.R. "Binet's Formula for the Tribonacci Seuqence." *The Fibonacci Quarterly*, Vol. *20.2* (1982): pp. 118-120.

[11] Waddill, M.E. "Using Matrix Techniques to Establish Properties of a Generalized Tribonacci Sequence", Applications of Fibonacci Numbers, Volume **4**. Ed. G.E. Bergum, A.N. Philippou and A.F. Horadam. Dordrecht: Kluwer, 1990, pp. 299-308.

[12] Waddill, M.E. "The Tetranacci Sequence and Generalizations." *The Fibonacci Quarterly*, Vol. *30.1* (1992): pp. 9-20.

[13] Waddill, M.E. and Sacks, L. "Another Generalized Fibonacci Sequence." *The Fibonacci Quarterly*, Vol. *5.3* (1967): pp. 209-222.

[14] Yalavigi, C.C. "Properties of Tribonacci Numbers." *The Fibonacci Quarterly*, Vol. *10.3* (1972): pp. 231-246.

AMS Classification number: 11B37, 11B83

A TYPE OF SEQUENCE CONSTRUCTED FROM FIBONACCI NUMBERS

Aihua Li and Sindhu Unnithan

1. RF-SEQUENCES

Consider the Fibonacci sequence $(F_n)_{n\geq 1}$. We define a sequence $(c_n)_{n\geq 1}$ below:

Definition 1.1: *A sequence $(c_n)_{n\geq 1}$ of positive integers is called an RF-sequence if there exist a sequence of associated integers $(q_n)_{n\geq 1}$ such that*

$$c_n = c_{n-1}q_{n-1} + F_{n-1}$$

$$c_{n-1} = c_{n-2}q_{n-2} + F_{n-2}$$

$$\dots$$

$$c_i = c_{i-1}q_{i-1} + F_{i-1}$$

$$\dots$$

$$c_2 = c_1q_1 + F_1.$$

We call the sequence $(q_n)_{n\geq 1}$ the quotient sequence associated to $(c_n)_{n\geq 1}$.

Example 1.2: *The sequence $(F_n)_{n\geq 2}$ is an RF-sequence but $(F_n)_{n\geq 1}$ is not.*

Note that the first sequence starts from F_2 and for $n \geq 2, F_{n+1} = F_n \cdot 1 + F_{n-1}$. So it is an RF-sequence associated with the quotient sequence $(1)_{n\geq 1}$. The second sequence starts from F_1. If it forms an RF-sequence, then by the defintion of RF-sequence, $F_4 = F_3q_3 + F_3$ for some $q_3 \in \mathbb{Z}$. It implies $3 = 2q_3 + 2$, that is, 2 divides 3, a contradiction.

The first author was supported by the BORSF under grant LEQSF(2000-03)RD-A-24.

This paper is in final form and no version of it will be submitted for publication elsewhere.

F.T. Howard (ed.), Applications of Fibonacci Numbers, Volume 9: Proceedings of the Tenth International Research Conference on Fibonacci Numbers and their Applications, 159–166.

Let us give an explicit formula for the nth term of an RF-sequence that can be proved by mathematical induction.

Lemma 1.3: *Let $(c_n)_{n\geq 1}$ be an RF-sequence. Then for $n \geq 2$,*

$$c_n = c_1 q_1 q_2 \cdots q_{n-1} + F_1 q_2 q_3 \cdots q_{n-1} + F_2 q_3 q_4 \cdots q_{n-1} + \cdots + F_{n-2} q_{n-1} + F_{n-1}.$$

Theorem 1.4: *For all $n \geq 2$, we have $(c_{n+1}, c_n, c_{n-1}) = 1$.*

Proof: Let $(c_n)_{n\geq 1}$ be an RF-sequence with $n > 2$. By definition $c_{n+1} = c_n q_n + F_n$, $c_n = c_{n-1} q_{n-1} + F_{n-1}$, and $c_{n-1} = c_{n-2} q_{n-2} + F_{n-2}$, where $q_{n+1}, q_n, q_{n-1} \in \mathbb{Z}$. Assume d is the greatest common divisor of c_{n+1}, c_n, and c_{n-1}. Then d divides both F_n and F_{n-1}. Therefore $d = 1$ since F_n and F_{n-1} are relatively prime.

This theorem gives a way to identify a class of non-RF-sequences.

Example 1.5: *For a given positive integer k, the sequence $(c_n)_{n\geq 1} = (kn)_{n\geq 1}$ is not an RF-sequence.*

Now let $\alpha = \frac{1+\sqrt{5}}{2}$. By [2, Lemma 9.1], Fibonacci numbers satisfy the inequalities:

$$\alpha^{n-2} < F_n < \alpha^{n-1} \quad (n > 2).$$

This gives a boundary for any RF-sequence.

Proposition 1.6: *Suppose $(c_n)_{n\geq 1}$ is an RF-sequence associated with the quotient sequence $(q_n)_{n\geq 1}$ and $q_n \neq 0$ for each $n \in \mathbb{Z}$. Then*

$$\left|\frac{c_n}{q_1 q_2 \cdots q_{n-1}}\right| \leq c_1 + \alpha^n + 1 \quad \textit{and} \quad \lim_{n\to\infty} \frac{c_n}{\alpha^n q_1 q_2 \cdots q_{n-1}} = 0.$$

Proof: By lemma 1.3,

$$\left|\frac{c_n}{q_1 q_2 \cdots q_{n-1}}\right| \leq c_1 + \frac{F_1}{|q_1|} + \frac{F_2}{|q_1 q_2|} + \cdots + \frac{F_{n-1}}{|q_1 q_2 \cdots q_{n-1}|}$$

$$\leq c_1 + F_1 + F_2 + \cdots + F_{n-1} = c_1 + \alpha^n - 1.$$

It is well known that the sum of the first n odd-subscripted Fibonacci numbers is $\sum_{k=1}^{n} F_{2k-1} = F_{2n}$ [2, Page 71]. A similar relation for an RF-sequence is given below:

Proposition 1.7: *Let $(c_n)_{n\geq 1}$ be an RF-sequence associated with the quotient sequence $(q_n)_{n\geq 1}$. Then*

$$\sum_{k=1}^{2n} \sigma_k c_k = F_{2n},$$

where

$$\sigma_k = \begin{cases} 1 & \text{if } k \text{ is even} \\ -q_k & \text{if } k \text{ is odd.} \end{cases}$$

Proof:

$$\sum_{k=1}^{2n} \sigma_k c_k = \sum_{k=1}^{n} \sigma_{2k} c_{2k} + \sum_{k=1}^{n} \sigma_{2k-1} c_{2k-1}$$

$$= \sum_{k=1}^{n} c_{2k} + \sum_{k=1}^{n} (-q_{2k-1} c_{2k-1})$$

$$= \sum_{k=1}^{n} (c_{2k} - q_{2k-1} c_{2k-1}) = \sum_{k=1}^{n} F_{2k-1} = F_{2n}.$$

Recall the formula $\sum_{k=1}^{n} (-1)^{k+1} F_{k+1} = (-1)^{n-1} F_n$ [2, Page 88 (20)] which results in the formula $\sum_{k=1}^{2n} (-1)^{k+1} F_{k+1} = -F_{2n}$. This can be derived from the above proposition as a special case, by taking $c_n = F_{n+1}$ and $q_n = 1$ for all $n \geq 1$. In this case, $\sigma_k = (-1)^k$ and this yields

Corollary 1.8:

$$\sum_{k=1}^{2n} (-1)^{k+1} F_{k+1} = -F_{2n}.$$

It is shown in [3, Theorem II, page 108] that every positive integer has a unique canonical representation as a sum of distinct non-consecutive Fibonacci numbers. Using this fact, we show that every positive integer is a special linear combination of finitely many numbers from a fixed RF-sequence.

Proposition 1.9: *For a given RF-sequence $(c_n)_{n \geq 1}$, every positive integer m can be expressed as a linear combination of finitely many c_n's in the form of*

$$m = \sum_{k=1}^{j} \tau_{i_k} c_{i_k},$$

where $\tau_{i_k} \in \{1, 1 - q_{i_k}, -q_{i_k} | k \in \mathbb{N}\}$.

Proof: The proof is straightforward. First we write m as a sum of finitely many Fibonacci numbers. Then we apply the formula $F_n = c_{n+1} - c_n q_n$.

A two by two matrix $\boldsymbol{Q}$ is well related to the Fibonacci numbers (see [2]). Let

$$\boldsymbol{Q} = \begin{bmatrix} 1 & 1 \\ 1 & 0 \end{bmatrix}.$$

It is well known that $\boldsymbol{Q}^n = \begin{bmatrix} F_{n+1} & F_n \\ F_n & F_{n-1} \end{bmatrix}$ $(n \geq 2)$. A recursive formula for the matrix $\begin{bmatrix} c_{n+1} & c_n \\ c_n & c_{n-1} \end{bmatrix}$ is given below

Proposition 1.10: *Suppose $(c_n)_{n\geq 1}$ is an RF-sequence associated with the quotient sequence $(q_n)_{n\geq 1}$. Then for $n \geq 2$,*

$$\begin{bmatrix} c_{n+2} & c_{n+1} \\ c_{n+1} & c_n \end{bmatrix} = q_n \begin{bmatrix} c_{n+1} & c_n \\ c_n & c_{n-1} \end{bmatrix} + \boldsymbol{Q}^n.$$

2. *RF*-SEQUENCES WITH A CONSTANT QUOTIENT

In this section we consider an *RF*-sequence $(c_n)_{n\geq 1}$ with the constant quotient sequence $(q)_{n\geq 1}$, that is, $q_n = q$ for all n.

Lemma 2.1: *Let $(c_n)_{n\geq 1}$ be an RF-sequence with a constant quotient q. Then*

(1) *$q \geq 0$ and $(c_n)_{n\geq 1}$ is an increasing sequence;*
(2) *If $q \neq 0$, then q divides both c_1 and $c_3 - c_2$;*
(3) *$c_{n+1} + c_n = (c_n + c_{n-1})q + F_{n+1}$ for $n \geq 2$;*
(4) *$c_n = c_1 q^{n-1} + F_1 q^{n-2} + F_2 q^{n-3} + \cdots + F_{n-2} q + F_{n-1}$ for $n \geq 2$.*

Note that if $q = 1$, $c_n = F_{n+1} + c_1 - 1$ for all n. In particular, if $q = c_1 = 1$, then $c_n = F_{n+1}$, so the sequence $(c_n)_{n\geq 1}$ is the Fibonacci sequence starting from F_2 (see example 1.2).

Proposition 2.2: *Let $(c_n)_{n\geq 1}$ be an RF-sequence with a constant quotient $q \neq 1$. Then*

$$c_1 + c_2 + \cdots + c_n = \sum_{i=1}^{n} (c_1 + F_{n-i+1} - 1) q^i$$

$$= \frac{1}{1-q}(c_1 + F_{n+2} - c_{n+1} - 1).$$

Proof: We list the formulas for c_n in the table below:

$$\begin{aligned}
c_1 &= c_1 \\
c_2 &= c_1q + F_1 \\
c_3 &= c_1q^2 + F_1q + F_2 \\
c_4 &= c_1q^3 + F_1q^2 + F_{2q} + F_3 \\
&\cdots \\
c_{n-1} &= c_1q^{n-2} + F_1q^{n-3} + F_2q^{n-4} + \cdots + F_{n-3}q + F_{n-2} \\
c_n &= c_1q^{n-1} + F_1q^{n-2} + F_2q^{n-3} + \cdots + F_{n-2}q + F_{n-1}.
\end{aligned}$$

Recall that $F_1 + F_2 + \cdots + F_{n-1} = F_{n+1} - 1$. If we add the right side items diagonally, the result is

$$\begin{aligned}
c_1 + c_2 + \ldots c_n &= c_1 + \sum_{i=1}^{n-1} F_i + \left(c_1 + \sum_{i=1}^{n-2} F_i\right) q + \left(c_1 + \sum_{i=1}^{n-3} F_i\right) q^2 + \\
&\cdots + (c_1 + F_1)q^{n-2} + c_1q^{n-1} \\
&= \sum_{i=0}^{n-1} (c_1 + F_{n-i+1} - 1)q^i.
\end{aligned}$$

If we add the right side along the columns, we have

$$\begin{aligned}
\sum_{i=1}^{n} c_i &= c_1 \sum_{i=1}^{n} q^{i-1} + F_1 \sum_{i=1}^{n-1} q^{i-1} + \cdots + F_{n-2}(1+q) + F_{n-1} \\
&= \frac{1}{1-q} \left(c_1(1-q^n) + F_1(1-q^{n-1}) + \cdots + F_{n-2}(1-q^2) + F_{n-1}(1-q)\right) \\
&= \frac{1}{1-q} \left(c_1 + \sum_{i=1}^{n-1} F_i - (c_1q^n + F_1q^{n-1} + \cdots + F_{n-2}q^2 + F_{n-1}q)\right) \\
&= \frac{1}{1-q} \left(c_1 + F_{n+1} - 1 - (c_1q^{n-1} + F_1q^{n-2} + \cdots + F_{n-2}q + F_{n-1})q\right) \\
&= \frac{1}{1-q}(c_1 + F_{n+2} - F_n - 1 - c_nq) \\
&= \frac{1}{1-q}(c_1 + F_{n+2} - c_{n+1} - 1).
\end{aligned}$$

It is obvious that for an RF-sequence $(c_n)_{n\geq 1}$ as above, $c_n + c_{n-1} = (c_{n-1} + c_{n-2})q + F_n$ for $n \geq 3$. A natural question then is: "Can the sum sequence be extended to an RF-sequence with the same constant quotient?" The answer is "not always". We give an extension theorem.

Theorem 2.3: *Let $(c_n)_{n\geq 1}$ be an RF-sequence with a constant quotient q, as above. Let $d_n = c_{n-1} + c_{n-2}$ for $n \geq 3$. Then the sum sequence $(d_n)_{n\geq 3}$ can be extended to an RF-sequence $(d_n)_{n\geq 1}$ if and only if $q^2|c_1 - q$.*

Proof:

"$\Longrightarrow$": Assume there exist $d_1, d_2 \in \mathbb{N}$ such that $(d_n)_{n\geq 1}$ forms an RF-sequence. By definition, $d_3 = c_1 + c_2 = c_1 + qc_1 + F_1$ and $d_3 = d_2q + F_1 \Longrightarrow (q+1)c_1 = qd_2$. Thus $q|c_1$ so that $c_1 = qr$ for some $r \in \mathbb{N}$. We can rewrite $d_3 = q^2r + qr + 1$. On the other hand, $d_3 = d_2q + 1$. These two forms of d_3 give $d_2 = qr + r$ since $q > 0$. Combining this with the definition, $d_2 = d_1q + 1$, we have $r - 1 = q(d_1 - r)$ and so $q|r - 1$. Further, $c_1 - q = qr - q = q(r-1) \Longrightarrow q^2|c_1 - q$.

For the other implication, assume $q^2|c_1 - q$. By defintion $d_n = d_{n-1} + F_{n-1}$ for $n \geq 3$. To extend the sequence $(d_n)_{n\geq 3}$ into an RF-sequence $(d_n)_{n\geq 1}$, we need to find two positive integers d_1 and d_2 such that $d_3 = d_2q + 1$ and $d_2 = d_1q + 1$. Let $c_1 - q = q^2m$ for some $m \in \mathbb{N}$. Thus $c_1 = q^2m + q$ and $c_2 = q^3m + q^2 + 1$. Define $d_1 = qm + m + 1 \in \mathbb{N}$ and $d_2 = qd_1 + 1 = q^2m + qm + q + 1 \in \mathbb{N}$. Then $d_3 = c_1 + c_2 = q^3m + q^2m + q^2 + q + 1 = qd_2 + 1$. Therefore $(d_n)_{n\geq 1}$ forms an RF-sequence.

Now let us investigate the condition $q^2|c_1 - q$.

Let c_1 be a positive integer. There are two positive factors q of c_1 satisfying $q^2|c_1 - q$, that is, $q = 1$ or $q = c_1$. We call them trivial solutions to the divisiblity $q^2|c_1 - q$. We are interested in non-trivial solutions to the divisibility.

Lemma 2.4: *Let c_1, q be positive integers such that $q^2|c_1 - q$. Then*

(1) *$q|c_1$, $q^2 \not| \, c_1$, and $c_1 - q$ is not a prime.*

(2) *If $c_1 = ab$ where $a, b \in \mathbb{N}$, then $a^2|c_1 - a \Longleftrightarrow a|b - 1$.*

Proposition 2.5: *In the following cases, we give all solutions for the divisibility*

$$(*) \quad q^2|c_1 - q.$$

(1) *If $c_1 = p^r$, where p is a prime and r is a positive integer greater than 1, then the divisibility $(*)$ has only trivial solutions;*

(2) *If $c_1 = 2p^r$, where p is an odd prime and $r \in \mathbb{N}$, then $q = 2$ is the only non-trivial solution to $(*)$;*

(3) *If $c_1 = p(p+2)$, where p and $p+2$ are two consecutive primes, then the divisibility $(*)$ has only trivial solutions.*

Proof: It is straight forward by testing all factors of c_1 using Lemma 2.4(3).

The solutions of the divisibility $(*)$ provide extensions of the sum sequence of an RF-sequence with constant quotient.

Corollary 2.6: *Let $(c_n)_{n\geq 1}$ be an RF-sequence with a constant quotient q, where $q \neq 1, 2$, or c_1. Then the sum sequence $(c_{n+1} + c_n)_{n\geq 2}$ can not be extended to an RF-sequence with the same constant quotient q if $c_1 = p^r (r \in \mathbb{N})$, $c_1 = 2p^r (r > 1)$, or $c_1 = p(p+2)$, where p and $p+2$ are both prime numbers.*

3. OTHER TYPES OF *RF*-SEQUENCES

First we consider an *RF*-sequence associated with the natural numbers.

Theorem 3.1: *Let* $(c_n)_{n\geq 1}$ *be an RF-sequence with the quotient sequence* $(q_n = n)_{n\geq 1}$. *Then*

$$\frac{c_n}{(n-1)!} - c_1 = \sum_{i=1}^{n-1} \frac{F_i}{i!}.$$

Proof:

It is easy to show the explicit formula for c_n by mathematical induction:

$$c_n = c_1(n-1)! + \frac{F_1(n-1)!}{1!} + \frac{F_2(n-1)!}{2!} + \frac{F_3(n-1)!}{3!} + \cdots + F_{n-2}(n-1) + F_{n-1}$$

$$= (n-1)!\left(c_1 + \frac{F_1}{1!} + \frac{F_2}{2!} + \frac{F_3}{3!} + \cdots + \frac{F_{n-1}}{(n-1)!}\right).$$

Thus the theorem follows.

Similarly, we develop a sum formula for *RF*-sequences with quotients that are powers of a constant. Our focus is on an *RF*-sequence $(s_n)_{n\geq 1}$ such that $s_{n+1} = s_n \cdot q^n + F_n$. It is obvious that $q \geq 0$. Let us consider the situation $q > 0$.

Proposition 3.2: *Let* $(s_n)_{n\geq 1}$ *be an RF-sequence such that* $s_{n+1} = s_n \cdot q^n + F_n$ *for each* $n \geq 2$. *Suppose* $q > 0$. *Then for* $n \geq 2$,

$$s_n = s_1 q^{n(n-1)/2} + \sum_{i=1}^{n-1} F_i q^{(n+i)(n-i-1)/2}.$$

Theorem 3.3: *Consider the sequence* $(s_n)_{n\geq 1}$ *as above. Then*

$$\sum_{i=1}^{n} s_i = s_2 q^{(n^3-n)/6} + \sum_{j=1}^{n-1}\left(\sum_{i=j+1}^{n-1} q^{(i+j)(i-j-1)/2}\right) F_j.$$

Proof: By Proposition 3.2,

$$\sum_{i=1}^{n} s_i = \sum_{i=1}^{n}\left(s_1 q^{i(i-1)/2} + \sum_{j=1}^{i-1} F_j q^{(i+j)(i-j-1)/2}\right)$$

$$= s_1 \sum_{i=1}^{n} q^{i(i-1)/2} + \sum_{i=1}^{n}\left(\sum_{j=1}^{i-1} F_j q^{(i+j)(i-j-1)/2}\right)$$

$$= s_1 q^{n(n-1)(n+1)/6} + \sum_{j=1}^{n-1}\left(\sum_{i=j+1}^{n-1} q^{(i+j)(i-j-1)/2}\right) F_j.$$

FUTURE DIRECTIONS

It is shown in our paper that the Fibonacci sequence is a special kind of RF-sequence. An RF-sequence has a few similar properties that the Fibonacci sequence has. In the future, we would like to develop more properties for a general RF-sequence using the properties of the Fibonacci sequence. Another direction is to study the RF-sequences associated with a quotient sequence of a specific interest. Similar sequences can also be constructed from the Lucas sequence.

REFERENCES

[1] Bicknell, Marjorie and Hoggatt, Verner E., Jr. A Primer for the Fibonacci Numbers. A publication of Fibonacci Association, 1972.

[2] Koshy, Thomas. _Fibonacci and Lucas Numbers with Applications, Pure & Applied Mathematics. John Wiley & Sons, Inc., 2001.

[3] Vajda, S. Fibonacci and Lucas Numbers, and the Golden Section: Theory and Applications. Ellis Horwood Limited, New York, 1989.

[4] Vorob'ev, N.N. Fibonacci Numbers. Translated from Russian by Halina Moss, Blaisdell Publication Company, New York · london, 1951.

AMS Classification Numbers: 11B39, 11B38, 11Y55

CULLEN NUMBERS IN BINARY RECURRENT SEQUENCES

Florian Luca and Pantelimon Stănică

1. INTRODUCTION

Mentioned in the excellent book of R. Guy [5], the *Cullen numbers* are elements of the sequence $C_n := n \cdot 2^n + 1$ (see Section **B20** of [5]). They happen to be composite (see [2] and [8]) for all $1 \le n < 412000$, except for $n = 1$, 141, 4713, 5795, 6611, 18496, 32292, 32469, 59656, 90825, 262419, 361275. John Conway (cited in [5]) observes that the Cullen number C_n is divisible by $p = 2n - 1$ if p is a prime of the form $8k \pm 3$. Hooley [6] showed that almost all Cullen numbers are composite. In spite of the fact that the primes in this sequence are rare, it is still believed that there are infinitely many Cullen primes. Related to these are the *Woodall numbers* (or *Cullen numbers of the second kind*) given by $W_n := n \cdot 2^n - 1$. The number W_n is prime for $n = 2$, 3, 6, 30, 75, 81, 115, 123, 249, 362, 384, 462, 512, 751, 822, 5312, 7755, 9531, 12379, 15822, 18885, 22971, 23005, 98726, 143018, 151023 and for no other n with $n < 416000$ (see [1] for details). We also mention that in [9] it is shown that $\log \gcd(C_n, C_m) \ll \sqrt{m \log m}$ holds for all $m > n > 0$ and a similar result holds for the Woodall numbers. Here, and elsewhere throughout this paper, we use the Vinogradov symbols $\gg$ and $\ll$, as well as the Landau symbols O and o with their usual meanings, and for a real number $x \ge 1$ we use $\log x$ for the natural logarithm of x.

2. THE RESULTS

We fix a nonzero integer l and a sequence of integers $s := (s_n)_{n \ge 0}$. We define the (s,l)-*Cullen numbers* as

$$C_n(s,l) := s_n \cdot 2^n + l \quad \text{for all } n \ge 0.$$

This paper is in final form and no version of it will be submitted for publication elsewhere.

F.T. Howard (ed.), Applications of Fibonacci Numbers, Volume 9: Proceedings of the Tenth International Research Conference on Fibonacci Numbers and their Applications, 167–175.

Notice that when $s_n := n$ for all n, then the $(s,1)$-Cullen numbers are simply the regular Cullen numbers, while the $(s,-1)$-Cullen numbers are simply the regular Woodall numbers. Throughout this paper, we let $(u_n)_{n\geq 0}$ be a nondegenerate binary recurrent sequence of integers. That is, $(u_n)_{n\geq 0}$ is a sequence of integers such that there exist two integers a and b such that the recurrence formula

$$u_{n+2} = au_{n+1} + bu_n \quad \text{holds for all } n \geq 0.$$

We also assume that $\Delta := a^2+4b > 0$. It is then well known that there exist two constants A and B so that the formula $u_k = A\alpha^k - B\beta^k$ holds for all $k \geq 0$, where α and β are the roots of the *characteristic equation* $x^2 - ax - b = 0$. This formula is sometimes referred to as *Binet's formula*. Notice that since $\Delta > 0$, it follows that α and β are real. The sequence $(u_n)_{n\geq 0}$ is called *nondegenerate* if $AB\alpha\beta \neq 0$ and $\alpha \neq -\beta$. We shall also assume that $a > 0$. Notice that this is not a real obstruction, for if $a < 0$, then we may replace the sequence $(u_n)_{n\geq 0}$ by the sequence $((-1)^n u_n)_{n\geq 0}$ which has the same arithmetic properties as our initial sequence $(u_n)_{n\geq 0}$, and which satisfies the same recurrence relation as $(u_n)_{n\geq 0}$ does with a replaced by $-a$. The case $a = 0$ is not allowed because this leads to $\alpha = -\beta$. We shall also adopt the convention that α is the largest root of the characteristic equation. Notice that $\alpha > \max\{|\beta|, 1\}$. Specifically, $\alpha = \frac{1}{2}\left(a + \sqrt{a^2+4b}\right)$, $\beta = \frac{1}{2}\left(a - \sqrt{a^2+4b}\right)$, and the values of A and B are given by $A = \frac{u_1 - u_0\beta}{\alpha-\beta}$, $B = \frac{u_1-u_0\alpha}{\alpha-\beta}$. The binary recurrent sequence $(v_n)_{n\geq 0}$ whose starting values are $v_0 := 0$ and $v_1 := 1$ is called *the Lucas sequence* with parameters a and b. We reserve the notation $(v_n)_{n\geq 0}$ for the Lucas sequence. When $a = b = 1$, the Lucas sequence is precisely the *Fibonacci sequence* $(F_n)_{n\geq 0}$, while when $a = 2$ and $b = 1$, the Lucas sequence is precisely the *Pell sequence* $(P_n)_{n\geq 0}$. Some of our results address also *almost Lucas sequences*, i.e., binary recurrent sequences $(u_n)_{n\geq 0}$ for which $u_0 := 0$ and $u_1 \geq 1$. Notice that in this case the formula $u_n = u_1 v_n$ holds for all $n \geq 0$, and the numbers A and B satisfy $A = B = \frac{u_1}{\alpha-\beta}$.

In the present paper, we investigate the occurrence of (s,l)-Cullen numbers in such binary recurrent sequences. As applications, we find all regular Cullen and Woodall numbers that are also either Fibonacci or Pell numbers. We also give a slightly more general result pertaining to the arithmetic structure of binary recurrent sequences having $b := \pm 1$.

Theorem 1: *Let $(v_n)_{n\geq 0}$ be a Lucas sequence for which $a > 0$ is odd and $b \equiv 1 \pmod{16}$. Assume also, that $(s_n)_{n\geq 0}$ is a sequence of positive integers satisfying $s_n = O(\alpha^{2^{n-1}})$, and assume that N is a nonnegative integer such that $v_N \equiv 3 \pmod 4$ or $v_N \equiv 0 \pmod 8$. Then, there exists an effectively computable constant $c_1 := c_1(\alpha, s, N)$, depending only on α, s and N, such that all positive integer solutions of the diophantine equation $C_n(s, v_N) = v_k$ satisfy $\max\{n, k\} < c_1$.*

The proof of our Theorem 1 is entirely elementary. Notice that the sequence $(s_n)_{n\geq 0}$ itself is not that important, rather what matters about it is that it does not grow too fast, that is, that it satisfies $\log s_n \leq 2^{n-1}\log\alpha + O(1)$. So, the above theorem can be reformulated by saying that if N is a fixed positive integer and if m is a large positive integer such that $2^n | v_m - v_N$, then $\log\frac{v_m - v_N}{2^n} - 2^{n-1}\log\alpha \gg 1$. In fact, it will be plain from our argument that the result continues to hold in a somewhat larger range for s, namely when s satisfies $\log s_n < c\cdot 2^{n-1} + O(1)$ with any constant c strictly smaller than $\frac{3}{2}\cdot\log\alpha \geq \frac{3}{2}\log\left(\frac{1+\sqrt{5}}{2}\right) \approx 1.12573$. Here, we used the well-known fact that $\alpha \geq \frac{1+\sqrt{5}}{2}$. In particular, c can be chosen to be $c := 1$, and under this form we obtain a result without a dependence on α in the upper bound for $\log s_n$. By nonelementary methods, we prove the following theorem.

Theorem 2: *Let $(u_n)_{n\geq 0}$ by any nondegenerate binary recurrent sequence of integers with $b := \pm 1$. Let l be any integer and $p_1 < \cdots < p_t$ be any fixed prime numbers. For any positive integer m write*

$$u_m - l = PQ,$$

where P is the largest divisor of $u_m - l$ composed only from the primes $p_1, \ldots, p_t$ and Q is coprime to $p_1 \cdot \ldots \cdot p_t$. Then there exist two computable constants c_1 and c_2 depending only on the sequence $(u_n)_{n\geq 0}$, the number l, and on the prime numbers $p_1, \ldots, p_t$, such that for $m > c_1$ we have $\log |Q| > |P|^{c_2}$.

The proof of Theorem 2 uses the theory of lower bounds for linear forms in logarithms of algebraic numbers, and a very good introduction to this topic is [11]. Since such lower bounds usually involve some astronomical constants, the constant c_1 turns out to be very large, while the constant c_2 turns out to be very small. In a certain sense, Theorem 2 is an extension of Theorem 1 when $b := \pm 1$. Indeed, when $b := \pm 1$, take $t := 1$ and $p_1 := 2$ in the statement of Theorem 2. The conclusion of Theorem 2 asserts that if l is any fixed integer (like $l := u_N$ for some fixed N, for example), then there exists a computable constant c_2 which depends only on the sequence $(u_n)_{n\geq 0}$ and the number l, such that the diophantine equation of the form $u_m = s2^n + l$ with m and n positive integers and s an integer such that $|s| \leq \exp(2^{n/c_2}) = 2^{(1/\log 2)2^n/c_2}$ has only finitely many effectively computable positive integer solutions m and, in fact, all such solutions have $m < c_1$, where c_1 is the constant appearing in the statement of Theorem 2. Thus, Theorem 2 is somewhat similar to Theorem 1 when $b := \pm 1$. Of course, even in this case Theorem 1 is better, but there l has a very particular value, $t := 1$, $p_1 := 2$, and a and b satisfy some restrictive congruence conditions. In the general case, i.e., with an arbitrary value of b, it seems to be hard to obtain good lower bounds on Q in terms of P comparable to the ones in the case $b := \pm 1$, where P and Q are defined in the statement of Theorem 2. However, at least when the two roots of the characteristic equation of $(u_n)_{n\geq 0}$ are real, or when the coefficients a and b of the recurrence relation for $(u_n)_{n\geq 0}$ are not coprime and $l \neq 0$, one can use similar methods as in the proof of Theorem 2 to show that there exist two effectively computable constants c_1 and c_2, depending again only on $(u_n)_{n\geq 0}$, l, and $\mathcal{P}$, such that the inequality $\log |Q| > c_2 \left(\frac{m}{\log m}\right)$ holds for all $m > c_1$. Finally, in the worst case, in which the two roots of the characteristic equation of $(u_n)_{n\geq 0}$ are complex conjugates and a and b are coprime, it is an immediate application of the Subspace Theorem (see [10]) that for every $\varepsilon > 0$ the inequality $|Q| > |u_m|^{1-\varepsilon}$ holds for all but finitely many values of the positive integer m. We do not give further details in this direction and restrict ourselves to presenting the proof of Theorem 2 as stated above.

Returning to elementary arguments, by using the method of proof of Theorem 1, we have the following result for the case in which $(v_n)_{n\geq 0}$ is the Fibonacci or Pell sequence.

Theorem 3: *(i) There are only three Fibonacci numbers that are also Cullen numbers, namely $F_1 = F_2 = 1$ and $F_4 = 3$. There are only two Fibonacci numbers that are also Woodall numbers, namely $F_1 = F_2 = 1$.*

(ii) There is only one Pell number that is also a Cullen number, namely $P_1 = 1$. There is only one Pell number that is also a Woodall number, namely $P_1 = 1$.

3. THE PROOF OF THE THEOREMS

We start with the nonelementary proof of Theorem 2.

Proof of Theorem 2: Throughout this proof, all constants which appear are positive, effectively computable, and labelled increasingly as c_3, $c_4, \ldots$. We reserve the notation c_1 and c_2 for the final constants asserted in the statement of Theorem 2.

Write

$$u_n = A\alpha^n - B\beta^n.$$

Since $b := \pm 1$, we get that $\beta = \pm\alpha^{-1}$. Recall that we are assuming that $a > 0$, for otherwise we may replace a by $-a$, l by $\pm l$, and $(u_n)_{n\geq 0}$ by $((-1)^n u_n)_{n\geq 0}$. In particular, $\alpha > 1 > |\beta|$. We treat only the case of the parameter $l \neq 0$ since the other case is even easier. Let m be a large positive integer and write $z := \alpha^m$, therefore

$$u_m = Az \pm Bz^{-1}.$$

In particular,

$$u_m - l = Az \pm Bz^{-1} - l = Az^{-1}(z - z_1)(z - z_2),$$

where $z_{1,2}$ are the roots of the equation

$$x^2 - \frac{l}{A}x \pm \frac{B}{A} = 0.$$

Let $\mathbf{K}$ be the smallest number field containing all the numbers α, β and z_1, z_2 for both choices of signs $\pm$ above. Let p be any of the prime numbers $p_1, \ldots, p_t$, and let π be any prime ideal of $\mathcal{O}_{\mathbf{K}}$ sitting above p, where we use $\mathcal{O}_{\mathbf{K}}$ for the ring of algebraic integers inside $\mathbf{K}$. It is easy to see that the inequality

$$|u_m| > c_3\alpha^m \tag{1}$$

holds for sufficiently large values of m, where one can take $c_3 := |A|/2$. In particular, for large m, we have

$$|u_m| > 2|l|, \tag{2}$$

and therefore $u_m - l$ is nonzero. Thus, none of the factors $z - z_1$ and $z - z_2$ is zero either. For any algebraic number γ in $\mathbf{K}$ and any prime ideal π in $\mathcal{O}_{\mathbf{K}}$ we write $\text{ord}_\pi(\gamma)$ for the order at which π appears in the factorization in prime ideals of the factional ideal $[\gamma]$ generated by γ inside $\mathbf{K}$. Using a linear form in p-adic logarithms (immediate application of Theorem 1 in [12]), we get that both inequalities

$$\text{ord}_\pi(z - z_1) = \text{ord}_\pi(\alpha^m - z_1) < c_4 \log m,$$

and

$$\text{ord}_\pi(z - z_2) = \text{ord}_\pi(\alpha^m - z_2) < c_4 \log m$$

hold with some constant c_4 depending on the sequence $(u_n)_{n\geq 0}$, the number l, as well as the prime number p. The constant c_4 can be taken to be of the form $c_5 \cdot p^2$, where c_5 is absolute. Since p can take only finitely many values, we get that if we write

$$P := p_1^{\alpha_1} \cdot \ldots \cdot p_t^{\alpha_t},$$

then the inequality

$$\max\{\alpha_i | i = 1, \ldots, t\} < c_6 \log m$$

holds with some computable constant c_6, which can be taken to be $c_6 := c_5 \cdot p_t^2$. Now

$$u_m - l = PQ,$$

therefore

$$\log |u_m - l| = \log |P| + \log |Q|.$$

Since for large m we have

$$|u_m - l| > c_7 \alpha^m \tag{3}$$

with $c_7 := |A|/2$, we get

$$c_8 m - c_9 < \log |u_m - l| = \log |P| + \log |Q| < c_{10} \log m + \log |Q|.$$

Here, one can take $c_8 := \log \alpha > 0, c_9 := |\log(|A|/2)| > 0$, and $c_{10} := c_6 \log(p_1 \cdot \ldots \cdot p_t)$. The above inequality implies that

$$\log |Q| > c_8 m - c_9 - c_{10} \log m.$$

However, the inequality

$$c_8 m - c_9 - c_{10} \log m > c_{11} m \tag{4}$$

clearly holds for large values of m, where c_{11} can be chosen to be any fixed constant strictly smaller than c_8. So, let us just take $c_{11} := c_8/2$, and let us write c_1 for the computable constant so that all the inequalities (1) - (4) hold for $m > c_1$. Thus, on the one hand we have that $\log |Q| > c_{11} m$, while on the other hand we have that $|P| < \exp(c_{10} \log m) = m^{c_{10}}$, which imply that $\log |Q| > c_{11} |P|^{1/c_{10}}$. It is now clear that this last inequality implies the conclusion of Theorem 2 with any constant c_2 strictly smaller than $1/c_{10}$. □

To prove Theorem 1, we need to recall some known facts about the distribution of the Lucas sequence $(v_n)_{k \geq 0}$ modulo 2^n.

Lemma 4: *If $a > 0$ is odd and $b \equiv 1 \pmod{16}$ then the Lucas sequence $(v_k)_{k \geq 0}$ satisfies the following properties:*

(i) if $v_N \equiv 3 \pmod 4$ and $v_k \equiv v_N \pmod{2^n}$, $n \geq 6$, $k > N$, then $k \equiv N \pmod{3 \cdot 2^{n-1}}$,

(ii) if $v_N \equiv 0 \pmod 8$ and $v_k \equiv v_N \pmod{2^n}$, $n \geq 6$, $k > N$, then $k \equiv N \pmod{3 \cdot 2^{n-2}}$.

Proof: It is well known (see [3], for instance), that under the assumptions of the lemma, the period of $(v_k)_{k \geq 0}$ modulo 2^n is $3 \cdot 2^{n-1}$. Theorem 1.1 of [3] shows that every residue r modulo 2^n satisfying $r \equiv 3 \pmod 4$ appears only once in every period of $(v_k)_{k \geq 0}$ modulo 2^n, which implies (i). To see (ii), we use again the same Theorem 1.1 of [3] which says that every residue r modulo 2^n satisfying $r \equiv 0 \pmod 8$ appears exactly twice in every period of $(v_k)_{k \geq 0}$ modulo 2^n. From the proof of that same theorem (p. 303), we deduce that the distance between two such *consecutive* residues is $3 \cdot 2^{n-2}$, which completes the proof of our lemma. □

Remark: The previous lemma implies that if $k > N \geq 0$ are consecutive indices satisfying $v_k \equiv v_N \pmod{2^n}$ and $v_N \equiv 0 \pmod 8$, then $k - N = 3 \cdot 2^{n-2}$.

For the purpose of the next lemma, we assume that $(u_n)_{n \geq 0}$ is an almost Lucas sequence with $a > 0$. Write $\gamma := \sqrt{\Delta} = \sqrt{a^2 + 4b} = \alpha - \beta$. Observe that $1 \leq \gamma = \alpha - \beta$. We will not be using all inequalities from the next lemma, but we thought they might have an interest of their own.

Lemma 5: *The inequalities*

$$A\alpha^{k-1} \leq u_k = A(\alpha^k - \beta^k) \leq u_1\alpha^{k-1}, \ if\ b > 0 \tag{5}$$

$$u_1\alpha^{k-1} \leq u_k = A(\alpha^k - \beta^k) \leq \frac{au_1}{\gamma}\alpha^{k-1}, \ if\ b < 0 \tag{6}$$

hold for all positive integers k.

Proof: Since $u_1 \geq 1$ and $u_n = u_1 v_n$ holds for all $n \geq 0$, we may divide both inequalities (5) and (6) across by u_1 and restrict our attention to proving these inequalities for the sequence $(v_n)_{n\geq 0}$

If $b > 0$, then since $b = -\alpha\beta$, we get that $\beta < 0$. Thus,

$$A(\alpha^k - \beta^k) = \frac{\alpha^k - \beta^k}{\alpha - \beta} = \frac{\alpha^k - \beta^k}{\alpha + |\beta|}.$$

Clearly,

$$\frac{\alpha^k - \beta^k}{\alpha + |\beta|} \leq \frac{\alpha^k + |\beta|^k}{\alpha + |\beta|} \leq \alpha^{k-1},$$

with the last inequality holding because $\alpha > |\beta|$, which takes care of the inequality from the right hand side of (5). The inequality from the left hand side of (5) is implied by

$$\alpha^k - |\beta|^k > \alpha^{k-1},$$

which is equivalent to $\alpha^{k-1}(\alpha - 1) > |\beta|^k$. The previous inequality is implied by $\alpha^{k-1} > |\beta|^{k-1}$ and $\alpha - 1 > |\beta|$, with the last inequality being true because $\alpha - |\beta| = \alpha + \beta = a \geq 1$. When $b < 0$, then $\alpha > \beta > 0$, therefore the inequality from the right hand side of (6) is simply

$$\frac{\alpha^k - \beta^k}{\alpha - \beta} < \frac{\alpha^{k-1}(\alpha + \beta)}{\alpha - \beta},$$

which is obvious because $\alpha^{k-1}(\alpha + \beta) > \alpha^k > \alpha^k - \beta^k$. The inequality from the left hand side of (6) is simply

$$\alpha^{k-1} < \frac{\alpha^k - \beta^k}{\alpha - \beta},$$

which is also obvious because it is equivalent to $\alpha^{k-1}\beta > \beta^k$. □

Proof of Theorem 1: Let k and n be nonnegative integers such that $v_k = C_n(s, v_N)$. Thus,

$$v_k = s_n \cdot 2^n + v_N. \tag{7}$$

We may certainly assume that $n \geq 6$ and that $k \geq 1$. Equation (7) implies that $v_k \equiv v_N \pmod{2^n}$. By Lemma 4, we get that if $k > N$, then

$$k \geq N + 3 \cdot 2^{n-1}, \text{ if } v_N \equiv 3 \pmod 4, \tag{8}$$

$$k \geq N + 3 \cdot 2^{n-2}, \text{ if } v_N \equiv 0 \pmod 8. \tag{9}$$

Using Lemma 5, inequalities (8) and (9), together with the fact that $A = \frac{1}{\alpha-\beta} \geq \frac{1}{2\alpha}$ when $b > 0$ (clearly, this last inequality holds when $b < 0$ as well, but we shall need it only in the case $b > 0$), we deduce that

$$s_n \cdot 2^n + v_N = v_k \geq A\alpha^{k-1} \geq \frac{1}{2} \cdot \alpha^{3 \cdot 2^{n-1}+N-2}, \text{ if } b > 0 \text{ and } v_N \equiv 3 \pmod 4, \tag{10}$$

$$s_n \cdot 2^n + v_N = v_k \geq A\alpha^{k-1} \geq \frac{1}{2} \cdot \alpha^{3 \cdot 2^{n-2}+N-2}, \text{ if } b > 0 \text{ and } v_N \equiv 0 \pmod 8, \tag{11}$$

$$s_n \cdot 2^n + v_N = v_k \geq v_1\alpha^{k-1} \geq \alpha^{3 \cdot 2^{n-1}+N-1}, \text{ if } b < 0 \text{ and } v_N \equiv 3 \pmod 4, \tag{12}$$

$$s_n \cdot 2^n + v_N = v_k \geq v_1\alpha^{k-1} \geq \alpha^{3 \cdot 2^{n-2}+N-1}, \text{ if } b < 0 \text{ and } v_N \equiv 0 \pmod 8. \tag{13}$$

But $\alpha > 1$, and $s_n = O(\alpha^{2^{n-1}})$, therefore the previous inequalities are false, for $n > c_1$ where c_1 is some computable constant depending only on α, s and N. In fact, it is clear that the dependence on s is encrypted only in the constant understood in the inequality $s_n = O(\alpha^{2^{n-1}})$. Theorem 1 is therefore proved. □

Remark: Any polynomial with integer coefficients in the variable n is an example of a sequence $(s_n)_{n \geq 0}$.

Although Theorem 3 is not a direct application of Theorem 1, the proof of this result can be achieved along the same lines. We need the following lemma.

Lemma 6:

(i) If $n \geq 3$, then $F_k \equiv \pm 1 \pmod{2^n}$ if and only if $k \equiv \pm 1, \pm 2 \pmod{3 \cdot 2^{n-1}}$.

(ii) If $n \geq 2$, then $P_k \equiv 1 \pmod{2^n}$ if and only if $k \equiv \pm 1 \pmod{2^n}$.

Proof: (*i*) This is well known (see [7]). (*ii*) By Theorem 3.1 (*a*) of [4], the sequence $(P_k)_{k \geq 0}$ has period 2^n modulo 2^n. Thus, if $n = 2$, then looking at the first four terms of the Pell sequence, we get that the sequence $(P_k)_{k \geq 0}$ is congruent to 0, 1, 2, 1 modulo 4. Thus, $P_k \equiv 1 \pmod 4$ if and only if $k \equiv \pm 1 \pmod 4$. Assume now that $n > 2$ and proceed by induction. By Lemma 4.3 of [4] and the induction hypothesis, the residue 1 appears exactly twice in one period of $(P_k)_{k\geq 0}$ modulo 2^{n+1}, so it is sufficient to prove that $P_1 \equiv P_{2^{n+1}-1} \equiv 1 \pmod{2^{n+1}}$. Since $P_1 = 1$, it suffices to show that $P_{2^{n+1}-1} \equiv 1 \pmod{2^{n+1}}$. Using the relation

$$P_{m+n} = P_{m-1}P_n + P_mP_{n+1},$$

we obtain $P_{2^n+2^n-1} = (P_{2^n-1})^2 + (P_{2^n})^2$. By the induction hypothesis, $P_{2^n-1} \equiv 1 \pmod{2^n}$, therefore $(P_{2^n-1})^2 \equiv 1 (\text{mod } 2^{n+1})$. By Proposition 2.4($a$) of [4], $P_{2^n} \equiv 0 \pmod{2^{n+1}}$. Thus, $P_{2^{n+1}-1} \equiv 1 \pmod{2^{n+1}}$. □

Proof of Theorem 3: We first look at the case of the Fibonacci numbers. Assume that k and n are nonnegative integers such that $F_k = n \cdot 2^n + 1$. When $n = 0$, 1, we obtain the obvious solutions $(k, n) = (1, 0), (2, 0), (4, 1)$. The case $n = 2$ does not render a solution, so, from here on, we assume that $n \geq 3$, and therefore that $k \geq 5$. Thus

$$F_k \equiv 1 \pmod{2^n},$$

which implies, by Lemma 6, that $k \equiv \pm 1, \pm 2 \pmod{3 \cdot 2^{n-1}}$. Since $k \geq 5$, it follows that

$$k + 2 \geq 3 \cdot 2^{n-1}. \tag{14}$$

Since the inequality

$$\frac{\alpha^k}{\sqrt{5}} - \frac{1}{2} < F_k < \frac{\alpha^k}{\sqrt{5}} + \frac{1}{2},$$

holds, where $\alpha := (1 + \sqrt{5})/2$ is the golden section, we get

$$n \cdot 2^n + 1 = F_k > \frac{\alpha^k}{\sqrt{5}} - \frac{1}{2} \geq \frac{\alpha^{3\cdot 2^{n-1}-2}}{\sqrt{5}} - \frac{1}{2}.$$

Therefore, n must satisfy the inequality

$$\sqrt{5} \cdot n \cdot 2^{n+1} + 3\sqrt{5} > 2 \cdot \alpha^{3\cdot 2^{n-1}-2}, \tag{15}$$

which is impossible because the function

$$x \mapsto 2 \cdot \alpha^{3\cdot 2^{x-1}-2} - \sqrt{5} \cdot x \cdot 2^{x+1} - 3\sqrt{5}$$

is positive for all $x \geq 3$ (in fact, the largest positive zero of the above function is $x_0 \approx 2.71031$).

The argument for the case of the Pell sequence is entirely similar. Let again k and n be nonnegative integers satisfying the equation $P_k = n \cdot 2^n + 1$. When $n = 0$, 1, 2 we only get the solution $(k, n) = (1, 0)$. Assume now that $n \geq 3$. If $P_k = n \cdot 2^n + 1$, then $P_k \equiv 1 \pmod{2^n}$. This implies, by Lemma 6 (ii), that $k + 1 \geq 2^n$, and employing Lemma 5, we get

$$n \cdot 2^n + 1 = P_k \geq \frac{\sqrt{2}}{4}(1 + \sqrt{2})^{2^n-2}, \tag{16}$$

which is impossible for $n \geq 3$ because the largest positive zero of the function

$$x \mapsto \frac{\sqrt{2}}{4}(1+\sqrt{2})^{2^x-2} - x \cdot 2^x - 1$$

is $x_1 \approx 2.69847$.

The analysis of the diophantine equations involving Woodall rather than Cullen numbers and Fibonacci or Pell numbers is entirely similar. □

ACKNOWLEDGMENTS

Our deepest thanks go to the referee whose detailed and helpful comments significantly improved both the quality as well as the presentation of the present paper. The first author was partly supported by grants ECOS-ANUIES M02-M01, SEP-CONACYT 37259-E and SEP-CONACYT 37260-E. The second author was partially supported by a research grant from the School of Sciences at his institution. The second author is also associated with the Institute of Mathematics of Romanian Academy, Bucharest, Romania.

REFERENCES

[1] Ballinger, R. Proth Search Page. http://www.prothsearch.net/woodall.html.
[2] Caldwell, C. The Prime Pages. http://primes.utm.edu/.
[3] Carlip, W. and Jacobson, E. "On the Stability of Certain Lucas Sequences Modulo 2^k." *The Fiboancci Quarterly*, Vol. *34.4* (1996): pp. 298-305.
[4] Carlip, W. and Jacobson, E. "Stability of Two-Term Recurrence Sequences with Even Parameter." *Finite Fields Appl.*, Vol. *3* (1997): pp. 70-83.
[5] Guy, R. Unsolved Problems in Number Theory, (2nd ed.) Springer-Verlag, New York, 1994.
[6] Hooley, C. Applications of the Sieve Methods to the Theory of Numbers. Cambridge University Press, Cambridge, 1976.
[7] Jacobson, E. "Distribution of the Fibonacci Numbers Modulo 2^k." *The Fibonacci Quarterly*, Vol. *30.3* (1992): pp. 211-215.
[8] Keller, W. "New Cullen Primes." *Math. Comput.*, Vol. *64* (1995): pp. 1733-1741.
[9] Luca, F. On the Greatest Common Divisor of Two Cullen Numbers to appear in *Abh. Math. Sem. Univ. Hamburg.*
[10] Schlickewei, H.P. "S-Unit Equations Over Number Fields." *Inventiones Math.*, Vol. *102* (1990): pp. 95-107.
[11] Shorey, T.N. and Tijdeman, R. Exponential Diophantine Equations. Cambridge University Press, Cambridge, 1986.
[12] Yu, K. "Linear Forms in p-adic Logarithms." III, *Compositio Math*, Vol. *91.3* (1994): pp. 241-276.

AMS Classification Numbers: 11B37, 11B39, 11B83, 11J86, 11K31

A GENERALIZATION OF EULER'S FORMULA AND ITS CONNECTION TO FIBONACCI NUMBERS

Jonathan F. Mason and Richard H. Hudson

1. INTRODUCTION

This paper began as a simple proof generalizing Euler's well-known formula for the vertices, faces, and edges of a cube in 3 dimensions, to a tessaract, and to higher dimensions. Let an n-cube with n-dimensional volume 1 consist of all n-tuples $(x_1, x_2, \ldots, x_n)$ where each x_i, $i = 1, \ldots, n$ satisfies $0 \le x_i \le 1$. The boundary points of the n-cube are the vertices, which we will call 0-cubes to indicate that they are 0-dimensional. For each such vertex, we clearly have x_i fixed to be 0 or 1. A 1-cube will be an edge of the n-cube. For an edge, we have exactly one of the x_i free to take on values between 0 and 1 (inclusive) and the other x_i fixed to be 0 or 1 for each $i = 1, \ldots, n$. Similarly, a k-cube, $k \le n$, will have exactly k of the x_i free to take on values between 0 and 1 (inclusive) and $n - k$ fixed to be 0 or 1.

By representing each vertex in this way, it is clear that there are 2^n vertices in an n-cube. For a k-cube, since $n - k$ of the x_i are fixed, and k are not fixed, we must have exactly

$$\binom{n}{k} * 2^{n-k} \tag{1.1}$$

k-cubes in an n-cube. In particular, there are $n \cdot (2^{n-1})$ edges and $\binom{n}{2} \cdot (2^{n-2})$ faces. Thus,

$$\text{Vertices} + \text{Faces} - \text{Edges} = 2^n - n \cdot (2^{n-1}) + \binom{n}{2} \cdot (2^{n-2}) = 2^{n-3}(n^2 - 5n + 8), \tag{1.2}$$

This paper is in final form and no version of it will be submitted for publication elsewhere.

F.T. Howard (ed.), *Applications of Fibonacci Numbers, Volume 9: Proceedings of the Tenth International Research Conference on Fibonacci Numbers and their Applications*, 177–185.

which is a natural generalization of the well-known formula of Euler when $n = 3$, namely $V + F = E + 2$.

Note that it is an easy consequence of the binomial theorem (see, e.g., [1, p. 9]), that

$$\sum_{k=0}^{n} \binom{n}{k} \cdot (2^{n-k}) = (1+2)^n = 3^n \tag{1.3}$$

This gives the following table, which appears in [2, p. 89] when $n \geq 5$:

1	0	0	0	0	0	0	0	...	0
2	1	0	0	0	0	0	0	...	0
4	4	1	0	0	0	0	0	...	0
8	12	6	1	0	0	0	0	...	0
16	32	24	8	1	0	0	0	...	0
32	80	80	40	10	1	0	0	...	0
64	192	240	160	60	12	1	0	...	0
128	448	672	560	280	84	14	1	...	0
$\vdots$	$\vdots$	$\vdots$	$\vdots$	$\vdots$	$\vdots$	$\vdots$	$\vdots$	$\ddots$	$\vdots$
$\binom{n}{0} \cdot 2^n$	$\binom{n}{1} \cdot 2^{n-1}$	$\binom{n}{2} \cdot 2^{n-2}$	$\binom{n}{3} \cdot 2^{n-3}$	$\binom{n}{4} \cdot 2^{n-4}$	$\binom{n}{5} \cdot 2^{n-5}$	...	$\binom{n}{k} \cdot 2^{n-k}$	...	1

Table 1.1

Looking carefully at Table 1.1, we note that there is a one-to-one correspondence between entries in the table and the sequence of Fibonacci numbers. In Section 2, we will show how to prove this correspondence, but it is a somewhat more complicated derivation than the similar well-known correspondence between the Fibonacci numbers and the diagonal's of Pascal's Triangle, so we will first illustrate it pictorially for small n in Tables 1.2 and 1.3 below:

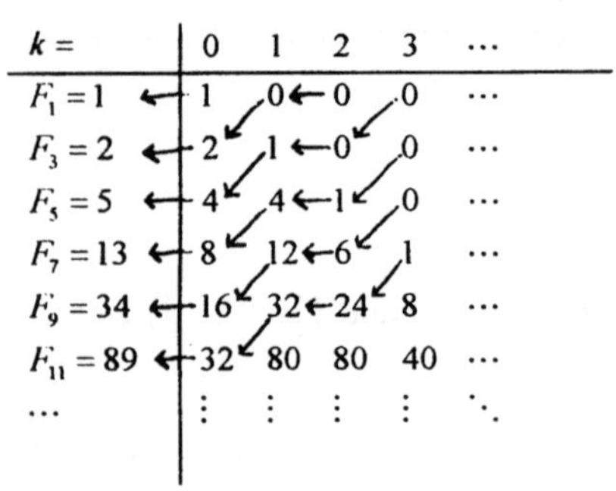

Table 1.2

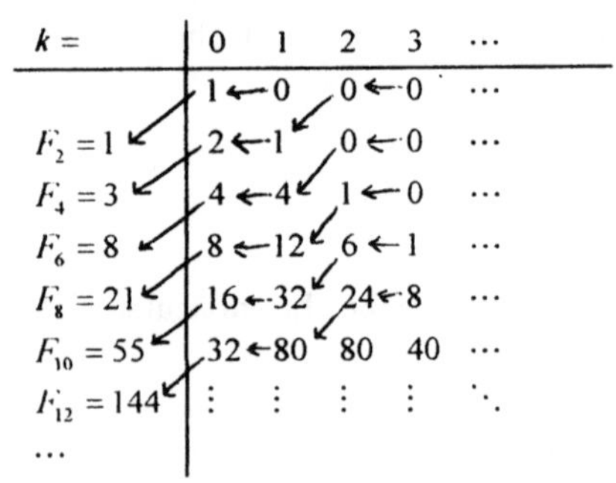

Table 1.3

Following the arrows and adding we obtain each Fibonacci number F_i exactly once.

2. THE PROOF OF THE FIBONACCI CORRESPONDENCE ILLUSTRATED IN TABLES 1.2 AND 1.3

We begin by defining A_n to be the sum of the terms starting with $\binom{n}{0} \cdot 2^n$ in the first column plus $\binom{n-1}{2} \cdot 2^{n-3}$ in the third column. We continue summing by moving up one row and over two columns each time. Note that we will encounter 0's when $2k > n-k$ or $3k > n$. Thus, there will only be $\lfloor \frac{n}{3} \rfloor + 1$ elements in the summation of A_n, and

$$A_n = \sum_{k=0}^{\lfloor \frac{n}{3} \rfloor} \binom{n-k}{2k} \cdot 2^{n-3k}, \; n \geq 0. \tag{2.1}$$

Similarly, we define B_n to be the same sequences as A_n but starting in the second column. Therefore,

$$B_n \sum_{k=1}^{\lfloor \frac{n+1}{3} \rfloor} \binom{n-k}{2k-1} \cdot 2^{n-3k+1}, \; n \geq 2 \text{ where } B_0 = B_1 = 0. \tag{2.2}$$

With these definitions now in place, we will show that $A_n + B_n = F_{2n+1}$, $n \geq 0$, and $A_{n-1} + B_n = F_{2n}$, $n \geq 1$. A bit more generally, $A_{\lfloor \frac{n}{2} \rfloor} + B_{\lfloor \frac{n+1}{2} \rfloor} = F_{\frac{n+1}{2}}$, $n \geq 1$, and therefore,

$$\sum_{k=0}^{\lfloor \frac{n}{6} \rfloor} \binom{\lfloor \frac{n}{2} \rfloor - k}{2k} \cdot 2^{\lfloor \frac{n}{2} \rfloor - 3k} + \sum_{k=1}^{\lfloor \frac{n+3}{6} \rfloor} \binom{\lfloor \frac{n+1}{2} \rfloor - k}{2k-1} \cdot 2^{\lfloor \frac{n+1}{2} \rfloor - 3k+1} = F_{n+1}. \tag{2.3}$$

To prove this we will argue by induction.

Initial Cases: $n = 0$ and $n = 1$: It is trivial to show by substitution that for $n = 0$ we get F_1 and for $n = 1$ we get F_2. Hence, the base cases both hold.

Now, we will assume that the result is true for both $n = m-1$ and $n = m-2$, and we will show that it is true for $n = m$. In other words, we will assume that $A_{\lfloor \frac{m-2}{2} \rfloor} + B_{\lfloor \frac{m-1}{2} \rfloor} = F_{m-1}$ and $A_{\lfloor \frac{m-1}{2} \rfloor} + B_{\lfloor \frac{m}{2} \rfloor} = F_m$, and we will show that $A_{\lfloor \frac{m}{2} \rfloor} + B_{\lfloor \frac{m+1}{2} \rfloor} = F_{m+1}$. By our assumptions we know that

$$A_{\lfloor \frac{m-2}{2} \rfloor} + B_{\lfloor \frac{m-1}{2} \rfloor} + A_{\lfloor \frac{m-1}{2} \rfloor} + B_{\lfloor \frac{m}{2} \rfloor} = F_{m-1} + F_m = F_{m+1}. \tag{2.4}$$

We must now break this problem into two cases:

Case 1: $\quad m \equiv 0 \pmod 2$

If $m \equiv 0(\mod 2)$, then $\lfloor \frac{m-2}{2} \rfloor = \lfloor \frac{m-1}{2} \rfloor = \lfloor \frac{m}{2} \rfloor - 1 = \lfloor \frac{m+1}{2} \rfloor - 1$. Therefore, by (2.4), we know that $A_{\lfloor \frac{m-2}{2} \rfloor} + B_{\lfloor \frac{m-2}{2} \rfloor} + A_{\lfloor \frac{m-2}{2} \rfloor} + B_{\lfloor \frac{m}{2} \rfloor} = F_{m+1}$. Thus,

$$2 \cdot \sum_{k=0}^{\lfloor \frac{m-2}{6} \rfloor} \binom{\lfloor \frac{m}{2} \rfloor - k - 1}{2k} \cdot 2^{\lfloor \frac{m}{2} \rfloor - 3k - 1} + \sum_{k=1}^{\lfloor \frac{m}{6} \rfloor} \binom{\lfloor \frac{m}{2} \rfloor - k - 1}{2k-1} \cdot 2^{\lfloor \frac{m}{2} \rfloor - 3k} + B_{\lfloor \frac{m}{2} \rfloor} = F_{m+1}. \quad (2.5)$$

It will help us later to move the 2 into the first summation and then bring out the first term of that summation. We are then left with

$$\binom{\lfloor \frac{m}{2} \rfloor - 1}{0} * 2^{\lfloor \frac{m}{2} \rfloor} + \sum_{k=1}^{\lfloor \frac{m-2}{6} \rfloor} \binom{\lfloor \frac{m}{2} \rfloor - k - 1}{2k} \cdot 2^{\lfloor \frac{m}{2} \rfloor - 3k}$$
$$+ \sum_{k=1}^{\lfloor \frac{m}{6} \rfloor} \binom{\lfloor \frac{m}{2} \rfloor - k - 1}{2k-1} \cdot 2^{\lfloor \frac{m}{2} \rfloor - 3k} + B_{\lfloor \frac{m}{2} \rfloor} = F_{m+1}. \quad (2.6)$$

We can make the substitution $\binom{\lfloor \frac{m}{2} \rfloor - 1}{0} \cdot 2^{\lfloor \frac{m}{2} \rfloor} = \binom{\lfloor \frac{m}{2} \rfloor}{0} \cdot 2^{\lfloor \frac{m}{2} \rfloor}$, and obtain

$$\binom{\lfloor \frac{m}{2} \rfloor}{0} * 2^{\lfloor \frac{m}{2} \rfloor} + \sum_{k=1}^{\lfloor \frac{m-2}{6} \rfloor} \binom{\lfloor \frac{m}{2} \rfloor - k - 1}{2k} \cdot 2^{\lfloor \frac{m}{2} \rfloor - 3k}$$
$$+ \sum_{k=1}^{\lfloor \frac{m}{6} \rfloor} \binom{\lfloor \frac{m}{2} \rfloor - k - 1}{2k-1} \cdot 2^{\lfloor \frac{m}{2} \rfloor - 3k} + B_{\lfloor \frac{m}{2} \rfloor} = F_{m+1}. \quad (2.7)$$

To progress further, we have three more cases to consider.

Case 1(a): $m \equiv 0 \pmod 6$, so that $\lfloor \frac{m-2}{6} \rfloor = \lfloor \frac{m}{6} \rfloor - 1$. From (2.7), we have that

$$\binom{\lfloor \frac{m}{2} \rfloor}{0} * 2^{\lfloor \frac{m}{2} \rfloor} + \sum_{k=1}^{\lfloor \frac{m}{6} \rfloor - 1} \binom{\lfloor \frac{m}{2} \rfloor - k - 1}{2k} \cdot 2^{\lfloor \frac{m}{2} \rfloor - 3k}$$
$$+ \sum_{k=1}^{\lfloor \frac{m}{6} \rfloor} \binom{\lfloor \frac{m}{2} \rfloor - k - 1}{2k-1} \cdot 2^{\lfloor \frac{m}{2} \rfloor - 3k} + B_{\lfloor \frac{m}{2} \rfloor} = F_{m+1}. \quad (2.8)$$

Then, after we pull out the last term of the second sum, our two sums have the same indices and we are free to combine them as follows

$$\binom{\lfloor\frac{m}{2}\rfloor}{0} * 2^{\lfloor\frac{m}{2}\rfloor} + \sum_{k=1}^{\lfloor\frac{m}{6}\rfloor-1}\left[\binom{\lfloor\frac{m}{2}\rfloor-k-1}{2k}+\binom{\lfloor\frac{m}{2}\rfloor-k-1}{2k-1}\right]\cdot 2^{\lfloor\frac{m}{2}\rfloor-3k}$$
$$+\binom{\lfloor\frac{m}{2}\rfloor-\lfloor\frac{m}{6}\rfloor-1}{2*\lfloor\frac{m}{6}\rfloor-1}+B_{\lfloor\frac{m}{2}\rfloor}=F_{m+1} \tag{2.9}$$

Again, using a result of Pascal, see [1, p. 8], we can simplify this to

$$\binom{\lfloor\frac{m}{2}\rfloor}{0}\cdot 2^{\lfloor\frac{m}{2}\rfloor}+\sum_{k=1}^{\lfloor\frac{m}{6}\rfloor-1}\binom{\lfloor\frac{m}{2}\rfloor-k}{2k}\cdot 2^{\lfloor\frac{m}{2}\rfloor-3k}+\binom{\lfloor\frac{m}{2}\rfloor-\lfloor\frac{m}{6}\rfloor-1}{2*\lfloor\frac{m}{6}\rfloor-1}+B_{\lfloor\frac{m}{2}\rfloor}=F_{m+1}. \tag{2.10}$$

Now, because

$$\left\lfloor\frac{m}{2}\right\rfloor-\left\lfloor\frac{m}{6}\right\rfloor-1=2*\left\lfloor\frac{m}{6}\right\rfloor-1,\ \binom{\lfloor\frac{m}{2}\rfloor-\lfloor\frac{m}{6}\rfloor-1}{2*\lfloor\frac{m}{6}\rfloor-1}=1=\binom{\lfloor\frac{m}{2}\rfloor-\lfloor\frac{m}{6}\rfloor}{2*\lfloor\frac{m}{6}\rfloor}, \tag{2.11}$$

it becomes evident that we can add the two terms on each side of the summation to the ends of the summation. Then (2.10) becomes

$$\sum_{k=0}^{\lfloor\frac{m}{6}\rfloor}\binom{\lfloor\frac{m}{2}\rfloor-k}{2k}\cdot 2^{\lfloor\frac{m}{2}\rfloor-3k}+B_{\lfloor\frac{m}{2}\rfloor}=F_{m+1}. \tag{2.12}$$

But, $\sum_{k=0}^{\lfloor\frac{m}{6}\rfloor}\binom{\lfloor\frac{m}{2}\rfloor-k}{2k}\cdot 2^{\lfloor\frac{m}{2}\rfloor-3k}=A_{\lfloor\frac{m}{2}\rfloor}$. Therefore, $A_{\lfloor\frac{m}{2}\rfloor}+B_{\lfloor\frac{m+1}{2}\rfloor}=F_{m+1}$, and our theorem is proven when $m\equiv 0 \pmod 6$.

Case 1(b): $m\not\equiv 0 \pmod 6$, but $m\equiv 0 \pmod 2$, so that $m\equiv 2 \pmod 6$ or $m\equiv 4 \pmod 6$.

Equation (2.7) will still hold, so

$$\binom{\lfloor\frac{m}{2}\rfloor}{0} * 2^{\lfloor\frac{m}{2}\rfloor}+\sum_{k=1}^{\lfloor\frac{m-2}{6}\rfloor}\binom{\lfloor\frac{m}{2}\rfloor-k-1}{2k}\cdot 2^{\lfloor\frac{m}{2}\rfloor-3k}$$
$$+\sum_{k=1}^{\lfloor\frac{m}{6}\rfloor}\binom{\lfloor\frac{m}{2}\rfloor-k-1}{2k-1}\cdot 2^{\lfloor\frac{m}{2}\rfloor-3k}+B_{\lfloor\frac{m}{2}\rfloor}=F_{m+1}. \tag{2.13}$$

However, this time $\lfloor \frac{m-2}{6} \rfloor = \lfloor \frac{m}{6} \rfloor$, so our first step will be to combine the summations.

$$\binom{\lfloor \frac{m}{2} \rfloor}{0} \cdot 2^{\lfloor \frac{m}{2} \rfloor} + \sum_{k=1}^{\lfloor \frac{m}{6} \rfloor} \left[\binom{\lfloor \frac{m}{2} \rfloor - k - 1}{2k} + \binom{\lfloor \frac{m}{2} \rfloor - k - 1}{2k-1} \right] \cdot 2^{\lfloor \frac{m}{2} \rfloor - 3k} + B_{\lfloor \frac{m}{2} \rfloor} = F_{m+1}. \quad (2.14)$$

As before, we may combine the two terms of the summand and add the first term of the equation to the sum. This leaves us with

$$\sum_{k=0}^{\lfloor \frac{m}{6} \rfloor} \binom{\lfloor \frac{m}{2} \rfloor - k}{2k} \cdot 2^{\lfloor \frac{m}{2} \rfloor - 3k} + B_{\lfloor \frac{m}{2} \rfloor} = F_{m+1}. \quad (2.15)$$

Again, this just means $A_{\lfloor \frac{m}{2} \rfloor} + B_{\lfloor \frac{m+1}{2} \rfloor} = F_{m+1}$, and the formula is proven for $m \equiv 0$ (mod 2).

Case 2: $m \equiv 1 \pmod 2$

If $m \equiv 1 \pmod 2$, then $\lfloor \frac{m-2}{2} \rfloor + 1 = \lfloor \frac{m-1}{2} \rfloor = \lfloor \frac{m}{2} \rfloor = \lfloor \frac{m+1}{2} \rfloor - 1$. Therefore, by (2.4), we know that $A_{\lfloor \frac{m-2}{2} \rfloor} + B_{\lfloor \frac{m}{2} \rfloor} + A_{\lfloor \frac{m}{2} \rfloor} + B_{\lfloor \frac{m}{2} \rfloor} = F_{m+1}$. Thus,

$$\sum_{k=0}^{\lfloor \frac{m-2}{6} \rfloor} \binom{\lfloor \frac{m}{2} \rfloor - k - 1}{2k} \cdot 2^{\lfloor \frac{m}{2} \rfloor - 3k - 1} + 2^{*} \sum_{k=1}^{\lfloor \frac{m+2}{6} \rfloor} \binom{\lfloor \frac{m}{2} \rfloor - k}{2k-1} \cdot 2^{\lfloor \frac{m}{2} \rfloor - 3k + 1} + A_{\lfloor \frac{m}{2} \rfloor} = F_{m+1}. \quad (2.16)$$

Now, we will move the 2 inside the second summation. Then,

$$\sum_{k=0}^{\lfloor \frac{m-2}{6} \rfloor} \binom{\lfloor \frac{m}{2} \rfloor - k - 1}{2k} \cdot 2^{\lfloor \frac{m}{2} \rfloor - 3k - 1} + \sum_{k=1}^{\lfloor \frac{m+2}{6} \rfloor} \binom{\lfloor \frac{m}{2} \rfloor - k}{2k-1} \cdot 2^{\lfloor \frac{m}{2} \rfloor - 3k + 2} + A_{\lfloor \frac{m}{2} \rfloor} = F_{m+1}, \quad (2.17)$$

which becomes

$$\sum_{k=0}^{\lfloor \frac{m-2}{6} \rfloor} \binom{\lfloor \frac{m}{2} \rfloor - k - 1}{2k} \cdot 2^{\lfloor \frac{m}{2} \rfloor - 3k - 1} + \sum_{k=0}^{\lfloor \frac{m-4}{6} \rfloor} \binom{\lfloor \frac{m}{2} \rfloor - k - 1}{2k+1} \cdot 2^{\lfloor \frac{m}{2} \rfloor - 3k - 1} + A_{\lfloor \frac{m}{2} \rfloor} = F_{m+1}. \quad (2.18)$$

Case 2(a): $m \equiv 3 \pmod 6$.

Then, $\lfloor \frac{m-4}{6} \rfloor = \lfloor \frac{m-2}{6} \rfloor - 1$, so our first summation can be rewritten to produce the equation

$$\binom{\lfloor \frac{m}{2} \rfloor - \lfloor \frac{m-2}{6} \rfloor - 1}{2^* \lfloor \frac{m-2}{6} \rfloor} \cdot 2^{\lfloor \frac{m}{2} \rfloor - 3^* \lfloor \frac{m-2}{6} \rfloor - 1} + \sum_{k=0}^{\lfloor \frac{m-4}{6} \rfloor} \binom{\lfloor \frac{m}{2} \rfloor - k - 1}{2k} \cdot 2^{\lfloor \frac{m}{2} \rfloor - 3k - 1} +$$

$$\sum_{k=0}^{\lfloor \frac{m-4}{6} \rfloor} \binom{\lfloor \frac{m}{2} \rfloor - k - 1}{2k+1} \cdot 2^{\lfloor \frac{m}{2} \rfloor - 3k - 1} + A_{\lfloor \frac{m}{2} \rfloor} = F_{m+1}. \tag{2.19}$$

However, $\lfloor \frac{m}{6} \rfloor = \lfloor \frac{m-2}{6} \rfloor$, and since $\binom{\lfloor \frac{m}{2} \rfloor - \lfloor \frac{m-2}{6} \rfloor - 1}{2^* \lfloor \frac{m}{6} \rfloor} \cdot 2^{\lfloor \frac{m}{2} \rfloor - 3^* \lfloor \frac{m-2}{6} \rfloor - 1} = 1$, we may now write (2.19) as

$$\begin{aligned} 1 + \sum_{k=0}^{\lfloor \frac{m-4}{6} \rfloor} \binom{\lfloor \frac{m}{2} \rfloor - k - 1}{2k} \cdot 2^{\lfloor \frac{m}{2} \rfloor - 3k - 1} \\ + \sum_{k=0}^{\lfloor \frac{m-4}{6} \rfloor} \binom{\lfloor \frac{m}{2} \rfloor - k - 1}{2k+1} \cdot 2^{\lfloor \frac{m}{2} \rfloor - 3k - 1} + A_{\lfloor \frac{m}{2} \rfloor} = F_{m+1}. \end{aligned} \tag{2.20}$$

We may also combine the two summations to produce

$$1 + \sum_{k=0}^{\lfloor \frac{m-4}{6} \rfloor} \left[\binom{\lfloor \frac{m}{2} \rfloor - k - 1}{2k} + \binom{\lfloor \frac{m}{2} \rfloor - k - 1}{2k+1} \right] \cdot 2^{\lfloor \frac{m}{2} \rfloor - 3k - 1} + A_{\lfloor \frac{m}{2} \rfloor} = F_{m+1}. \tag{2.21}$$

Again, we may combine the two combinations as follows

$$1 + \sum_{k=0}^{\lfloor \frac{m-4}{6} \rfloor} \binom{\lfloor \frac{m}{2} \rfloor - k}{2k+1} \cdot 2^{\lfloor \frac{m}{2} \rfloor - 3k - 1} + A_{\lfloor \frac{m}{2} \rfloor} = F_{m+1}. \tag{2.22}$$

Finally, we can add the one into the summation because

$$\binom{\lfloor \frac{m}{2} \rfloor - \lfloor \frac{m+2}{6} \rfloor}{2 \cdot \lfloor \frac{m+2}{6} \rfloor + 1} \cdot 2^{\lfloor \frac{m}{2} \rfloor - 3^* \lfloor \frac{m+2}{6} \rfloor - 1} = 1 \tag{2.23}$$

when $m \equiv 3 \pmod 6$. Therefore,

$$\sum_{k=0}^{\lfloor\frac{m+2}{6}\rfloor} \binom{\lfloor\frac{m}{2}\rfloor - k}{2k+1} \cdot 2^{\lfloor\frac{m}{2}\rfloor-3k-1} + A_{\lfloor\frac{m}{2}\rfloor} = F_{m+1}. \tag{2.24}$$

Now, we must rewrite the summation as

$$\sum_{k=1}^{\lfloor\frac{m+2}{6}\rfloor+1} \binom{\lfloor\frac{m}{2}\rfloor - k + 1}{2k-1} \cdot 2^{\lfloor\frac{m}{2}\rfloor-3k+2} + A_{\lfloor\frac{m}{2}\rfloor} = F_{m+1}. \tag{2.25}$$

However, $\lfloor\frac{m+3}{6}\rfloor = \lfloor\frac{m+2}{6}\rfloor + 1$ and $\lfloor\frac{m+1}{2}\rfloor = \lfloor\frac{m}{2}\rfloor + 1$, so

$$\sum_{k=1}^{\lfloor\frac{m+3}{6}\rfloor} \binom{\lfloor\frac{m+1}{2}\rfloor - k}{2k-1} \cdot 2^{\lfloor\frac{m+1}{2}\rfloor-3k+1} + A_{\lfloor\frac{m}{2}\rfloor} = F_{m+1}. \tag{2.26}$$

Therefore,

$$B_{\lfloor\frac{m+1}{2}\rfloor} + A_{\lfloor\frac{m}{2}\rfloor} = F_{m+1}. \tag{2.27}$$

and our theorem is proven for $m \equiv 3 \pmod 6$.

Case 2(b): $m \equiv 1, 5 \pmod 6$

Then, $\lfloor\frac{m-4}{6}\rfloor = \lfloor\frac{m-2}{6}\rfloor$, so, by (2.18),

$$\sum_{k=0}^{\lfloor\frac{m-2}{6}\rfloor} \left[\binom{\lfloor\frac{m}{2}\rfloor - k - 1}{2k} + \binom{\lfloor\frac{m}{2}\rfloor - k - 1}{2k+1}\right] \cdot 2^{\lfloor\frac{m}{2}\rfloor-3k-1} + A_{\lfloor\frac{m}{2}\rfloor} = F_{m+1}. \tag{2.28}$$

The two terms of this summation can be combined into

$$\sum_{k=0}^{\lfloor\frac{m-2}{6}\rfloor} \binom{\lfloor\frac{m}{2}\rfloor - k}{2k+1} \cdot 2^{\lfloor\frac{m}{2}\rfloor-3k-1} + A_{\lfloor\frac{m}{2}\rfloor} = F_{m+1}. \tag{2.29}$$

This equation can be easily transformed into (2.25). Therefore, our equation holds when $m \equiv 1 \pmod 6$ and when $m \equiv 5 \pmod 6$. Thus it is true for $m \equiv 1 \pmod 2$, and this completes the proof.

REFERENCES

[1] Burton, David M. Elementary Number Theory. New York, NY: The McGraw-Hill Companies, Inc., 1998.
[2] Pickover, Clifford A. Surfing through Hyperspace. New York, NY: Oxford University Press, 1999.

AMS Classification Numbers: 11B39, 11A99

EXTENSIONS OF GENERALIZED BINOMIAL COEFFICIENTS

Richard L. Ollerton and Anthony G. Shannon

1. INTRODUCTION

Bondarenko gives an excellent account of the history and properties of the generalized binomial coefficient [2]. He describes this coefficient, written $\binom{n}{m}_s$ for $n, m \geq 0$ and $s \geq 1$, as the number of ways m objects can be placed in n cells, each of which holds a maximum of $s - 1$ objects. The ordinary binomial coefficients are obtained when $s = 2$. This description tacitly assumes that empty cells produce distinct arrangements, that the objects are placed in the cells in a given order and that the order of the objects within the cells is not important. So, for instance, $\binom{3}{2}_2 = 3$ from the arrangements

1	2	
1		2
	1	2

and $\binom{2}{2}_3 = 3$ from the arrangements

1,2	
1	2
	1,2

This paper is in final form and no version of it will be submitted for publication elsewhere.

F.T. Howard (ed.), Applications of Fibonacci Numbers, Volume 9: Proceedings of the Tenth International Research Conference on Fibonacci Numbers and their Applications, 187–199.

Some properties, reminiscent of the ordinary binomial coefficient, include

$$\binom{n}{m}_s = \sum_{i=0}^{s-1} \binom{n-1}{m-i}_s, \tag{1.1}$$

$$\binom{n}{m}_s = \begin{cases} 0 & n < 0,\ m < 0 \text{ or } m > (s-1)n \\ 1 & n \geq 0 \ \&\ m = 0, \end{cases} \tag{1.2}$$

$$\binom{n}{0}_s = \binom{n}{(s-1)n}_s = 1, \tag{1.3}$$

$$\sum_{m=0}^{(s-1)n} \binom{n}{m}_s x^m = (1 + x + \cdots + x^{s-1})^n, \tag{1.4}$$

$$\sum_{m=0}^{(s-1)n} \binom{n}{m}_s = s^n \text{ and} \tag{1.5}$$

$$\sum_{m=0}^{(s-1)n} (-1)^m \binom{n}{m}_s = \begin{cases} 0 & s \text{ even} \\ 1 & s \text{ odd} \end{cases}. \tag{1.6}$$

To extend this generalization we consider three possibilities:

a. whether empty cells give new arrangements (if not then empty cells may be placed at the end of the arrangement),

b. whether the order of objects across the cells is free (if not then each cell must contain objects all of which are less than those in the next non-blank cell to the right), and

c. whether the order of objects within the cells gives new arrrangements (if not then they may be listed in increasing order).

Taking values of the corresponding binary variables a, b and c as $1 \equiv$ Yes and $0 \equiv$ No, combinations of these possibilities lead to eight cases. Setting $k = abc_2$, we shall denote the number of ways m objects can be placed in n cells, each of which holds a maximum of $q = s-1$ objects, according to case k as

$$\binom{n}{m}_q^k \text{ with } n, m, q \geq 0 \text{ and } 0 \leq k = abc_2 \leq 7,$$

and call these the k-extensions of the generalized binomial coefficient. The usual generalized binomial coefficient described by Bondarenko has $k = 4$. (Whilst we recognize that there is the possibility of a slight confusion, the use of q instead of s is more natural in terms of our derivation and results.) The essential ideas are illustrated in Table 1 which shows the arrangements that give the k-extensions for $q = 2$ and $m = n = 3$.

We note that the 7-extension includes all possible permutations across the cells and that different cell orders are achieved by the combination of properties a. and c. The other cases can be viewed either as subsets of the 7-extension or as invoking equivalence classes amongst its elements. Interrelationships can be illustrated by the following schematic, where extensions below are subsets of the linked extensions above.

```
       7
     / | \
   3   5   6
   | X   X |
   1   2   4
     \ | /
       0
```

In the following sections we develop recurrence relations for these extensions, present a unifying recurrence relation and explore generalizations of the Fibonacci sequence by considering diagonal sums of the various arrays produced.

2. RECURRENCE RELATIONS AND OTHER PROPERTIES

To develop recurrence relations for each k-extension we consider how each arrangement is formed using the m objects and n cells. In each case, the total number of arrangements equals the number of ways zero objects can be placed in the first cell multiplied by the number of arrangements with one less cell, plus the number of ways one object can be placed in the first cell multiplied by the number of arrangements with one less cell and one less object, and so on up to the number of ways of placing q objects in the first cell. Referring to the three possibilities noted previously, for n & $m > 0$ we have:

a. Empty cells will be ignored if all arrangements are left-justified with respect to the empty cells (see Table 1a). Thus, if empty cells do not give new arrangements then there are no ways of placing zero objects in the first cell. Otherwise there will be one way.

b. If the order of objects across the cells is free then a factor of $C_i^m = \binom{m}{i}$ is introduced at each step, where i is the current number of objects being placed in a cell.

c. If the order within cells is important then a factor of $i!$ is introduced at each step.

This leads to the recurrence relations shown in Table 2. Consideration of the form of each recurrence relation leads to the unifying recurrence relation

$$\binom{n}{m}_q^k = \sum_{i=1-a}^{q} C_{ib}^{m} i!^{c} \binom{n-1}{m-i}_q^k \tag{2.1}$$

for $0 \le k = abc_2 \le 7$, n & $m > 0$ and $q \ge 0$ (c.f. (1.1)). In each case we assume that there is precisely one way of placing $m = 0$ objects in $n \ge 0$ cells, which motivates the boundary conditions (c.f. (1.2))

$$\binom{n}{m}_q^k = \begin{cases} 0\begin{cases} n < 0 \text{ or } m < 0 \\ n = 0 \;\&\; m > 0 \end{cases} \\ 1 \; n \ge 0 \;\&\; m = 0. \end{cases}$$

For each k and q, arrays can be formed with rows given by n and columns given by m. Examples are shown in Table 3 for $q = 2$.

Properties:

i. Consideration of the boundary conditions and the summation limits of (2.1) immediately gives $\binom{n}{m}_q^k = 0$ for $m > qn$ (c.f. (1.2)).

ii. Extending (1.3),

$$\binom{n}{qn}_q^k = \sum_{i=1-a}^{q} C_{ib}^{qn} i!^c \binom{n-1}{qn-i}_q^k \qquad (n, q > 0)$$

$$= C_{qb}^{qn} q!^c \binom{n-1}{q(n-1)}_q^k \qquad \text{(using the boundary conditions)}$$

$$= q!^{nc} \prod_{i=1}^{n} C_{qb}^{qi} \qquad \text{(in fact for } n, q \ge 0\text{)}$$

which reverts to the value 1 for $k = 0, 4$ and also indicates that arrays with $k = 0, 1, 2, 3$ have the same final row values as those with $k = 4, 5, 6, 7$ respectively.

iii. For each value of k and q, a generating function may be derived in the following manner. Let

$$g(k, n, q; x) = \sum_{m=0}^{qn} m!^{-b} \binom{n}{m}_q^k x^m. \tag{2.2}$$

Substituting (2.1) for $n, m > 0$ gives

$$g(k,n,q;x) = \binom{n}{0}_q^k + \sum_{m=1}^{qn} m!^{-b} \sum_{i=1-a}^{q} C_{ib}^m i!^c \binom{n-1}{m-i}_q^k x^m$$

$$= 1 - a + \sum_{i=1-a}^{q} i!^c \sum_{m=0}^{qn} C_{ib}^m m!^{-b} \binom{n-1}{m-i}_q^k x^m$$

$$= 1 - a + \sum_{i=1-a}^{q} i!^{c-b} x^i \sum_{m=0}^{qn} (m-i)!^{-b} \binom{n-1}{m-i}_q^k x^{m-i}$$

$$= 1 - a + \sum_{i=1-a}^{q} i!^{c-b} x^i \sum_{m=-i}^{qn-i} m!^{-b} \binom{n-1}{m}_q^k x^m$$

(changing the summation index to $j = m - i$ then reverting to m)

$$= 1 - a + \sum_{i=1-a}^{q} i!^{c-b} x^i \sum_{m=0}^{q(n-1)} m!^{-b} \binom{n-1}{m}_q^k x^m \text{ (using the boundary conditions)}$$

$$= 1 - a + g(k,n-1,q;x) \sum_{i=1-a}^{q} i!^{c-b} x^i \tag{2.3}$$

with $g(k,0,q;x) = \sum_{m=0}^{0} m!^{-b} \binom{0}{m}_q^k x^m = \binom{0}{0}_q^k = 1$.

If $T_q^k(x) = \sum_{i=1-a}^{q} i!^{c-b} x^i$ then (2.3) may be solved to give

$$\sum_{m=0}^{qn} m!^{-b} \binom{n}{m}_q^k x^m = \begin{cases} \frac{T_q^k(x)^{n+1}-1}{T_q^k(x)-1} (= n+1 \text{ for } T_q^k(x) = 1) & k = 0 \text{ to } 3 \\ T_q^k(x)^n & k = 4 \text{ to } 7. \end{cases} \tag{2.4}$$

(1.4) follows immediately by setting $k = 4$.

iv. Comparison of the ways in which the arrangements are formed leads to the simple relationships

$$\binom{n}{m}_q^3 = m! \binom{n}{m}_q^0 \text{ and } \binom{n}{m}_q^7 = m! \binom{n}{m}_q^4. \tag{2.5}$$

This may also be shown by mathematical induction using the relevant recurrence relations. Thus, the exponential generating functions of the $k = 3, 7$ arrays are just the ordinary generating functions of the $k = 0, 4$ arrays respectively (as may also be seen directly from (2.2)). Further interrelationships between coefficients with different values of k could be investigated, as could relationships between coefficients of the same k-extension but smaller values of q.

v. As well as producing various new sequences, the k-extension arrays link a number of results which can be discerned readily from the relevant arrays:

$(k, q) = (4, 1)$ gives the Pascal triangle and $(4, q)$ gives the higher order Pascal arrays described by Bondarenko.
$(0, 2)$ gives the Fibonacci sequence as a limiting row.
$(1, 2)$ gives the Jacobsthal sequence as a limiting row (Sloane [11] A001045).
$(3, 2)$ has a limiting row of $a_m = m!F_{m+1}$ (Sloane A005442).
$(0, 3)$ gives the Tribonacci sequence as a limiting row, higher order Fibonacci sequences occur similarly for larger values of q.
$(0, \infty)$ has a limiting row of increasing integer powers of 2.
$(1, \infty)$ has a limiting row of Sloane A051296.
$(2, \infty)$ gives the ordered Bell numbers as a limiting row (Sloane A000670).
$(4, \infty)$ gives the Pascal triangle as diagonals (i.e. the Pascal square).
$(6, \infty)$ has increasing integer powers of n for the nth row.
$(7, \infty)$ has nth row $\{1, n, n(n+1), n(n+1)(n+2)\ldots\}$.

vi. Non-binary values of a, b, c could also be considered although the combinatorial meaning becomes less clear.

3. ROW SUMS

To extend the row sums (1.5) and (1.6), set $x = \pm 1$ respectively in (2.4). When $b = 1$ $(k = 2, 3, 6, 7)$ these become weighted row sums.

Properties:

i. For example, $c = b$ $(k = 0, 3, 4, 7)$ leads to $T_q^k(1) = \sum_{i=1-a}^{q} i!^{c-b} = q + a$ so that, in particular, $k = 4$ gives the generalized binomial coefficient row sum result (1.5) in the form

$$\sum_{m=0}^{qn} \binom{n}{m}_q^4 = (q+1)^n.$$

For $k = 0$,

$$\sum_{m=0}^{qn} \binom{n}{m}_q^0 = \frac{q^{n+1} - 1}{q - 1}.$$

It may be verified that the rows of the $(k = 0, q = 2)$ array in Table 3 do in fact sum to the value $2^{n+1} - 1$.

ii. In the case of the alternating-sign row sum, $T_q^k(-1) = \sum_{i=1-a}^{q} i!^{c-b}(-1)^i$. For example, when $k = 4$,

$$T_q^4(-1) = \begin{cases} 0 & q \text{ odd} \\ 1 & q \text{ even} \end{cases} \quad \text{so that}$$

$$\sum_{m=0}^{qn} \binom{n}{m}_q^4 (-1)^m = T_q^4(-1)^n = \begin{cases} 0 & q \text{ odd} \\ 1 & q \text{ even} \end{cases}$$

and (1.6) is effectively recovered.

When k = 0,

$$T_q^0(-1) = \begin{cases} -1 & q \text{ odd} \\ 0 & q \text{ even} \end{cases}$$

and the alternating-sign row sum takes the values

$$\sum_{m=0}^{qn} \binom{n}{m}_q^0 (-1)^m = \frac{T_q^0(-1)^{n+1} - 1}{T_q^0(-1) - 1} = \begin{cases} \left.\begin{matrix} 0 & n \text{ odd} \\ 1 & n \text{ even} \end{matrix}\right\} & q \text{ odd} \\ \quad 1 & q \text{ even} \end{cases}$$

which may also be confirmed in Table 3 for $q = 2$.

iii. Unweighted sums could be investigated further, as could diagonal sums of both the weighted and unweighted row sum arrays (with n as rows, q as columns).

4. DIAGONAL SUMS AND FIBONACCI GENERALIZATIONS

We define the following (weighted) diagonal sums of the extended binomial coefficients as

$$d(k,n,q) = \sum_{m=0}^{n} m!^{-b} \binom{n-m}{m}_q^k = \sum_{m=0}^{\lfloor qn/(q+1) \rfloor} m!^{-b} \binom{n-m}{m}_q^k \quad \text{for } n \geq 0 \tag{4.1}$$

and $d(k,n,q) = 0$ otherwise. The steps involved in finding the recurrence relations mimic those for the generating function. Substituting (2.1) for $n, m > 0$ leads to

$$d(k,n,q) = 1 - a + \sum_{i=1-a}^{q} i!^{c-b} d(k, n-i-1, q) \tag{4.2}$$

with

$$d(k,0,q) = \sum_{m=0}^{0} m!^{-b} \binom{0-m}{m}_q^k = \binom{0}{0}_q^k = 1.$$

Properties:

i. Columns of both these and the unweighted diagonal sum arrays, with n as rows and q as columns, may be thought of as generalized Fibonacci sequences (although the weighted sums are in general non-integer when $b = 1$ and $c = 0$ $(k = 2, 6)$).

ii. When $a = 1$ and $b = c$ $(k = 4$ or $7)$ we recover the generalized Fibonacci recurrence relation of arbitrary order with $d(k, n-1, q) = U_n$ defined by

$$U_n = \sum_{i=0}^{q} U_{n-i-1} \text{ for } n > 1$$

and $U_{-n} = 0$ for $n = 0, \ldots, q-1, U_1 = 1$. Thus, the weighted diagonal sums are

$$U_{n+1} = \sum_{m=0}^{n} \binom{n-m}{m}_q^4$$

$$= \sum_{m=0}^{n} m!^{-1} \binom{n-m}{m}_q^7 \quad \text{(as expected from (2.5))}$$

where, as in (4.1), the upper summation limit may be reduced to $\lfloor qn/(q+1) \rfloor$.

iii. Similarly, $a = 0$ and $b = c$ $(k = 0$ or $3)$ gives $d(k, n-1, q) = V_n$ where

$$V_n = 1 + \sum_{i=1}^{q} V_{n-i-1} \text{ for } n > 1$$

and $V_{-n} = 0$ for $n = 0, \ldots, q-1, V_1 = 1$.

When $q = 2, \ldots, 10$, we observe Sloane's A023434-42 (the "dying rabbit" sequences). This provides alternative recurrence relations of lower order (but generally more terms) than those given by Sloane. For $q = 2, V_n = 1 + V_{n-2} + V_{n-3}$ rather than $V_n = V_{n-1} + V_{n-2} - V_{n-4}$, which can be demonstrated readily by equating them.

It may also be observed that the columns of the $d(0, n, q)$ array converge to the standard Fibonacci sequence for $n \geq 0$ (i.e. $F_n = d(0, n-1, \infty)$) which thus has the alternative recurrence relation (given by Hoggatt [4])

$$V_n = 1 + \sum_{i=1}^{\infty} V_{n-i-1} = 1 + \sum_{i=1}^{n-2} V_{n-i-1}.$$

It is of interest to consider this Fibonacci limiting column of the 0-extension diagonal array in light of the higher order Fibonacci limiting rows of the $\binom{n}{m}_q^0$ arrays noted earlier.

The recurrence relations for other values of k and q could also be explored as could alternating-sign diagonal sums.

iv. Further simple properties which may be observed include the limiting columns of $d(4, n, q)$ (increasing integer powers of 2) and $d(5, n, q)$ (Sloane A051295).

v. Finally, the arrays determined by either the weighted or unweighted diagonal sums themselves yield diagonal sums which may be explored.

5. CONCLUSION

Extending the generalized binomial coefficients leads to a generalized recurrence relation, the various cases of which can readily be given concrete interpretations. Applications of these extensions to earlier results are topics for further research. This could include interrelationships between the k-extensions, explicit and asymptotic formulae, expansion of the work of Bollinger & Burchard [1], column generating functions and further Fibonacci sequence generalizations as exemplified by Hoggatt & Bicknell [6], generalized Pascal pyramids with analogues of the results of Hoggatt & Alexanderson [5], distributions modulo prime p, fractal triangles and determinants along the lines of Bondarenko [2] and Ollerton & Shannon [9]. Their combinatorial applications could also be considered as in the work of Letac & Takács [7] who, in effect, related the permutations associated with Bondarenko's $\binom{3}{m}_{s=3}$ to random walks along the edges of a dodecahedron, or in the connections of combinatorial matrices to planar networks [3]. Such research should lead to further generalizations of the Fibonacci sequence which would be different from those discussed by Shannon [10] and the standard generalizations of Philippou and his colleagues [8].

APPENDIX

For further exploration, arrays of these coefficients can be generated by the following Mathematica code (Mathematica 4, Wolfram Research, Inc., 1988-2000).
karray[a,b,c,n,q] gives an array of the ($k = abc_2$)-extended binomial coefficients for rows -1 to n and columns -1 to $qn + 1$.

```
k[a_,b_,c_,n_,m_,q_] := k[a,b,c,n,m,q] =
    Which [m<0 ||n<0,0,m==0&&n>=0,1,n==0&&m>0,0,n>0,
        Sum[k[a,b,c,n-1,m-i,q]Binomial[m,b i]i!^c, {i,1-a,q}]]
karray [a_,b_,c_,n_,q_] :=
    MatrixForm[Table[k[a,b,c,i,j,q], {i,-1,n}, {j,-1,q n+1}]]
```

REFERENCES

[1] Bollinger, R.C. & Burchard, C.L. "Lucas's Theorem and Some Related Results for Extended Pascal Triangles." *American Mathematical Monthly*, Vol. *97.3* (1990): pp. 198-204.

[2] Bondarenko, B.A. Generalized Pascal Triangles and Pyramids. (Translated by R.C. Bollinger.) Santa Clara, Calif.: The Fibonacci Association, 1993.
[3] Fomin, S. & Zelevinsky, A. "Total Positivity: Tests and Parameterizations." *Mathematical Intelligencer*, Vol. *22.1* (2000): pp. 23-33.
[4] Hoggatt, V.E., Jr. Fibonacci and Lucas Numbers. Santa Clara, California: The Fibonacci Association, 1972.
[5] Hoggatt, V.E. Jr & Alexanderson, G.L. "A Property of Multinomial Coefficients." *The Fibonacci Quarterly*, Vol. *9.4* (1971): pp. 420-421.
[6] Hoggatt, V.E., Jr. & Bicknell, M. "Diagonal Sums of Generalized Pascal Triangles." *The Fibonacci Quarterly*, Vol. *7.4* (1969): pp. 341-358, 393.
[7] Letac, G. & Takács, L. "Random Walks on a Dodecahedron." *Journal of Applied Probability*, Vol. *17.2* (1980): pp. 373-384.
[8] Philippou, A.N. & Muwafi, A.A. "Waiting for the kth Consecutive Success and the Fibonacci Sequence of Order k." *The Fibonacci Quarterly*, Vol. *20.1* (1982): pp. 28-32.
[9] Ollerton, R.L. & Shannon, A.G. "Some Properties of Generalized Pascal Squares and Triangles." *The Fibonacci Quarterly*, Vol. *36.2* (1998): pp. 98-109.
[10] Shannon, A.G. "Iterative Formulas Associated with Generalized Third Order Recurrence Relations." *SIAM Journal of Applied Mathematics*, Vol. *23.3* (1972): pp. 364-368.
[11] Sloane, N.J.A. The On-Line Encyclopedia of Integer Sequences, http://www.research.att.com/~njas/sequences/, (2001).

${}^{0}\binom{3}{3}_2 = 3$			${}^{1}\binom{3}{3}_2 = 5$			${}^{2}\binom{3}{3}_2 = 12$			${}^{3}\binom{3}{3}_2 = 18$		
1,2	3		1,2	3		1,2	3		1,2	3	
			2,1	3					2,1	3	
						1,3	2		1,3	2	
									3,1	2	
						2,3	1		2,3	1	
									3,2	1	
1	2,3		1	2,3		1	2,3		1	2,3	
			1	3,2					1	3,2	
1	2	3	1	2	3	1	2	3	1	2	3
						1	3	2	1	3	2
						2	1,3		2	1,3	
									2	3,1	
						2	1	3	2	1	3
						2	3	1	2	3	1
						3	1,2		3	1,2	
									3	2,1	
						3	1	2	3	1	2
						3	2	1	3	2	1

Table 1a. Arrangements giving the k-extensions of the generalized binomial coefficient $\binom{n}{m}_q^k$ for $q = 2, m = n = 3$ and $k = 0$ to 3.

${}^{4}\binom{3}{3}_2 = 7$			${}^{5}\binom{3}{3}_2 = 13$			${}^{6}\binom{3}{3}_2 = 24$			${}^{7}\binom{3}{3}_2 = 42$		
1,2	3		1,2	3		1,2	3		1,2	3	
			2,1	3					2,1	3	
1,2		3	1,2	3		1,2		3	1,2		3
			2,1	3					2,1		3
						1,3	2		1,3	2	
									3,1	2	
						1,3		2	1,3		2
									3,1		2
						2,3	1		2,3	1	
									3,2	1	
						2,3		1	2,3		1
									3,2		1
1	2,3		1	2,3		1	2,3		1	2,3	
			1	3,2					1	3,2	
1	2	3	1	2	3	1	2	3	1	2	3
						1	3	2	1	3	2
1		2,3	1		2,3	1		2,3	1		2,3
			1		3,2				1		3,2
						2	1,3		2	1,3	
									2	3,1	
						2	1	3	2	1	3
						2	3	1	2	3	1
						2		1,3	2		1,3
									2		3,1
						3	1,2		3	1,2	
									3	2,1	
						3	1	2	3	1	2
						3	2	1	3	2	1
						3		1,2	3		1,2
									3		2,1
	1,2	3		1,2	3		1,2	3		1,2	3
				2,1	3					2,1	3
							1,3	2		1,3	2
										3,1	2
							2,3	1		2,3	1
										3,2	1
	1	2,3		1	2,3		1	2,3		1	2,3
				1	3,2					1	3,2
							2	1,3		2	1,3
										2	3,1
							3	1,2		3	1,2
										3	2,1

Table 1b. Arrangements giving the k-extensions of the generalized binomial coefficient $\binom{n}{m}_q^k$ for $q = 2, n = m = 3$ and $k = 4$ to 7.

k	Recurrence relation	a	b	c
0	${}^{0}\binom{n}{m}_q = \sum_{i=1}^{q} {}^{0}\binom{n-1}{m-i}_q$	0	0	0
1	${}^{1}\binom{n}{m}_q = \sum_{i=1}^{q} i!\, {}^{1}\binom{n-1}{m-i}_q$	0	0	1
2	${}^{2}\binom{n}{m}_q = \sum_{i=1}^{q} {}^{m}C_i\, {}^{2}\binom{n-1}{m-i}_q$	0	1	0
3	${}^{3}\binom{n}{m}_q = \sum_{i=1}^{q} {}^{m}C_i\, i!\, {}^{3}\binom{n-1}{m-i}_q$	0	1	1
4	${}^{4}\binom{n}{m}_q = \sum_{i=0}^{q} {}^{4}\binom{n-1}{m-i}_q$	1	0	0
5	${}^{5}\binom{n}{m}_q = \sum_{i=0}^{q} i!\, {}^{5}\binom{n-1}{m-i}_q$	1	0	1
6	${}^{6}\binom{n}{m}_q = \sum_{i=0}^{q} {}^{m}C_i\, {}^{6}\binom{n-1}{m-i}_q$	1	1	0
7	${}^{7}\binom{n}{m}_q = \sum_{i=0}^{q} {}^{m}C_i\, i!\, {}^{7}\binom{n-1}{m-i}_q$	1	1	1

Table 2. Recurrence relations for the k-extensions of the generalized binomial coefficient.
a: empty cells important, b: order across cells free, c: order within cells important
$1 \equiv$ Yes, $0 \equiv$ No

{0, 0, 0}

1	0	0	0	0	0	0	0	0	0	0	0	0
1	1	1	0	0	0	0	0	0	0	0	0	0
1	1	2	2	1	0	0	0	0	0	0	0	0
1	1	2	3	4	3	1	0	0	0	0	0	0
1	1	2	3	5	7	7	4	1	0	0	0	0
1	1	2	3	5	8	12	14	11	5	1	0	0
1	1	2	3	5	8	13	20	26	25	16	6	1

{0, 0, 1}

1	0	0	0	0	0	0	0	0	0	0	0	0
1	1	2	0	0	0	0	0	0	0	0	0	0
1	1	3	4	4	0	0	0	0	0	0	0	0
1	1	3	5	10	12	8	0	0	0	0	0	0
1	1	3	5	11	20	32	32	16	0	0	0	0
1	1	3	5	11	21	42	72	96	80	32	0	0
1	1	3	5	11	21	43	84	156	240	272	192	64

{0, 1, 0}

1	0	0	0	0	0	0	0	0	0	0	0	0
1	1	1	0	0	0	0	0	0	0	0	0	0
1	1	3	6	6	0	0	0	0	0	0	0	0
1	1	3	12	42	90	90	0	0	0	0	0	0
1	1	3	12	66	330	1170	2520	2520	0	0	0	0
1	1	3	12	66	450	2970	15120	52920	113400	113400	0	0
1	1	3	12	66	450	3690	30240	204120	1020600	3515400	7484400	7484400

{0, 1, 1}

1	0	0	0	0	0	0	0	0	0	0	0	0
1	1	2	0	0	0	0	0	0	0	0	0	0
1	1	4	12	24	0	0	0	0	0	0	0	0
1	1	4	18	96	360	720	0	0	0	0	0	0
1	1	4	18	120	840	5040	20160	40320	0	0	0	0
1	1	4	18	120	960	8640	70560	443520	1814400	3628800	0	0
1	1	4	18	120	960	9360	100800	1048320	9072000	58060800	239500800	479001600

{1, 0, 0}

1	0	0	0	0	0	0	0	0	0	0	0	0
1	1	1	0	0	0	0	0	0	0	0	0	0
1	2	3	2	1	0	0	0	0	0	0	0	0
1	3	6	7	6	3	1	0	0	0	0	0	0
1	4	10	16	19	16	10	4	1	0	0	0	0
1	5	15	30	45	51	45	30	15	5	1	0	0
1	6	21	50	90	126	141	126	90	50	21	6	1

{1, 0, 1}

1	0	0	0	0	0	0	0	0	0	0	0	0
1	1	2	0	0	0	0	0	0	0	0	0	0
1	2	5	4	4	0	0	0	0	0	0	0	0
1	3	9	13	18	12	8	0	0	0	0	0	0
1	4	14	28	49	56	56	32	16	0	0	0	0
1	5	20	50	105	161	210	200	160	80	32	0	0
1	6	27	80	195	366	581	732	780	640	432	192	64

{1, 1, 0}

1	0	0	0	0	0	0	0	0	0	0	0	0
1	1	1	0	0	0	0	0	0	0	0	0	0
1	2	4	6	6	0	0	0	0	0	0	0	0
1	3	9	24	54	90	90	0	0	0	0	0	0
1	4	16	60	204	600	1440	2520	2520	0	0	0	0
1	5	25	120	540	2220	8100	25200	63000	113400	113400	0	0
1	6	36	210	1170	6120	29520	128520	491400	1587600	4082400	7484400	7484400

{1, 1, 1}

1	0	0	0	0	0	0	0	0	0	0	0	0
1	1	2	0	0	0	0	0	0	0	0	0	0
1	2	6	12	24	0	0	0	0	0	0	0	0
1	3	12	42	144	360	720	0	0	0	0	0	0
1	4	20	96	456	1920	7200	20160	40320	0	0	0	0
1	5	30	180	1080	6120	32400	151200	604800	1814400	3628800	0	0
1	6	42	300	2160	15120	101520	635040	3628800	18144000	76204800	239500800	479001600

Table 3. Example arrays of the $\binom{n}{m}_q^k$ extended binomial coefficients for $q = 2$ ($n \geq 0$ gives rows, $m \geq 0$ gives columns). Values of $\{a, b, c\}$ are indicated above each array.

AMS Classification Numbers: 11B39, 11B65, 15A36

SOME PARITY RESULTS REGARDING t-CORE PARTITIONS

Neville Robbins and M.V. Subbarao*

1. INTRODUCTION

If n is a natural number, then a *partition* of n is a representation of n as the sum of one or more natural numbers. For example, the equation:

$$10 = 5 + 3 + 2 \tag{1}$$

defines a partition of 10. It is customary to use the following notation: a partition of n will be given by

$$n = \lambda_1 + \lambda_2 + \cdots + \lambda_r \tag{2}$$

where the λ_i are natrual numbers such that

$$\lambda_1 \geq \lambda_2 \geq \cdots \geq \lambda_r. \tag{3}$$

The summands λ_i are called the *parts* of the partition, and the integer r is the number of parts. A good reference on partitions is [1].

Corresponding to every partition of n, there is a *Ferrers-Young diagram*, consisting of r left-justified rows, and having λ_i equally-spaced nodes in row i, where $1 \leq i \leq r$. For example, the Ferrers-Young diagram that corresponds to the partition of 10 given in (1) is:

*The second author was supported by a grant for NSERC.

This paper is in final form and no version of it will be submitted for publication elsewhere.

F.T. Howard (ed.), Applications of Fibonacci Numbers, Volume 9: Proceedings of the Tenth International Research Conference on Fibonacci Numbers and their Applications, 201–211.

○ ○ ○ ○ ○
○ ○ ○ (*)
○ ○

Referring to the Ferrers-Young diagram of a partition, if (i,j) denotes the node in row i and column j, then the (i,j)-*hook*, denoted $H(i,j)$, is defined as the union of $\{(i,j)\}$ and the set of all nodes directly to the right of, or directly below (i,j). Formally, we write:

$$H(i,j) = \{(i,k) : k \geq j\} \cup \{(m,j) : m \geq i\}. \tag{4}$$

Note that $H(i,j)$ is never empty, since $(i,j) \in H(i,j)$. For example, referring to $(*)$, we have:

$$H(1,2) = \{(1,2),(1,3),(1,4),(1,5),(2,2),(3,2)\}. \tag{5}$$

Corresponding to each hook, $H(i,j)$, there is a *hook number*, denoted $|H(i,j)|$, which is simply the number of nodes in the hook. For example, referring to (5), we have $|H(1,2)| = 6$. (This choice of notation is more elaborate than that used in [5], but is hopefully more clear.)

Let us compute the hook numbers for each hook in $(*)$. Corresponding to the nodes in row 1, we have hook numbers: 7, 6, 4, 2, 1. Corresponding to the nodes in row 2, we have hook numbers: 4,3,1. Corresponding to the nodes in row 3, we have hook numbers: 2, 1.

If the integer $t \geq 2$, then a partition of n is said to be *t-core* if no hook number is a multiple of t. For example, the above hook number computations show that the partition of 10 given by (1) is 5-core. Let $c_t(n)$ denote the number of t-core partitions of n (with $c_t(0) = 1$). (In the literature, $c_t(n)$ has also been denoted $C_t(n), a_t(n)$.) Using an identity attributed to Gauss (see [1], (2.2.13)), it is easily shown that

$$c_2(n) = \begin{cases} 1 & \text{if } n = k(k+1)/2 \\ 0 & \text{otherwise} \end{cases}. \tag{6}$$

In the last two decades of the twentieth century, numerous researchers investigated t-core partitions, which arise in the representation theory of the symmetric group. According to Klyachko [9], if p is prime, then $c_p(n)$ is the number of blocks of defect zero of the group S_n over a field of characteristic p. Let (a/p) denote the Legendre symbol. The formula:

$$c_3(n) = \sum_{d|(3n+1)} (d/3) \tag{7}$$

appears in [9], was subsequently obtained by Granville and Ono [5] using the theory of modular forms, and was independently obtained by Robbins [16] using combinatorial methods. One of the most significant results regarding t-core partitions is due to Granville and Ono [5], namely:

$$c_t(n) > 0 \ \ \forall n \in N \ \ \forall t \geq 4. \tag{8}$$

There are some remarkable links between t-core partitions and algebraic number theory. Let $h(-m)$ denote the class number of the imaginary quadratic field obtained by adjoining $\sqrt{-m}$ to the rationals. Ono and Sze [14] proved that if $8n+5$ is square-free, then

$$c_4(n) = \frac{1}{2}h(-32n-20). \tag{9}$$

As a consequence of a theorem of Ono and Wilson [15], it follows that $c_t(n)$ changes parity infinitely often as n tends to infinity for all $t \geq 2$. In [7], Hirschhorn and Sellers proved that

$$c_4(n) \equiv 0 \pmod 2 \;\; if \;\; n \equiv 2, 8 \pmod 9. \tag{10}$$

In this note, we study the parity of the coefficients $c_t(n)$ when $t = 2^k$, $k \geq 2$. In Theorem 3 below, we generalize (10). Furthermore, we obtain an asymptotic estimate for the number of integers n such that $n \leq x$ and $c_4(n)$ is odd. We also mention that Boylan [2] has studied $c_{2^t}(n)$ using Hecke algebra methods and obtained some strong results. Our results do not overlap with Boylan's, which are in a different direction. This will be spelled out in our concluding remarks.

The symbol $p(n)$ will denote the unrestricted partition function.

2. PRELIMINARIES

To make some of our results easier to state, we introduce the set

$$W = \{m \in N : m = p^\alpha \nu^2, p\ prime,\ p \equiv 5 (\bmod 8),\ \alpha \equiv 1 \pmod 4,\ (\nu, 2p) = 1\}.$$

Definition 1: Let $\sum_{n=0}^{\infty} B(n)x^n = \prod_{n=1}^{\infty}(1-x^n)^{15}$, where $|x| < 1$.

Proposition 1: For all $n \geq 1$ we have

$$p(n) \equiv p^*(n) \pmod 2$$

where

$$p^*(n) = \sum \left\{ p\left(\frac{n-j}{16}\right) : j \equiv n \pmod{16},\ 0 \leq j \leq n,\ 8j+5 \in W \right\}.$$

Remark: Proposition 1 was proved by Majumdar in [11]. It is easily verified that if $1 \leq j \leq 100$, then $8j+5 \in W$ iff $j \in \{1, 3, 4, 5, 6, 7, 12, 13, 14, 18, 19, 21, 22, 24, 28, 30, 33, 34, 36, 39, 40, 41, 43, 46, 48, 49, 50, 52, 57, 59, 63, 67, 68, 69, 75, 76, 79, 81, 82, 84, 87, 88, 90, 91, 94, 96, 99\}$.

Proposition 2: $B(n)$ is odd if and only if $8n+5 \in W$.

Remark: The method used in [11] to prove Proposition 1 also serves to prove Proposition 2.

In this paper, we will use the following identities:

$$\sum_{n=0}^{\infty} c_t(n)x^n = \prod_{n=1}^{\infty} \frac{(1-x^{tn})^t}{1-x^n} \tag{11}$$

$$\prod_{n=1}^{\infty}(1-x^n)^3=\sum_{n=0}^{\infty}(-1)^n(2n+1)x^{\frac{n(n+1)}{2}} \tag{12}$$

$$\prod_{n=1}^{\infty}(1-x^n)^3\equiv\sum_{n=0}^{\infty}x^{\frac{n(n+1)}{2}} \pmod 2. \tag{13}$$

Remark: (11) is the generating function identity for t-core partitions; it may be found in [8]; (12) is due to Jacobi; (13) follows from (12).

3. RESULTS ABOUT THE PARITY OF THE COEFFICIENTS $c_{2^t}(n)$

Theorem 1: For all $t \geq 1$ and all $n \geq 1$, we have

$$c_{4^t}(n) \equiv \sum\left\{\prod_{i=1}^{t} B(m_i) : n=\sum_{i=1}^{t} 16^{i-1}m_i\right\} \pmod 2.$$

Proof: (11) implies

$$\sum_{n=0}^{\infty} c_{4^t}(n)x^n=\prod_{n=1}^{\infty}\frac{(1-x^{4^t n})^{4^t}}{1-x^n}\equiv\prod_{n=1}^{\infty}(1-x^n)^{16^t-1}$$

$$\equiv\left(\prod_{n=1}^{\infty}(1-x^n)^{15}\right)^{\sum_{i=1}^{t}16^{i-1}}\equiv\prod_{i=1}^{t}\prod_{n=1}^{\infty}\left(1-x^{16^{i-1}n}\right)^{15}$$

$$\equiv\prod_{i=1}^{t}\left(\sum_{n=0}^{\infty}B(n)x^{16^{i-1}n}\right)\equiv\prod_{i=1}^{t}\left(\sum_{n=0}^{\infty}B\left(\frac{n}{16^{i-1}}\right)x^n\right)\equiv$$

$$\sum_{n=0}^{\infty}\left(\sum\left\{\prod_{i=1}^{t}B(m_i) : n=\sum_{i=1}^{t}16^{i-1}m_i\right\}\right)x^n \pmod 2.$$

The conclusion now follows by matching coefficients of like powers of x.

Corollary 1: $c_4(n)$ is odd if and only if $8n+5 \in W$.

Proof: Appealing to Theorem 1 with $t=1$, we have $c_4(n)\equiv B(n) \pmod 2$. The conclusion now follows from Proposition 2.

Corollary 1, stated more obliquely, is Theorem 3 in [7]. To put Corollary 1 into perspective, it should be noted that $c_4(n)$ is even for almost all n. Indeed, by Corollary 1, we have $c_4(n)$ is odd if and only if $\frac{8n+5}{p}$ is a square, a condition not satisfied by almost all integers.

In [7], Theorems 1 and 2, Hirschhorn and Sellers proved that if $n \equiv 2, 8 \pmod 9$, then $c_4(n)$ is even. The following theorems generalize their results.

Theorem 2: If p is a prime such that $p \equiv 3 \pmod 4$, let j be the unique integer such that $0 < j < p$ and $p \parallel (8j + 5)$. Then $c_4(n)$ is even for any $n \equiv j \pmod{p^2}$.

Proof: By Corollary 1, it suffices to show that $8n + 5 \notin W$. Since $(8, p) = 1$ by hypothesis, the congruence $8j + 5 \equiv 0 \pmod p$ has a unique solution in the range $0 < j < p$. If $p = 3$ or 7, then $j = 2$, and $8j + 5 \not\equiv 0 \pmod{p^2}$. If $p \geq 11$, then $p^2 > 8j + 5$, so $8j + 5 \not\equiv 0 \pmod{p^2}$. By hypothesis, $n \equiv j \pmod{p^2}$, so $p|(8n + 5)$, $p^2 \not| (8n + 5)$. Therefore $8n + 5 \notin W$, so we are done.

Theorem 3: If p is a prime such that $p \equiv 3 \pmod 4$, then there are $p - 1$ integers, j, such that $0 < j < p^2$ and $p \parallel (8j + 5)$. If $n \equiv j \pmod{p^2}$ for any such j, then $c_4(n)$ is even.

Proof: By Theorem 2, there exists a j such that $0 < j < p$ and $p|(8j + 5)$. If $p^2|(8k + 5)$, then $k = j + rp$ where $8r \equiv -\frac{8j+5}{p} \pmod p$. This congruence has a unique solution $\pmod p$. The conclusion now follows from Theorem 2.

Example: Applying Theorem 3, with $p = 7$, we have $7 \parallel (8j + 5)$ for $j = 2, 9, 16, 23, 37, 44$, so $c_4(n)$ is even if $n \equiv 2, 9, 16, 23, 37, 44 \pmod{49}$. If we apply Theorem 3 with $p = 3$, we obtain the previously quoted result of Hirschhorn and Sellers.

Theorem 4: $c_8(n) \equiv \sum_{j \geq 0} B(n - 8j(j + 1)) \pmod 2$.

Proof: Using (11), and subsequently (13), we have

$$\sum_{n=0}^{\infty} c_8(n)x^n = \prod_{n=1}^{\infty} \frac{(1 - x^{8n})^8}{1 - x^n} \equiv \prod_{n=1}^{\infty} \frac{(1 - x^n)^{64}}{1 - x^n}$$

$$\equiv \prod_{n=1}^{\infty} (1 - x^n)^{63} \equiv \left(\prod_{n=1}^{\infty} (1 - x^n)^{15}\right) \left(\prod_{n=1}^{\infty} (1 - x^n)^{16}\right)^3$$

$$\equiv \left(\prod_{n=1}^{\infty} (1 - x^n)^{15}\right) \left(\prod_{n=1}^{\infty} (1 - x^{16n})\right)^3 \equiv \left(\sum_{n=0}^{\infty} B(n)x^n\right) \left(\sum_{n=0}^{\infty} x^{8n(n+1)}\right)$$

$$\equiv \sum_{n=0}^{\infty} \left(\sum_{j \geq 0} B(n - 8j(j + 1))\right) x^n \pmod 2.$$

The conclusion now follows by matching coefficients of like powers of x.

An alternate way of computing the parity of $c_8(n)$ is given by Theorem 5 below.

Theorem 5: Let $m = \left[\frac{-1+\sqrt{1+8n}}{2}\right]$. Then

$$c_8(n) \equiv \sum_{j=0}^{m} B\left(\frac{2n - j(j + 1)}{8}\right) \pmod 2.$$

Proof: As in the proof of Theorem 4 above, we have:

$$\sum_{n=0}^{\infty} c_8(n)x^n \equiv \prod_{n=1}^{\infty}(1-x^n)^{63} \equiv \left(\prod_{n=1}^{\infty}(1-x^n)^{60}\right)\left(\prod_{n=1}^{\infty}(1-x^n)^{3}\right)$$

$$\equiv \left(\prod_{n=1}^{\infty}(1-x^{4n})^{15}\right)\left(\prod_{n=1}^{\infty}(1-x^n)^{3}\right) \equiv \left(\sum_{n=0}^{\infty} B(n)x^{4n}\right)\left(\sum_{n=0}^{\infty} x^{n(n+1)/2}\right)$$

$$\equiv \left(\sum_{n=0}^{\infty} B\left(\frac{n}{4}\right)x^n\right)\left(\sum_{n=0}^{\infty} x^{n(n+1)/2}\right) \equiv \sum_{n=0}^{\infty}\left(\sum_{j=0}^{m} B\left(\frac{n-\frac{j(j+1)}{2}}{4}\right)x^n\right) \pmod 2.$$

The conclusion now follows by simplifying and matching coefficients of like powers of x. The result of Theorem 5 reminds us of the parity criterion of P.J. MacMahon [10], namely

$$p(n) \equiv \sum_{j\geq 0} p\left(\frac{2n-j(j+1)}{8}\right) \pmod 2$$

which G. H. Hardy remarked, (see [6], p. 100) can be used to determine the parity of huge numbers such as $p(200)$ quite rapidly. There are many other parity criteria of $p(n)$ in the literature; see e.g. Subbarao [18]. The two criteria for parity of $c_8(n)$ given in Theorems 4 and 5 each require the computation of about $\sqrt{\frac{n}{8}}$ terms. The terms that arise using Theorem 5 will involve smaller arguments than those that arise using Theorem 4. For example, using Theorem 4 and Proposition 2, we have:

$$c_8(1000) \equiv B(1000)+B(984)+B(952)+B(904)+B(840)+$$
$$B(760)+B(664)+B(552)+B(424)+B(280)+B(120) \equiv$$
$$0+1+1+1+1+0+0+1+0+0+0 \equiv 1 \pmod 2,$$

while Theorem 5 yields:

$$c_8(1000) \equiv B(250)+B(243)+B(241)+B(220)+B(216)+$$
$$B(181)+B(175)+B(126)+B(118)+B(55)+B(45) \equiv$$
$$0+1+1+0+1+1+0+1+0+0+0 \equiv 1 \pmod 2.$$

Theorem 6:

$$c_{16}(n) \equiv \sum\left\{B\left(\frac{n-j}{16}\right)B(j) : 0\leq j\leq n, j\equiv n \pmod{16}\right\} \pmod 2.$$

Proof: This follows from Theorem 1, with $t = 2$.

In the next two theorems, we will use the following notation: $\delta(k,n)$ will represent the number of ways of representing n as a sum of k triangular numbers, including 0 as a summand, and where representations differing only by order of summands are considered distinct. Representations of integers as sums of triangular numbers have recently been studied by Ono, Robins, and Wahl [13], and by Milne [12]. We are using a slight modification of the notation used in [13], in order to avoid the necessity of using complicated subscripts.

Theorem 7: $c_{2^t}(n) \equiv \delta\left(\frac{4^t-1}{3}, n\right) \pmod 2$.

Proof: Using (11), and subsequently (13), we have

$$\sum_{n=0}^{\infty} c_{2^t}(n)x^n = \prod_{n=1}^{\infty} \frac{(1-x^{2^t n})^{2^t}}{1-x^n} \equiv \prod_{n=1}^{\infty} \frac{(1-x^n)^{4^t}}{1-x^n} \equiv$$

$$\prod_{n=1}^{\infty}(1-x^n)^{4^t-1} \equiv \left(\prod_{n=1}^{\infty}(1-x^n)^3\right)^{\frac{4^t-1}{3}} \equiv \left(\sum_{n=0}^{\infty} x^{\frac{n(n+1)}{2}}\right)^{\frac{4^t-1}{3}} \equiv$$

$$\sum_{n=0}^{\infty} \delta\left(\frac{4^t-1}{3}, n\right) x^n \pmod 2.$$

The conclusion now follows by matching coefficients of like powers of x.

Theorem 8: $\delta(5,n)$ is odd if and only if $8n+5 \in W$.

Proof: This follows from Theorem 7, with $k = 2$, and from Corollary 1.

Remarks: Explicit formulas for $\delta(k,n)$ are given for $k \in \{2,3,4,6,8,24\}$ in [13], and for $k = 4m^2, 4m(m+1)$ in [12]. Therefore Theorem 8 is not a consequence of results in either [12] or [13].

4. ASYMPTOTIC ESTIMATES CONCERNING THE PARITY OF $c_4(n)$

The results stated in Corollaries 1 and 2 can be sharpened as follows:

Definition 2: Let $W(x)$ denote the number of positive integers $n \le x$ such that $n \in W$.

Definition 3: Let $C_4(x)$ denote the number of positive integers $n \le x$ such that $c_4(n)$ is odd.

We will prove two theorems, namely:

Theorem 9:

$$W(x) = \frac{\pi^2}{32}\left(\frac{x}{\log x}\right) + O\left(\frac{x}{\log^2 x}\right).$$

Theorem 10:

$$C_4(x) = \frac{\pi^2}{4}\left(\frac{x}{\log x}\right) + O\left(\frac{x}{\log^2 x}\right).$$

We first prove Theorem 9 and then utilize it to prove Theorem 10. We make use of the results of the paper [19], which we state as a lemma below. First, some notation. Let k and r be fixed integers such that $1 \le r \le k$. Let $a \ge 0$, $d \ge 1$ be integers with $(a,d) = 1$, and $P(a,d)$ the arithmetic progression given by $\{a + nd,\ n = 1, 2, \ldots\}$. Let $A_{k,r}$ denote the set of positive integers n that are representable in the form $n = p^r m^k$ where m is a positive integer and the prime $p \in P(a,d)$. Let $A^*_{k,r}$ denote the subset of $A_{k,r}$ that arises with the restriction $(p,m) = 1$. Let $A_{k,r}(x)$ and $A^*_{k,r}(x)$ denote respectively the enumerative functions of these sets, so that

$$A_{k,r}(x) = \sum\{1 : n \le x, n \in A_{k,r}\}$$

and similarly for $A^*_{k,r}(x)$. Let $\zeta(s)$ denote the usual Riemann zeta function, while $\phi(n)$ denotes Euler's totient function. Then we have:

Lemma 1: As $x \to \infty$, we have

$$A_{k,r}(x) = \frac{r\zeta(\frac{k}{r})}{\phi(d)}\left(\frac{x^{\frac{1}{r}}}{\log x}\right) + O\left(\frac{x^{\frac{1}{r}}}{\log^2 x}\right)$$

and

$$A^*_{k,r}(x) = \frac{r\zeta(\frac{k}{r})}{\phi(d)}\left(\frac{x^{\frac{1}{r}}}{\log x}\right) + O\left(\frac{x^{\frac{1}{r}}}{\log^2 x}\right)$$

the O-term depending on at most k.

Remark: The special case of the lemma corresponding to $r = 1,\ k = 2$ was proved earlier by Cohen [3], [4], and improved by Schwarz [17].

Corresponding to the choice $a = 5, d = 8, r = 1, k = 2$, Lemma 1 gives

Proposition 3: Let V denote the set of integers n that are of the form $n = pm^2$, where p is a prime of the form $p \equiv 5 \pmod 8$ and $(m,p) = 1$. Let $V(x)$ denote the number of integers, n, such that $n \le x$ and $n \in V$. Then we have:

$$V(x) = \frac{\pi^2}{24}\left(\frac{x}{\log x}\right) + O\left(\frac{x}{\log^2 x}\right). \tag{14}$$

Let U denote the set of all squarefull numbers, that is, integers n in whose canonical factorization every prime factor carries an exponent ≥ 2. A trivial estimate yields

$$U(x) = O(x^{\frac{1}{2}}). \tag{15}$$

We split W into the two mutually exclusive subsets W_1 and W_2, where

$$W_1 = \{pm^2 : p \equiv 5 \pmod 8, \quad (p, 2m) = 1\}$$

$$W_2 = \{p^\alpha m^2 : p \equiv 5 \pmod 8, \quad (p, 2m) = 1, \quad 5 \le \alpha \equiv 1 \pmod 4)\}.$$

Clearly $W_2 \subset U$, so

$$W_2(x) \le U(x) = O(x^{\frac{1}{2}}) \tag{16}$$

using (15). Let $2^i || m$ mean, as usual, that $m \equiv 0 \pmod{2^i}$ but $m \not\equiv 0 \pmod{2^{i+1}}$. Recalling the definition of V given above, we shall partition V into mutually disjoint subsets $V_i, i = 0, 1, 2, \ldots$ such that $n \in V_i$ iff $2^i || m$. Thus $V_0 = W$ and

$$V(x) = V_0(x) + V_1(x) + V_2(x) + \ldots. \tag{17}$$

Clearly, for a given x, $V_i(x)$ is null for all sufficiently large i. A simple argument shows that $V_i(x) - W_1(\frac{x}{4^i})$ for all $i \le 0$. Hence (18) gives

$$V(x) = W_1(x) + W_1(\frac{x}{4}) + W_1(\frac{x}{4^2}) + \cdots$$

so that

$$V(\frac{x}{4}) = W_1(\frac{x}{4}) + W_1(\frac{x}{16}) + \cdots$$

hence

$$W_1(x) = V(x) - V(\frac{x}{4}) = \frac{\pi^2}{32}\left(\frac{x}{\log x}\right) + O\left(\frac{x}{\log^2 x}\right).$$

Thus Theorem 9 follows in view of (17).

To prove Theorem 10, let w denote an arbitrary element of the set W. Then

$$C_4(x) = \sum\{1 : n \le x, c_4(n) odd\} = \sum\{1 : n \le x, 8n + 5 = w \in W\}$$

$$= \sum\{1 : 1 \le n = \frac{w-5}{8} \le x, w \in W\} = \sum\{1 : w \in W, w \le 8x + 5\} = W(8x + 5).$$

Using Theorem 9, we get

$$C_4(x) = \frac{\pi^2}{32}\left(\frac{8x+5}{\log(8x+5)}\right) + O\left(\frac{8x+5}{\log^2(8x+5)}\right)$$

from which Theorem 10 follows.

Remark: As in [17], by using a well-known form of the prime number theorem for arithmetic progressions, we can obtain sharper versions of Theorems 9 and 10. For example, a sharper version of Theorem 10 would be:

Theorem 10a: For any positive integer t, we have

$$C_4(x) = \frac{\pi^2 x}{4}\left(\frac{1}{\log x} + \frac{C_1}{\log^2 x} + \cdots + \frac{C_{t-1}}{\log^t x}\right) + O_t\left(\frac{x}{\log^{t+1} x}\right)$$

as $x \to \infty$, where for each i such that $1 \leq i \leq t-1, C_i$ can be expressed as a linear combination of $\zeta(s)$ and its derivatives at $s = 2$. (See [17] for details.) The constant in the O term may depend on t.

CONCLUDING REMARKS

The results of Theorem 1, 4, and 5 can be extended to the case $c_t(n)$ where $t = 4^a m$ and m is odd. Using the binary representation of m, we can express $c_t(n)$ as a sum of multiple series of terms that are multiple products involving $B(n)$. Details will be given in separate note.

The techniques used above may also be used to obtain parity results regarding the coefficients in the power series expansions of certain powers of the Euler product $\prod_{n=1}^{\infty}(1-x^n)$. We intend to pursue this goal in the future. Finally, we wish to thank M. Boylan for sending us a preprint of his paper [2]. His results concern the parity of the functions $c_{2^t}(n)$ when n belongs to *certain arithmetic progressions.* (See [2], Theorems 1.1 and 1.2.) Thus his results do not overlap with ours.

ACKNOWLEDGMENT

We wish to thank the referee for his/her many valuable suggestions and meticulous attention to detail.

REFERENCES

[1] Andrews, G.E. The Theory of Partitions. Addison-Wesley (1976).
[2] Boylan, M. "Congruences for 2^t-core partitions." *J. Number Theory*, Vol. *92.1* (2002): pp. 131-138.
[3] Cohen, E. "Arithmetical Notes IX: On the Set of Integers Representable as a Product of a Prime and a Square." *Acta Arith.*, Vol. *7* (1962): pp. 417-420.
[4] Cohen, E. "On the Distribution of Certain Sequences of Integers." *Amer. Math. Monthly*, Vol. *70* (1963): pp. 516-521.
[5] Granville A. and Ono, K. "Defect Zero p-groups for Finite Simple Groups." *Trans. Amer. Math. Soc.*, Vol. *348* (1996): pp. 331-347.
[6] Hardy, G.H. Ramanujan. Chelsea Publishing Company, New York (1940).
[7] Hirschhorn, M. and Sellers, J.D. "Two Congruences Involving 4-cores." *Electronic J. Combinatorics*, Vol. *3.2* (1996).
[8] James, G.D. and Kerber, A. The Representation Theory of the Symmetric Group. Addison-Wesley, New York (1981).

[9] Klyachko, A.A. "Modular Forms and Representations of Finite Simple Groups." *J. Soviet Math.*, Vol. *26* (1984): pp. 1879-1887.

[10] MacMahon, P.A. "Note on the Parity of the Number Which Enumerates the Partitions of a Number." *Proc. Cambridge Phil. Soc.*, Vol. *20* (1921): pp. 281-283.

[11] Majumdar, K.N. "On the Parity of the Partition Function $p(n)$." *J. Indian Math. Soc.*, Vol. *13* (1949): pp. 23-24.

[12] Milne, S.C. "Infinite Families of Exact Sums of Squares Formulas, Jacobi Elliptic Functions, Continued Fractions, and Schur Functions." *Ramanujan J.*, Vol. *6* (2002): pp. 7-149.

[13] Ono, K., Robins, S. and Wahl, P.T. "On the Representation of Integers as Sums of Triangular Numbers." *Aequationes Math.*, Vol. *50* (1995): pp. 73-94.

[14] Ono, K. and Sze, L. "4-Core Partitions and Class Numbers." *Acta Arith.*, Vol. *80.3* (1997): pp. 249-272.

[15] Ono, K. and Wilson, B. "Parity of Fourier Coefficients of Modular Forms." *Illinois J. Math.*, Vol. *41* (1997): pp. 142-149.

[16] Robbins, N. "On t-Core Partitions." *The Fibonacci Quarterly*, Vol. *38* (2000): pp. 39-48.

[17] Schwarz, W. "Eine Bemerkung ueber die Anzahl der Zahlen pr^2 unterhalb x." *Colloq. Math.*, Vol. *19* (1968): pp. 117-120.

[18] Subbarao, M.V. "On the Parity of the Partition Function." *Congr. Numer.*, Vol. *56* (1987): pp. 265-275.

[19] Subbarao, M.V. and Suryanarayana, D. "On the Set of Integers $p^r m^k$." *Indian J. Math.*, Vol. *15* (1975): pp. 179-184; corrections and additions, same journal *17* (1975): p. 172.

AMS Classification Numbers: 11P83

GENERALIZED PELL NUMBERS AND POLYNOMIALS

A.G. Shannon and A.F. Horadam

1. INTRODUCTION

We define sequences of *generalized Pell numbers* with the notation introduced by Horadam [6]

$$\{P_{r,n}\} \equiv \{P_{r,n}(1, 2^r; 2^r, -1)\} \tag{1.1}$$

and by the second order recurrence relation

$$P_{r,n} = 2^r P_{r,n-1} + P_{r,n-2}, \quad n > 2 \tag{1.2}$$

with initial conditions $P_{r,1} = 1, P_{r,2} = 2^r$, (though we can allow for $n \leq 0$). For instance,

$$\{P_{0,n}(1, 1; 1, -1\} \equiv \{F_n\}, \tag{1.3}$$

$$\{P_{1,n}(1, 2; 2, -1\} \equiv \{P_n\}, \tag{1.4}$$

the ordinary Fibonacci and Pell sequences respectively. We also define an allied sequence

$$\{Q_{r,n}\} \equiv \{P_{r,n}(2^r, 2^r + 2; 2^r, -1)\}, \tag{1.5}$$

so that

$$\{Q_{0,n}(1, 3; 1, -1\} \equiv \{L_n\}, \tag{1.6}$$

the ordinary Lucas numbers. Note that $P_{r,0} = 0, Q_{r,0} = 2^r + 2 - 2^{2r}$, if we extend the definition to $n = 0$.

This paper is in final form and no version of it will be submitted for publication elsewhere.

F.T. Howard (ed.), Applications of Fibonacci Numbers, Volume 9: Proceedings of the Tenth International Research Conference on Fibonacci Numbers and their Applications, 213–224.

It is the intention of this paper to explore the number theoretic and combinatorial properties of these numbers and related polynomials $p_r(x)$ and $p_{r,n}(x)$ defined below in (3.2) and (3.5). Further, it is shown that any polynomial can be expressed in terms of related generalized Pell polynomials.

The $\{P_{r,n}\}$ arose in [9] in the combinatorial matrix defined by

$$S_{p,q}(n;2) = [s_{i,j}(n)]_{n\times n} \tag{1.7}$$

where

$$s_{i,j}(n) = \binom{j-1}{n-i} p^{i+j-n-1} q^{n-i}, \tag{1.8}$$

and

$$S'_{2^r,-1} S_{2^r,-1} = S_{2^{r+1},-1}, r \geq 0, \tag{1.9}$$

where

$$S'_{2^r,-1} = S_{2^r,-1} E \tag{1.10}$$

in which E is the unit matrix with rows reversed, that is, the elementary (self-inverse) matrix

$$E = [e_{i,j}]_{n\times n}$$

$$e_{i,j} = \begin{cases} 1 & \text{if } j = n-i+1 \\ 0 & \text{otherwise.} \end{cases}$$

An example of (1.9) when $r = 1$ is

$$\begin{pmatrix} 1 & 0 & 0 & 0 \\ 6 & 1 & 0 & 0 \\ 12 & 4 & 1 & 0 \\ 8 & 4 & 2 & 1 \end{pmatrix} \begin{pmatrix} 0 & 0 & 0 & 1 \\ 0 & 0 & 1 & 6 \\ 0 & 1 & 4 & 12 \\ 1 & 2 & 4 & 8 \end{pmatrix} = \begin{pmatrix} 0 & 0 & 0 & 1 \\ 0 & 0 & 1 & 12 \\ 0 & 1 & 8 & 48 \\ 1 & 4 & 16 & 64 \end{pmatrix} \tag{1.11}$$

The falling (from left to right) diagonal sums, starting at the bottom, in the S matrices (when considered as infinite in extent) are generalized Pell numbers $\{P_{r,n}\}$. For instance, in (1.11) we have $\{1, 2, 5, 12, \ldots\}$, the ordinary Pell numbers, on the left, and $\{1, 4, 17, 72, \ldots\}$ on the right, which is $\{P_{2,n}\}$.

2. GENERAL TERMS

The auxiliary equation associated with the recurrence relation (1.2) is given by

$$\lambda^2 - 2^r\lambda - 1 = 0 \tag{2.1}$$

which has roots given by

$$\alpha = \frac{2^r + \Delta}{2}, \text{ and } \beta = \frac{2^r - \Delta}{2}, \tag{i}$$

in which

$$\Delta = \alpha - \beta = \surd(4 + 2^{2r}). \tag{ii}$$

We note that

$$\alpha + \beta = 2^r, \ \alpha\beta = -1. \tag{iii}$$

The Binet forms of the general terms are

$$P_{r,n} = \frac{\alpha^n - \beta^n}{\Delta}, \tag{iv}$$

$$Q_{r,n} = \alpha^n + \beta^n. \tag{v}$$

Using (i)-(v), we then get identities analogous to the well-known results for Fibonacci, Pell and Lucas numbers:

$$Q_{r,n} = P_{r,n-1} + P_{r,n+1}, \tag{2.2}$$
$$P_{r,2n} = P_{r,n} Q_{r,n}, \tag{2.3}$$
$$\Delta^2 P_{r,n} = Q_{r,n+1} + Q_{r,n-1}, \tag{2.4}$$
$$P_{r,n+1} P_{r,n-1} - P_{r,n}^2 = (-1)^n, \tag{2.5}$$

$$Q_{r,n+1} Q_{r,n-1} = (-1)^{n-1} \Delta^2. \tag{2.6}$$

Since the proofs are trivial, they will be omitted.

Combining (2.2) and (2.4), we may introduce the concept of interrelated associated sequences.

Definition: $P_{r,n}^{(k)}$ and $Q_{r,n}^{(k)}$, the kth *associated sequences* of $P_{r,n}$ and $Q_{r,n}$ respectively, are defined by

$$P_{r,n}^{(k)} = P_{r,n+1}^{(k-1)} + P_{r,n-1}^{(k-1)}, \tag{2.7}$$

$$Q_{r,n}^{(k)} = Q_{r,n+1}^{(k-1)} + Q_{r,n-1}^{(k-1)}, \tag{2.8}$$

with $P_{r,n}^{(0)} \equiv P_{r,n}$, $Q_{r,n}^{(0)} = Q_{r,n}$ so that

$$P_{r,n}^{(1)} = Q_{r,n} \quad \text{by (2.2)}, \tag{2.9}$$

$$Q_{r,n}^{(1)} = \Delta^2 P_{r,n} \quad \text{by (2.4).} \tag{2.10}$$

Some leisurely substitutions using (2.2) and (2.4) lead readily to the conclusions that

$$P_{r,n}^{(2m)} = \Delta^{2m} P_{r,n}, \quad P_{r,n}^{(2m+1)} = \Delta^{2m} Q_{r,n}, \tag{2.11}$$

$$Q_{r,n}^{(2m)} = \Delta^{2m} Q_{r,n}, \quad Q_{r,n}^{(2m+1)} = \Delta^{2m+2} P_{r,n}. \tag{2.12}$$

Succinctly, we write

$$P_{r,n}^{(2m+1)} = Q_{r,n}^{(2m)}, \tag{2.13}$$

$$Q_{r,n}^{(2m+1)} = \Delta^2 P_{r,n}^{(2m)}. \tag{2.14}$$

3. GENERATING FUNCTIONS

For notational convenience we let

$$p_{r,n} = P_{r,n+1}. \tag{3.1}$$

Define formally

$$p_r(x) = \sum_{n=0}^{\infty} p_{r,n} x^n. \tag{3.2}$$

Then it can be shown from (1.2) that the generating function for $p_r(x)$ is

$$p_r(x) = \frac{1}{1 - 2^r x - x^2}. \tag{3.3}$$

Theorem 1:

$$p_r(x) = \exp\left(\sum_{m=1}^{\infty} Q_{r,m} \frac{x^m}{m} \right). \tag{3.4}$$

Proof:

$$\begin{aligned} \ln p_r(x) &= -\ln(1-\alpha x)(1-\beta x) \text{ using (i), (iii)} \\ &= -\ln(1-\alpha x) - \ln(1-\beta x) \\ &= \sum_{m=1}^{\infty} \frac{\alpha^m x^m}{m} + \sum_{m=1}^{\infty} \frac{\beta^m x^m}{m} \\ &= \sum_{m=1}^{\infty} (\alpha^m + \beta^m) \frac{x^m}{m} \\ &= \sum_{m=1}^{\infty} Q_{r,m} \frac{x^m}{m}, \text{ as required.} \end{aligned}$$

We next define a type of *generalized Pell polynomial*, $p_{r,n}(x)$, by means of an exponential generating function which has the form of a Sheffer generating function [4]:

$$\sum_{n=0}^{\infty} p_{r,n}(x)\frac{t^n}{n!} = e^{xt}p_r(t). \tag{3.5}$$

So

$$p_{r,n} = p_{r,n}(0)/n!. \tag{3.6}$$

Then

$$\sum_{n=0}^{\infty} p_{r,n}(x)\frac{t^n}{n!} = e^{xt}\sum_{n=0}^{\infty} p_{r,n}t^n$$

$$= e^{xt}\sum_{n=0}^{\infty} p_{r,n}(0)\frac{t^n}{n!} \tag{3.7}$$

analogous to the classical polynomials

$$\sum_{n=0}^{\infty} H_n(x)\frac{t^n}{n!} = e^{2xt}\sum_{n=0}^{\infty} H_n(0)\frac{t^n}{n!}, \tag{3.8}$$

$$\sum_{n=0}^{\infty} B_n(x)\frac{t^n}{n!} = e^{xt}\sum_{n=0}^{\infty} B_n(0)\frac{t^n}{n!}, \tag{3.9}$$

$$\sum_{n=0}^{\infty} E_n(x)\frac{t^n}{n!} = e^{xt}\sum_{n=0}^{\infty} E_n(0)\frac{t^n}{n!}, \tag{3.10}$$

of Hermite, Bernoulli and Euler respectively (Andrews et al, 1999).

4. POLYNOMIAL PROPERTIES

The Bernoulli polynomials can be expressed in the *umbral calculus* [7] by

$$B_n(x) = (x + B(0))^n$$

in which, after expansion of the binomial, a superscript is replaced by a subscript (and where $B_n(0) = B_n$). Similarly,

$$p_{r,n}(x) = (x + p_r(0))^n. \tag{4.1}$$

Theorem 2:

$$p_{r,n}(x) = \sum_{k=0}^{n} \binom{n}{k} p_{r,n-k}(0)x^k. \tag{4.2}$$

Proof:

$$\sum_{n=0}^{\infty} p_{r,n}(x)\frac{t^n}{n!} = \sum_{k=0}^{\infty} \frac{x^k t^k}{k!} \sum_{j=0}^{\infty} p_{r,j}t^j \text{ from (3.2), (3.5)}$$

$$= \sum_{n=0}^{\infty}\sum_{k=0}^{n} \frac{n!}{k!} p_{r,n-k} \frac{x^k t^n}{n!},$$

so

$$p_{r,n}(x) = \sum_{k=0}^{n} \frac{n!}{k!(n-k)!}(n-k)!p_{r,n-k}(0)x^k$$

$$= \sum_{k=0}^{n} \binom{n}{k} p_{r,n-k}(0)x^k, \text{ by (3.6), (3.7) as required (as in (4.1)).}$$

The first few expressions for $p_{r,n}(x)$ are set out in Table 1. The coefficients for non-zero $p_{r,n}(x)$ are elements of sequences which have entries in Sloane and Plouffe [12].

n	$p_{r,n}(x)$
0	$p_{r,0}$
1	$p_{r,0}x + p_{r,1}$
2	$p_{r,0}x^2 + 2p_{r,1}x + 2p_{r,2}$
3	$p_{r,0}x^3 + 3p_{r,1}x^2 + 6p_{r,2}x + 6p_{r,3}.$

Table 1. The first few expressions for $p_{r,n}(x)$

On the assumption of continuity and uniform convergence in the appropriate closed intervals, $p_{r,n}(x)$ is an *Appell polynomial* because

$$\begin{aligned}\frac{\partial}{\partial x}\sum_{n=0}^{\infty} p_{r,n}(x)\frac{t^n}{n!} &= te^{xt}p_r(t)\\ &= t\sum_{n=0}^{\infty} p_{r,n}(x)\frac{t^n}{n!}\\ &= \sum_{n=1}^{\infty} np_{r,n-1}(x)\frac{t^n}{n!}\end{aligned}$$

which yields the Appel set criterion:

$$p'_{r,n}(x) = np_{r,n-1}(x),\ n = 1,2,3,\dots. \tag{4.3}$$

Differentiating t times, we obtain

$$p^{(t)}_{r,n}(x) = \frac{n!}{(n-t)!}p_{r,n-t}(x). \tag{4.4}$$

The *differential equation* for $p_{r,n}(x)$ is now readily obtained.

Theorem 3:

$$p''_{r,n}(x) - (n-1)p'_{r,n-1}(x) = (n-1)p_{r,n-2}(x) = 0. \tag{4.5}$$

Proof:

From (4.3) we have that

$$\begin{aligned}p''_{r,n}(x) &= n(n-1)p_{r,n-2}(x)\\ &= (n-1)[(n-1)p_{r,n-2}(x) + p_{r,n-2}(x)]\\ &= (n-1)p'_{r,n-1}(x) + (n-1)p_{r,n-2}(x) \text{ as required.}\end{aligned}$$

Similarly from (4.3) we can obtain an *integration formula*

$$\int_0^x p_{r,n}(x) = \frac{p_{r,n+1}(x) - p_{r,n+1}(0)}{n+1}. \tag{4.6}$$

The $p_{r,n}(x)$ are not orthogonal since Shohat [10] has proved that the only system of orthogonal polynomials which is an Appell polynomial sequence is that which is reducible to the Hermite polynomials by a linear transformation. The $p_{r,n}(x)$ are related to the Hermite polynomials by the result in (4.7).

Theorem 4:

$$\sum_{m=0}^{\infty} H_m(x)p_{r,n}(y)\frac{t^m}{m!} = \exp(2xyt - y^2t^2)\sum_{n=0}^{\infty} H_n(x-yt)p_{r,n}(0)\frac{t^n}{n!}. \tag{4.7}$$

Proof: We use the known result [1]:

$$\sum_{n=0}^{\infty} H_{m+n}(x)\frac{y^n}{n!} = \exp(2xy - y^2)H_m(x-y)$$

so that the right hand side of (4.2)

$$\begin{aligned} e^{2xyt-y^2t^2}\sum_{n=0}^{\infty} H_n(x-yt)p_{r,n}(0)\frac{t^n}{n!} &= \sum_{n=0}^{\infty}\sum_{m=0}^{\infty} H_{m+n}(x)p_{r,n}(0)\frac{y^m t^{m+n}}{m!n!} \\ &= \sum_{n=0}^{\infty}\sum_{m=0}^{\infty} H_{m+n}(x)p_{r,n}(0)\binom{m+n}{n}\frac{y^m t^{m+n}}{(m+n)!} \\ &= \sum_{m=0}^{\infty} H_m(x)\left(\sum_{n=0}^{m}\binom{m}{n}p_{r,n}(0)y^{m-n}\right)\frac{t^m}{m!} \\ &= \sum_{m=0}^{\infty} H_m(x)p_{r,m}(y)\frac{t^m}{m!}, \text{ as required (by Theorem 2).} \end{aligned}$$

The $p_{r,n}(x)$ are not of binomial type [8] because

$$\begin{aligned} p_{r,n}(x+y) &= (x+y+p_r(0))^n \\ &\neq \sum_{i=0}^{n}\binom{n}{i}(x+p_r(0))^i(y+p_r(0))^{n-i} \\ &= \sum_{i=0}^{n}\binom{n}{i}p_{r,i}(x)p_{r,n-i}(y). \end{aligned}$$

We can also obtain an *addition formula*.

Theorem 5:

$$p_{r,n}(x+y) = \sum_{k=0}^{n}\sum_{j=0}^{k} p_{r,n-k} n! \frac{x^j}{j!}\frac{y^{k-j}}{(k-j)!}. \tag{4.8}$$

Proof: From (3.2) and (3.5)

$$\sum_{n=0}^{\infty} p_{r,n}(x+y)\frac{t^n}{n!} = \left[1 + (x+y)\frac{t}{1!} + (x+y)^2\frac{t^2}{2!} + (x+y)^3\frac{t^3}{3!} + \dots\right]\sum_{n=0}^{\infty} p_{r,n}t^n$$

$$= \sum_{m=0}^{\infty}(x+y)^m\frac{t^m}{m!}\sum_{n=0}^{\infty} p_{r,n}t^n$$

$$= \sum_{n=0}^{\infty}\sum_{m=0}^{\infty}\sum_{k=0}^{m}\binom{m}{k}x^k y^{m-k} p_{r,n}\frac{t^{n+m}}{m!}$$

$$= \sum_{n=0}^{\infty}\sum_{m=0}^{\infty}\sum_{k=0}^{m}\binom{m}{k}x^k y^{m-k} p_{r,n-m}\frac{t^{n}}{m!}$$ by changing the order of summation,

so

$$p_{r,n}(x+y) = \sum_{m=0}^{n}\sum_{k=0}^{m}\frac{n!}{k!(m-k)!}x^k y^{m-k} p_{r,n-m},$$ as required, on equating coefficients of t.

Special Cases:

(*i*) $p_{r,n}(2x) = \sum_{m=0}^{n}\sum_{k=0}^{m} n! p_{r,n-m}\frac{x^k}{k!}\frac{y^{m-k}}{(m-k)!}$ (*duplication formula*),

(*ii*) $p_{r,n}(x+1) = \sum_{m=0}^{n}\sum_{k=0}^{m}\frac{n!}{(m-k)!} p_{r,n-m}\frac{x^k}{k!}$,

(*iii*) $p_{r,n}(0) = \sum_{m=0}^{n} n! p_{r,n-m}(x)\frac{x^m}{m!}$.

More generally,

$$p_{r,n}(tx) = \sum_{m=0}^{n}\sum_{k=0}^{m} n!\binom{m}{k} p_{r,n-m}(t-1)\frac{x^m}{m!} \text{ (\textit{multiplication formula}).}$$

Further investigations could be made of properties analogous to those of other classical polynomials, such as Jacobi and Laguerre polynomials [13].

5. COMBINATORIAL PROPERTIES

A composition of the positive integer n is a vector $(a_1, a_2, \ldots, a_k)$, the components of which are the positive integers such that $a_1 + a_2 + \cdots + a_k = n$ [3]. If the vector has order k, then the composition is a *k-part composition.* In what follows $\gamma(n)$ indicates summation over all the compositions $(a_1, a_2, \ldots, a_k)$ of n, the number of components being variable [5]. Let

$$R_n = \sum_{\gamma(n)} \frac{(-1)^{k-1}}{k} p_{r,a_1} \cdots p_{r,a_k}. \tag{5.1}$$

Then formally

$$\begin{aligned}\sum_{n=1}^{\infty} R_n x^n &= \sum_{n=1}^{\infty}\sum \gamma(n)\frac{(-1)^{k-1}}{k} p_{r,a_1}\cdots p_{r,a_k}x^n \\ &= \sum_{k=1}^{\infty} - \left(-\sum_{n=1}^{\infty} p_{r,n}x^n\right)^k /k \\ &= \ln\left(1 + \sum_{n=1}^{\infty} p_{r,n}x^n\right) \\ &= \ln\left(\sum_{n=1}^{\infty} p_{r,n}x^n\right),\end{aligned}$$

that is,

Theorem 6:

$$\sum_{n=1}^{\infty} p_{r,n}x^n = \exp\left(\sum_{n=1}^{\infty} R_n x^n\right). \tag{5.2}$$

This, from (3.2) and (3.4), is satisfied by

$$R_n = \frac{1}{n}Q_{r,n}. \tag{5.3}$$

Thus,

Theorem 7:

$$Q_{r,n} = \sum_{\gamma(n)}(-1)^{k-1}\frac{n}{k}p_{r,a_1}\cdots p_{r,a_k}. \tag{5.4}$$

When $r = 0$, we find for the Fibonacci and Lucas numbers that

$$L_n = \sum_{\gamma(n)}(-1)^{k-1}\frac{n}{k}f_{a_1}\cdots f_{a_k}, \text{ using (1.3), (1.4)} \tag{5.5}$$

in which $f_n = F_{n+1}$. For instance, when $n = 3$,

$$\sum_{\gamma(3)}(-1)^{k-1}\frac{3}{k}f_{a_1}\cdots f_{a_k} = -\frac{3}{2}f_1f_2 - \frac{3}{2}f_2f_1 + \frac{3}{1}f_3 + \frac{3}{3}f_1f_1f_1$$

$$= -3 - 3 + 9 + 1 = 4 = L_3.$$

6. CONCLUDING COMMENTS

Theorem 8: Any polynomial can be expressed in terms of the generalized Pell polynomials.

Proof: From (3.4) and (3.5) we have that

$$\exp xt = \exp\left(-\sum_{m=1}^{\infty} Q_{r,m}\frac{x^m}{m}\right)\sum_{n=0}^{\infty} p_{r,n}(x)\frac{t^n}{n!}$$

$$= \frac{1}{p_r(x)}\sum_{n=0}^{\infty} p_{r,n}(x)\frac{t^n}{n!}$$

$$= (1 - 2^r x - x^2)\sum_{n=0}^{\infty} p_{r,n}(x)\frac{t^n}{n!};$$

on equating coefficients of t^n we get

$$x^n - p_{r,n}(x) - 2^r n p_{r,n-1}(x) - n(n-1)p_{r,n-2}(x). \tag{6.1}$$

For example, when $n = 2$, from Table 1,

$$\begin{aligned} p_{r,2}(x) - 2^{r+1}p_{r,1}(x) - 2p_{r,0}(x) &= (x^2 + 2^{r+1}x + 2^{2r+1} + 2) - (2^{r+1}x + 2^{2r+1}) - 2 \\ &= x^2. \end{aligned}$$

Gratitude is expressed to an anonymous referee for detailed useful comments.

REFERENCES

[1] Al-Salam, W.A. and Carlitz, L. "On Certain Generating Functions for the Hermite Polynomials." *Portugaliae Mathematica*, Vol. *25* (1966): pp. 35-45.

[2] Andrews, G.E., Askey, R. and Roy, R. Special Functions. Encyclopedia of Mathematics & Its Applications. Vol. 71. Cambridge: Cambridge University Press, 1999

[3] Brualdi, R.A. and Ryser, H.J. Combinatorial Matrix Theory. Encyclopedia of Mathematics & Its Applications. Vol. 39. Cambridge: Cambridge University Press, 1991.

[4] Freeman, J.M. "A Strategy for Determining Polynomial Orthogonality." In B.E. Sagan & R.P. Stanley (eds). Mathematical Essays in Honor of Gian-Carlo Rota. Boston: Birkhäuser, 1998, pp. 239-244.

[5] Hoggatt, V.E., Jr. and Lind, D.A. "Fibonacci and Binomial Properties of Weighted Compositions." *Journal of Combinatorial Theory*, Vol. *4.2* (1968): pp. 121-124.

[6] Horadam, A.F. "Basic Properties of a Certain Generalized Sequence of Numbers." *The Fibonacci Quarterly*, Vol. *3* (1965): pp. 161-176.

[7] Riordan, J. An Introduction to Combinatorial Analysis. New York: Wiley, 1958.

[8] Rota, G-C and Mullin, R. On the Foundations of Combinatorial Theory, III: Theory of Binomial Enumerations. New York: Academic Press, 1970.

[9] Shannon, A.G. and Ollerton, R.L. "Combinatorial Matrices and Linear Recursive Sequences." *The Fibonacci Quarterly*, Vol. *40* (2002): pp. 417-423.

[10] Shohat, J. "The Relation of the Classical Orthogonal Polynomials to the Polynomials of Appell." *American Journal of Mathematics*, Vol. *58* (1936): pp. 453-464.

[11] Singhal, J.P. and Srivastava, H.M. "A Class of Bilateral Generating Functions for Certain Classical Polynomials." *Pacific Journal of Mathematics*, Vol. *42* (1972): pp. 755-762.

[12] Sloane, J.A. and Plouffe, S. The Encyclopedia of Integer Sequences. San Diego: Academic Press, 1995.

[13] Thakare, N.K. "Encounters with Problems from Algebra, Discrete Mathematics and Soft as Well as Hard Analysis." *The Mathematics Student*, Vol. *68* (1999): pp. 219-241.

AMS Classification Numbers: 11B39, 11B68

A FURTHER NOTE ON LUCASIAN NUMBERS

Lawrence Somer

1. INTRODUCTION

This paper will extend and unify the results in [4] by completely determining all Lucasian numbers which are terms in certain Lucas sequences. Our specification of all Lucasian numbers will be based on results obtained in [1] in which all terms in particular Lucas sequences which do not have any primitive prime divisors are found.

Before proceeding, we will recall some definitions and known results. Our notation will be the same as that in [4]. Let $u(r, s)$ and $v(r, s)$ be Lucas sequences which satisfy the same second-order recursion relation

$$w_{n+2} = rw_{n+1} + sw_n \tag{1}$$

and have initial terms $u_0 = 0$, $u_1 = 1$, $v_0 = 2$, $v_1 = r$ respectively, where r and s are integers. Associated with the sequences $u(r, s)$ and $v(r, s)$ is the characteristic polynomial

$$f(x) = x^2 - rx - s \tag{2}$$

with characteristic roots α and β. Let $D = r^2 + 4s = (\alpha - \beta)^2$ be the discriminant of both $u(r, s)$ and $v(r, s)$. By the Binet formulas

$$u_n = (\alpha^n - \beta^n)/(\alpha - \beta) \tag{3}$$

and

$$v_n = \alpha^n + \beta^n. \tag{4}$$

The recurrences $u(r, s)$ and $v(r, s)$ are said to be <u>degenerate</u> if $\alpha\beta = -s = 0$ or α/β is a root of unity. It follows from (3) and (4) that u_n or v_n can be equal to zero for $n \geq 1$ only if both $u(r, s)$ and $v(r, s)$ are degenerate.

This paper is in final form and no version of it will be submitted for publication elsewhere.

F.T. Howard (ed.), Applications of Fibonacci Numbers, Volume 9: Proceedings of the Tenth International Research Conference on Fibonacci Numbers and their Applications, 225–234.

The integer m is a divisor of the recurrence $w(r,s)$ satisfying the relation (1) if $m \mid w_n$ for some $n \geq 1$. The prime p is a primitive prime divisor of w_n, $n \geq 1$, if $p \mid w_n$ but $p \not| w_i$ for $1 \leq i < n$. Given the Lucas sequence $v(r,s)$, the integer m is called Lucasian if m is a divisor of $v(r,s)$. In our main theorem, Theorem 2.6, we will show that if $u(r,s)$ and $v(r,s)$ are nondegenerate and $gcd(r,s) = 1$, then u_n is not Lucasian if $n \geq 27$. Theorem 2.6 will also find all terms u_a and v_b such that $a \geq 1$, $b \geq 1$, and $u_a \mid v_b$.

A related question is to determine all a and b such that $a \geq 1$, $b \geq 1$, and $v_a \mid u_b$. We will answer this question completely in Theorem 2.8 when both $v(r,s)$ and $u(r,s)$ are nondegenerate and $gcd(r,s) = 1$.

2. THE MAIN RESULTS

For reference, we will give the main results of [4] in Theorems 2.1-2.4. Theorems 2.6 and 2.8 will then generalize Theorems 2.1, 2.2 and 2.4.

Theorem 2.1: Let $u(r,s)$ and $v(r,s)$ be nondegenerate Lucas sequences for which $gcd(r,s) = 1$ and $D > 0$. Let a and b be positive integers. Then $u_a \mid v_b$ if and only if one of the following conditions also holds. For convenience, the value of u_a is given in these conditions. The expressions $r = *$ given below means that r can be any integer.
(i) $a = 1,\ r = *,\ s = *,\ b = *,\ u = 1;$
(ii) $a = 2, \mid r \mid = 1$ or $2,\ s = *,\ b = *,\ u_2 = r = v_1;$
(iii) $a = 2,\ \mid r \mid \geq 3,\ s = *,\ b \equiv 1 \pmod 2), u_2 = r = v_1;$
(iv) $a = 3, \mid r \mid = 1,\ s = 1,\ b \equiv 0 \pmod 3),\ u_3 = 2;$
(v) $a = 4,\ \mid r \mid = 1, s = *,\ b \equiv 2 \pmod 4),\ u_4 = v_2 = r^2 + 2s.$
In particular, u_n is not Lucasian if $n \geq 5$.

Theorem 2.2: Let $u(r,s)$ and $v(r,s)$ be nondegenerate Lucas sequences for which $gcd(r,s) = 1$ and $D < 0$. Then u_n is not Lucasian for $n > e^{452}2^{68}$.

Theorem 2.3: Let $u(r,s)$ and $v(r,s)$ be nondegenerate Lucas sequences for which $gcd(r,s) > 1$. Then there exists a constant $N(r,s)$ dependent on r and s such that u_n is not Lucasian for $n \geq N(r,s)$.

Theorem 2.4: Let $u(r,s)$ and $v(r,s)$ be nondegenerate Lucas sequences for which $gcd(r,s) = 1$ and $D > 0$. Let a and b be positive integers. If $|v_a| \geq 3$, then $v_a | u_b$ if and only if $2a|b$. If $|v_a| \leq 2$, then $v_a | u_b$ if and only if one of the following two conditions also holds (the value of v_a is given for convenience):
(i) $a = 1,\ |r| = 1,\ s = *,\ b = *,\ v_1 = r;$
(ii) $a = 1,\ |r| = 2,\ s \equiv 1 \pmod 2),\ b \equiv 0 \pmod 2),\ v_1 = r = u_2.$

Remark 2.5: The proofs of Theorem 2.1 and 2.4 given in [4] depend partly on the fact that if $D > 0$, then $u(r,s)$ is strictly increasing for $n \geq 2$ and $v(r,s)$ is strictly increasing for $n \geq 1$.

Theorem 2.6: Let $u(r,s)$ and $v(r,s)$ be nondegenerate Lucas sequences for which $gcd(r,s) = 1$. Let a and b be positive integers. Then u_a is Lucasian and $u_a | v_b$ if and only if one of the following conditions also holds. For convenience, the value of u_a is given in these conditions.
(i) $a = 1,\ r = *,\ s = *,\ b = *,\ u_1 = 1;$
(ii) $a = 2,\ |r| = 1$ or $2,\ s = *,\ b = *,\ u_2 = v_1 = r;$
(iii) $a = 2,\ |r| \geq 3,\ s = *,\ b \equiv 1 \pmod 2),\ u_2 = v_1 = r;$
(iv) $a = 3,\ r = *,\ s = \pm 1 - r^2, b = *, u_3 = \pm 1;$
(v) $a = 3,\ r \equiv 1 \pmod 2),\ s = \pm 2 - r^2, b \equiv 0 \pmod 3),\ u_3 = \pm 2;$

(vi) $a = 4$, $|r| = 1$, $s = *$, $b \equiv 2 (\text{mod } 4)$, $u_4 = \pm v_2$;
(vii) $a = 4$, $r \equiv 1 (\text{mod } 2)$, $s = (\pm 1 - r^2)/2$, $b \equiv 1 (\text{mod } 2)$, $u_4 = \pm u_2 = \pm v_1 = \pm r$;
(viii) $a = 5$, $|r| = 1$, $s = -2$, $b = *$, $u_5 = -1$;
(ix) $a = 5$, $|r| = 1$, $s = -3$, $b = *$, $u_5 = 1$;
(x) $a = 5$, $|r| = 12$, $s = -55$, $b = *$, $u_5 = 1$;
(xi) $a = 5$, $|r| = 12$, $s = -377$, $b = *$, $u_5 = 1$;
(xii) $a = 6$, $r = *$, $s = \pm 1 - r^2$, $b \equiv 3 (\text{mod } 6)$, $u_6 = \pm v_3$;
(xiii) $a = 7$, $|r| = 1$, $s = -5$, $b = *$, $u_7 = 1$;
(xiv) $a = 8$, $|r| = 1$, $s = -2$, $b \equiv 2 (\text{mod } 4)$, $u_8 = u_4 = \pm v_2 = \pm 3$;
(xv) $a = 10$, $|r| = 1$, $s = -2$, $b \equiv 5 (\text{mod } 10)$, $u_{10} = -v_5 = \pm 11$;
(xvi) $a = 10$, $|r| = 1$, $s = -3$, $b \equiv 5 (\text{mod } 10)$, $u_{10} = v_5 = \pm 31$;
(xvii) $a = 10$, $|r| = 12$, $s = -55$, $b \equiv 5 (\text{mod } 10)$, $u_{10} = v_5 = \pm 44868$;
(xviii) $a = 10$, $|r| = 12$, $s = -377$, $b \equiv 5 (\text{mod } 10)$, $u_{10} = v_5 = \pm 5519292$;
(xix) $a = 13$, $|r| = 1$, $s = -2$, $b = *$, $u_{13} - 1$;
(xx) $a = 14, |r| = 1$, $s = -5, b \equiv 7 \ (\text{mod } 14), u_{14} = v_7 = \pm 559$;
(xxi) $a = 26$, $|r| = 1$, $s = -2$, $b \equiv 13 \ (\text{mod } 26)$, $u_{26} = -v_{13} = \pm 181$.

Remark 2.7: By Theorem 2.6, if $u(r, s)$ is nondegenerate and $gcd(r, s) = 1$, then there exist only 12 possible indices n for which u_n can be Lucasian. It is noteworthy that for the Lucas sequences $u(\pm 1, -2)$, u_n is Lucasian for 10 of these indices, namely $n = 1, 2, 3, 4, 5, 6, 8, 10, 13, 26$. The only other nondegenerate Lucas sequences $u(r, s)$ for which $gcd(r, s) = 1$ and u_n is Lucasian for 5 or more indices n are $u(\pm 1, -3)$ and $u(\pm 1, -5)$. For $u(\pm 1, -3)$, u_n is Lucasian for the 6 indices $n = 1, 2, 3, 4, 5, 10$, while for $u(\pm 1, -5)$, u_n is Lucasian for the 5 indices $n = 1, 2, 4, 7, 14$.

Theorem 2.8: Let $u(r, s)$ and $v(r, s)$ be nondegenerate Lucas sequences for which $gcd(r, s) = 1$. Let a and b be positive integers. Then $v_a | u_b$ if and only if one of the following conditions also holds. For convenience, the value of v_a is given in condition (ii)-(vii).

(i) $2a | b$, $r = *$, $s = *$;
(ii) $a = 1$, $|r| = 1$, $s = *$, $b = *$, $v_1 = r$;
(iii) $a = 2$, $r \equiv 1 \ (\text{mod } 2)$, $s = (\pm 1 - r^2)/2$, $b = *$, $v_2 = \pm 1$;
(iv) $a = 2$, $r \equiv 0 \ (\text{mod } 2)$, $s = (\pm 2 - r^2)/2$, $b \equiv 0 \ (\text{mod } 2)$, $v_2 = \pm 2$;
(v) $a = 4$, $|r| = 1$, $s = -2$, $b = *$, $v_4 = 1$;
(vi) $a = 4$, $|r| = 2$, $s = -7$, $b \equiv 0 \ (\text{mod } 2)$, $v_4 = 2$;
(vii) $a = 5$, $|r| = 2$, $s = -3$, $b \equiv 0 \ (\text{mod } 2)$, $v_5 = \pm 2$.

Remark 2.9: Theorem 2.6 generalizes Theorems 2.1 and 2.2, which were proven in [4]. Theorem 2.8 generalizes Theorem 2.4 which was also proved in [4]. As contrasted to the proofs of Theorems 2.1, 2.2, and 2.4, the proofs of Theorems 2.6 and 2.7, which will be given in Section 4, do not treat the cases $D > 0$ and $D < 0$ separately.

The key result in proving Theorems 2.6 and 2.8 is Theorem 2.10 given below which is the main theorem of [1].

Theorem 2.10: Let $u(r, s)$ be a nondegenerate Lucas sequence for which $gcd(r, s) = 1$. Then u_n has a primitive prime divisor if $n > 30$. Moreover u_n has no primitive prime divisor only if $n = 1, 2, 3, 4, 5, 6, 7, 8, 10, 12, 13, 18$, or 30.

Remark 2.11: Consider all nondegenerate Lucas sequences $u(r, s)$ for which $gcd(r, s) = 1$. Tables 1 and 3 of [1] list all terms $u_n, n \geq 1$, which have no primitive prime divisors. We note that in [1], the authors define a prime p to be a primitive prime divisor of u_n if $p | u_n$ but

$p \not| Du_1u_2\ldots u_{n-1}$. In contrast to this definition, we do not include p as a primitive prime divisor of u_n if $p|D$, but $p \not| u_k$ for $1 \leq k < n$.

For reference, Theorem 2.12 lists all degenerate Lucas sequences $u(r,s)$ and $v(r,s)$.

Theorem 2.12: The Lucas sequences $u(r,s)$ and $v(r,s)$ are degenerate if and only if one of the following conditions holds:

(i) $s = 0$.
(ii) $\alpha/\beta = 1$ and $D = r^2 + 4s = 0$.
(iii) $\alpha/\beta = -1$, $r = 0$ and $s = N$ for some non-zero integer N.
(iv) α/β is a primitive cube root of unity, $r = N$, and $s = -N^2$ for some non-zero integer N.
(v) α/β is a primitive fourth root of unity, $r = 2N$, and $s = -2N^2$ for some non-zero integer N.
(vi) α/β is a primitive sixth root of unity, $r = 3N$, and $s = -3N^2$ for some non-zero integer N.

Proof: This is proved in [9, p. 613]. □

3. NECESSARY LEMMAS AND DEFINITIONS

The following lemmas and definition will be needed for the proofs of Theorems 2.6 and 2.8. Lemmas 3.1, 3.2, 3.3, and 3.5 are well-known (and follow readily from (1), (3), and (4)).

Lemma 3.1: $u_{2n} = u_n v_n$.

Lemma 3.2:

$$u_n(-r,s) = (-1)^{n+1}u_n(r,s). \tag{5}$$

$$v_n(-r,s) = (-1)^n v_n(r,s). \tag{6}$$

It follows from Lemma 3.2 that, $u_n(-r,s)$ is Lucasian if and only if $u_n(r,s)$ is Lucasian.

Lemma 3.3: Consider the Lucas sequences $u(r,s)$ and $v(r,s)$. Then $u_n|u_{in}$ for all $i \geq 1$ and $v_n|v_{(2j+1)n}$ for all $j \geq 0$. □

Lemma 3.4: If u_a is not Lucasian, and $a|c$, then u_c is not Lucasian.

Proof: By Lemma 3.3, $u_a|u_c$. It is now evident that u_c is not Lucasian if u_a is not Lucasian. □

Lemma 3.5: Let $u(r,s)$ and $v(r,s)$ be Lucas sequences for which $2 \not| gcd(r,s)$. □

(i) Suppose r is odd and s is even. Then $2 \not| u_n$ and $2 \not| v_n$ for $n \geq 1$.
(ii) Suppose r and s are both odd. Then $2|u_n$ if and only if $3|n$, and $2|v_n$ if and only if $3|n$.
(iii) Suppose r is even and s is odd. Then $2|u_n$ if and only if $2|n$, and $2|v_n$ for all $n \geq 0$.

Lemma 3.6: Let $v(r,s)$ be a Lucas sequence for which $2 \not| gcd(r,s)$.

(i) If r and s are both odd and $2^k \parallel v_3$ for some positive integer k, then $2 \parallel v_n$ for $n \equiv 0 \pmod 6$ and $2^k \parallel v_n$ for $n \equiv 3 \pmod 6$. Recall that $2^k \parallel a$ if $2^k|a$, but $2^{k+1} \not| a$.
(ii) If r is even and s is odd and $2^k \parallel v_1 = r$, then $2 \parallel v_{2n}$ and $2^k \parallel v_{2n+1}$ for all $n \geq 0$.

Proof: This is proved in [8]. □

Definition 3.7: For the Lucas sequence $u(r,s)$ the rank of appearance of the positive integer m in $u(r,s)$, denoted by $\omega(m)$, is the least positive integer n, if it exists, such that $m|u_n$. The rank of appearance of m in $v(r,s)$, denoted by $\overline{\omega}(m)$, is defined similarly.

Lemma 3.8: Let $u(r,s)$ and $v(r,s)$ be Lucas sequences for which $gcd(r,s) = 1$. Let p be an odd prime. If $\omega(p)$ is odd, then $\overline{\omega}(p)$ does not exist and $u_{\omega(p)}$ is not Lucasian.

Proof: This was proved by Carmichael [2, p. 47]. □

Lemma 3.9: Let $u(r,s)$ and $v(r,s)$ be Lucas sequences for which $gcd(r,s)=1$. Suppose that p is an odd prime and $\omega(p)=2n$. Then $\overline{\omega}(p)=n$.

Proof: This is proved in Proposition 2(iv) of [7]. □

Definition 3.10: The 2-valuation of the integer n, denoted by $[n]_2$ is the largest integer k such that $2^k|n$.

Lemma 3.11: Let $v(r,s)$ be a Lucas sequence for which $gcd(r,s)=1$. Suppose that u_a is Lucasian and that p and q are distinct odd prime divisors of u_a. Then $[\overline{\omega}(p)]_2=[(\overline{\omega}(q)]_2$.

Proof: This is proved in Proposition 2(ix) of [7]. □

Lemma 3.12: Let $u(r,s)$ and $v(r,s)$ be nondegenerate Lucas sequences for which $gcd(r,s)=1$. Let a and b be positive integers and let $d=gcd(a,b)$.

(i) $gcd(u_a,u_b)=u_d$;

(ii) $gcd(v_a,v_b)=\begin{cases} v_d \text{ if } [a]_2=[b]_2, \\ 1 \text{ or } 2 \text{ otherwise;}\end{cases}$

(iii) $gcd(u_a,v_b)=\begin{cases} v_d \text{ if } [a]_2>[b]_2, \\ 1 \text{ or } 2 \text{ otherwise.}\end{cases}$

Proof: This is proved in [6] and [3, section 5]. □

Lemma 3.13: Let $u(r,s)$ and $v(r,s)$ be nondegenerate Lucas sequences for which $gcd(r,s)=1$. Then $m|u_n$ if and only if $\omega(m)|n$. Moreover, if $m\geq 3$, then $m|v_n$ if and only if $\overline{\omega}(m)|n$ and $[\overline{\omega}(m)]_2=[n]_2$.

Proof: The results follow from Lemmas 3.3. and 3.12. □

Lemma 3.14: Let $u(r,s)$ be nondegenerate Lucas sequence for which $2\not|\, gcd(r,s)$. If r is even, s is odd, and $4|a$, then u_a is not Lucasian. If r and s are both odd and $6|a$, then u_a is not Lucasian. If r and s are both odd and $6|a$, the u_a is not Lucasian. If r and s are both odd, $4|u_3$, and $3|a$, then u_a is not Lucasian.

Proof: First suppose that r is even and s is odd. By Lemma 3.5 (iii), $2|v_n$ for all $n\geq 0$. Moreover, $u_2=v_1=r$. Suppose that $2^k\parallel r$. By Lemma 3.6 (ii), $2^{k+1}\not|\, v_n$ for any $n\geq 0$. However, $2|v_2$, and hence by Lemma 3.1, $2^{k+1}|u_4=u_2v_2$. Thus, u_4 is not Lucasian, and conseuqently by Lemma 3.4, u_a is not Lucasian if $4|a$.

Now suppose that both r and s are odd. By Lemma 3.5(ii), $2|u_3$ and $2|v_3$. Suppose that $2^k\parallel v_3$. By Lemma 3.6 (i), $2^{k+1}\not|\, v_n$ for any $n\geq 0$. However, $2^{k+1}|u_6=u_3v_3$. Therefore, u_6 is not Lucasian, and hence by Lemma 3.4, u_a is not Lucasian if $6|a$.

Finally, suppose that in addition to r and s both being odd, $4|u_3$. Since $u_3=r^2+s$, this can occur only if $s\equiv 3 \pmod 4$. However, then $v_3=r(r^2+3s)\equiv 2 \pmod 4$, and $2\parallel v_3$. By Lemma 3.6 (i), it follows that 4 is not a divisor of $v(r,s)$. Since $u_3|u_{3t}$ for all $t\geq 1$ by Lemma 3.3, we see that $4|u_{3t}$ for all $t\geq 1$. Hence, u_a is not Lucasian if $3|a$. □

Lemma 3.15: Let $u(r,s)$ and $v(r,s)$ be nondegenerate Lucas sequences for which $gcd(r,s)=1$. Then u_a is Lucasian for $a\geq 1$ if and only if at least one of the following four conditions holds:

(i) a is odd and $u_a=\pm 1$;

(ii) a is even and $u_{a/2}=\pm 1$;

(iii) a is even, $u_{a/2}$ is Lucasian, and $v_{a/2}=\pm 1$;

(iv) $a = 3$ and $u_a = \pm 2$.

Proof: We first show sufficiency. The sufficiency of (i) is obvious. The sufficiency of (ii) and (iii) follow from the fact that $u_a = u_{a/2}v_{a/2}$ by Lemma 3.1. The sufficiency of (iv) follows from the fact that by Lemma 3.5, $2|u_a$ for $a \geq 1$ only if $2|v_a$.

We now show necessity. Suppose first that a is odd, $u_a \neq \pm 1$, and u_a is Lucasian. If u_a has an odd prime divisor p, then by Lemma 3.13, $\omega(p)|a$ and hence $\omega(p)$ is odd. It now follows from Lemmas 3.4 and 3.8 that u_a is not Lucasian.

Now assume that $u_a \geq 2$ and u_a is a power of 2. By Lemma 3.5, we must have that r and s are both odd and $3|a$. Since $2|u_3 = r^2 + s$, we see that u_a has no primitive prime divisor if $a > 3$. It now follows from Theorem 2.10 that a must equal 3. By Lemma 3.14, u_a is not Lucasian if $4|u_3$. Hence, $a = 3$ and $u_a = \pm 2$. By our above discussion, we see that if a is odd and u_a is Lucasian, then either $u_a = \pm 1$ or $a = 3$ and $u_a = \pm 2$.

At this point, we assume that a is even, $u_{a/2} \neq \pm 1$, and $v_{a/2} \neq \pm 1$. Suppose that $2|u_{a/2}$. By Lemma 3.5, either r and s are both odd, or r is even and s is odd. If r and s are both odd, then $3|(a/2)$ and $2|v_{a/2}$ by Lemma 3.5. Suppose that $2^k \parallel v_3$. Then $2^{k+1}|u_6 = u_3v_3$, and hence $2^{k+1}|u_a$ by Lemma 3.3. Thus, u_a is not Lucasian by Lemma 3.6 (i). If r is even and s is odd, then $a/2$ is even by Lemma 3.5. Moreover, by Lemma 3.5 (iii), $2|v_n$ for all $n \geq 0$. Suppose that $2^k \parallel v_1$. Then $2^k|u_2 = r = v_1$, and hence $2^k|u_{a/2}$ by Lemma 3.3. Then $2^{k+1}|u_a = u_{a/2}v_{a/2}$, and u_a is not Lucasian by Lemma 3.6 (ii).

Next suppose that $2|v_{a/2}$. Since $2 \nmid u_{a/2}$ by our above arguments, we see by Lemma 3.5 that r is even, s is odd, and $a/2$ is odd. Then $u_{a/2}$ has an odd prime divisor. By our earlier discussion and Lemma 3.4, it follows that both $u_{a/2}$ and u_a are not Lucasian.

The only remaining case to consider is the one in which $u_{a/2}$ has an odd prime divisor p and $v_{a/2}$ has an odd prime divisor q. Since $gcd(u_{a/2}, v_{a/2}) = 1$ or 2 by Lemma 3.12 (iii), $p \neq q$. Suppose that $2^t \parallel a$. If $t = 1$, then $\omega(p)|(a/2)$ and $\omega(p)$ is odd by Lemma 3.13. Thus, both $u_{a/2}$ and u_a are not Lucasian by Lemmas 3.4 and 3.8. Consequently we must have that $t \geq 2$. By Lemmas 3.13 and 3.9, it follows that

$$[\overline{\omega}(p)]_2 \leq t - 2. \tag{7}$$

Since $q|v_{a/2}$, we see by Lemma 3.13 that

$$[\overline{\omega}(q)]_2 = t - 1. \tag{8}$$

Since $pq|u_a$ it now follows from Lemma 3.11 and (7) and (8) that u_a is not Lucasian. Hence, if a is even and u_a is Lucasian, then $u_{a/2} = \pm 1$ or $v_{a/2} = \pm 1$. We note further that if $v_{a/2} = \pm 1$ and u_a is Lucasian, then $u_{a/2}$ is also Lucasian since $u_a = u_{a/2}v_{a/2}$. Necessity is thus shown and our result follows. □

Remark 3.16: Suppose that $u(r,s)$ and $v(r,s)$ are nondegenerate, $gcd(r,s) = 1$, and u_a is Lucasian. We note that if $v_{a/2} = \pm 1$, then $u_a = u_{a/2}v_{a/2} = \pm u_{a/2}$. It now follows from Lemma 3.15 that if it is not the case that $a = 3$ and $u_a = \pm 2$, then either u_a or $u_{a/2}$ has no primitive prime divisor. In what follows, we will apply the results of [1], which determine all n such that u_n has no primitive prime divisor to find all a for which u_a is not Lucasian and it is not the case that $a = 3$ and $u_a = \pm 2$.

Lemma 3.17: Let $u(r, s)$ be a nondegenerate Lucas sequence for which $gcd(r, s) = 1$. If a is odd, then u_a is not Lucasian if $a \geq 12$, $a \neq 14$, and $a \neq 26$.

Proof: Suppose that a is odd, $a \geq 9$ and $a \neq 13$. By Lemma 3.15, if u_a were to be Lucasian, then u_a must equal ± 1, and consequently u_a would have no primitive prime divisor. It now follows from Theorem 2.10 that u_a is not Lucasian.

We now assume that $a = 18$, $a = 30$, or $a = 60$. By our earlier discussion, u_9 and u_{15} are both not Lucasian. Thus, u_a is not Lucasian by Lemma 3.4. Now assume that a is even and either $a = 22$ or $a \geq 28$. By our previous argument, we can assume that $a \neq 30$, $a \neq 36$, and $a \neq 60$. If u_a were to be Lucasian, then by Remark 3.16, either $u_{a/2}$ or u_a would have no primitive prime divisors. Therefore, by Theorem 2.10, u_a is not Lucasian in this case also.

We next suppose that $a = 20$ and u_a is Lucasian. By Theorem 2.10, u_{20} has a primitive prime divisor. Thus by Lemma 3.15 and Remark 3.16, $u_{10} = \pm 1$ and u_{10} has no primitive prime divisor. By Table 1 of [1], the only recurrences $u(r, s)$ for which u_{10} has no primitive prime divisor are $u(\pm 2, -3)$, $u(\pm 5, -7)$, and $u(\pm 5, -18)$. We see by inspection that in each case, $u_{10} \neq \pm 1$. Hence, u_{20} is not Lucasian.

Now assume that $a = 16$ and u_a is Lucasian. Since u_{16} has a primitive prime divisor by Theorem 2.10, we see by Lemma 3.15 and Remark 3.16 that $u_8 = \pm 1$. By Table 1 of [1], the only recurrences $u(r, s)$ for which u_8 has no primitive prime divisor are $u(\pm 2, -7)$ and $u(\pm 1, -2)$. In each case, we see by inspection that $u_8 \neq \pm 1$. Thus u_{16} is not Lucasian.

We finally suppose that $a = 12$ and u_a is Lucasian. By Lemma 3.15 either $u_6 = \pm 1$ or $v_6 = \pm 1$. First assume that $u_6 = \pm 1$. Since $u_6 = u_3 v_3$, $v_3 = r(r^2 = 3s) = \pm 1$. Thus, $r = \pm 1$. Then $r^2 + 3s = 3s + 1 = \pm 1$. Hence, $3s = -2$, which is impossible, or $3s = 0$ and $s = 0$, which is excluded by the assumption that $u(r, s)$ is nondegenerate. Therefore, by Lemma 3.15 and Remark 3.16, $v_6 = \pm 1$ and u_{12} has no primitive prime divisor. Since $v_6 = \pm 1$, it follows from Lemma 3.5 that r is odd and s is even. By Table 1 of [1], the only recurrences $u(r, s)$ for which r is odd and s is even and u_{12} has no primitive prime divisor are $u(\pm 1, -2)$ and $u(\pm 1, -4)$. In both cases, we see that for the corresponding Lucas sequence, $v(r, s)$, $v_6 \neq \pm 1$. Thus, u_{12} is not Lucasian. By Lemma 3.4, u_{24} is also not Lucasian. □

Lemma 3.18: Suppose that $u(r, s)$ and $v(r, s)$ are nondegenerate Lucas sequences for which $gcd(r, s) = 1$. Suppose that a is even, $u_{a/2} = \pm 1$, $|u_a| \geq 3$, and $\overline{\omega}(v_{a/2}) = a/2$. Then $u_a | v_b$ if and only if $b \equiv a/2 \pmod{a}$.

Proof: Since $u_a = u_{a/2} v_{a/2}$ by Lemma 3.1, $u_a = \pm v_{a/2}$. Note that $b \equiv a/2 \pmod{a}$ if and only if $b = (2k + 1)(a/2)$ for some $k \geq 0$. The result now follows from Lemma 3.12 (ii). □

4. PROOFS OF THE MAIN THEOREMS

We are now ready for the proofs of Theorems 2.6 and 2.8.

Proof of Theorem 2.6: A substantial part of this proof deals with examining nondegenerate Lucas sequences $u(r, s)$ for which $gcd(r, s) = 1$ and either $u_{a/2}$ or u_a has no primitive prime divisor. Since there exist infinitely many such Lucas sequences when $a/2$ or $a = 1$, 2, 3, 4 or 6 by Table 3 of [1], we will treat these cases separately.

If $u_a = 1$, then clearly $u_a | v_n$ for all $n \geq 1$. The result for the case in which $a = 1$ now follows since $u_1 = 1$. Suppose that $a = 2$. Then $u_2 = r = v_1$. If $|r| = 1$ or 2, then $u_2 | v_n$ for all n by our above observation and by Lemma 3.5 (iii). If $|r| \geq 3$, then condition (iii) holds for Lemma 3.18.

Now assume that $a = 3$. By Lemma 3.15, u_3 is Lucasian only if $u_3 = \pm 1$ or $u_3 = \pm 2$. We note that $u_3 = r^2 + s = \pm 1$ if and only if $s = \pm 1 - r^2$, and condition (iv) is satisfied. If $u_3 = r^2 + s = \pm 2$, then $s = \pm 2 - r^2$. Hence, $s \equiv r \pmod 2$. Consequently, r and s are both odd since $gcd(r, s) = 1$. By Lemma 3.5 (ii), it follows that condition (v) holds.

Next suppose that $a = 4$. By Lemma 3.15 either $u_2 = r = \pm 1$ or $v_2 = r^2 + 2s = \pm 1$. If $u_2 = \pm 1$, then

$$u_4 = u_2 v_2 = \pm v_2 = \pm(r^2 + 2s) = \pm(2s + 1). \tag{9}$$

We claim that $|v_2| \geq 3$. Suppose that $v_2 = \pm 1$. By (9), either $2s + 1 = 1$ or $2s + 1 = -1$. If $2s + 1 = 1$, then $s = 0$, which is excluded since $u(r, s)$ and $v(r, s)$ are both nondegenerate. If $2s + 1 = -1$, then $u(r, s)$ and $v(r, s)$ would be degenerate by Theorem 2.12 (iv), contrary to assumption. Noting that $v_2 = 2s + 1$ is odd, we see that $|u_4| = |v_2| \geq 3$. Since $v_1 = r = \pm 1$, it follows from Lemma 3.18 that condition (vi) holds. If $v_2 = r^2 + 2s = \pm 1$, then

$$u_4 = u_2 v_2 = \pm u_2 = \pm r = \pm v_1. \tag{10}$$

Moreover, $v_2 = \pm 1$ if and only if $s = (\pm 1 - r^2)/2$ and $r \equiv 1 \pmod 2$. If $r = \pm 1$, then $s = 0$ which contradicts the fact that both $u(r, s)$ and $v(r, s)$ are nondegenerate. Hence, $|u_4| = |u_2| = |r| = |v_1| \geq 3$. We now see by Lemma 3.18 that condition (vii) holds.

We now assume that $a = 6$. By Lemma 3.15, either $u_3 = \pm 1$ or $v_3 = \pm 1$. However, by the treatment of the case $a = 12$ in the latter part of the proof of Lemma 3.17, we see that v_3 cannot equal ± 1. Thus $u_3 = \pm 1$. Then $u_6 = u_3 v_3 = \pm v_3$. We will show that $|v_3| \geq 3$ and $\overline{\omega}(v_3) = 3$. Since $u_3 = r^2 + s$, it will then follow from Lemma 3.18 that condition (xii) holds.

We note that $v_3 \neq 0$ since $v(r, s)$ is nondegenerate. Suppose that

$$0 < |v_3| = |r(r^2 + 3s)| \leq 2.$$

Then $r = \pm 1$ or $r = \pm 2$. Substituting these values into $s = 1 - r^2$, and noting that $s \neq 0$, we obtain a value for $|v_3| = |r(r^2 + 3s)| > 3$ in each case.

We now claim that $\overline{\omega}(v_3) = 3$. If $\overline{\omega}(v_3) < 3$, then by Lemmas 3.3 and 3.12, $\overline{\omega}(v_3) = 1$ and $|v_1| = |v_3|$. If $|v_1| = |v_3|$, then, since $r \neq 0$, $r^2 + 3s = \pm 1$. Since, also, $r^2 + s = \pm 1$, we have $s = 0$ or ± 1; but $s \neq 0$ by Theorem 2.12 (i), and $r^2 + s = \pm 1$ is not possible for an integer $r \neq 0$ if $s = \pm 1$. Hence, $|v_1| \neq |v_3|$, and it follows that condition (xii) holds.

We next suppose that $a = 8$. By Lemma 3.15, $u_4 = \pm 1$ or $v_4 = \pm 1$. First suppose that

$$u_4 = u_2 v_2 = r(r^2 + 2s) = \pm 1.$$

Then $r = \pm 1$ and $r^2 + 2s = 2s + 1 = \pm 1$. If $2s + 1 = 1$, then $s = 0$, which is excluded since $u(r, s)$ is nondegenerate. If $2s + 1 = -1$, then $r = \pm 1$ and $s = -1$, which is also excluded by Lemma 2.12 (iv) since $u(r, s)$ is nondegenerate. Now assume that $v_4 = \pm 1$. Then $u_8 = u_4 v_4 = \pm u_4$, and u_8 has no primitive prime divisor. By Table 1 of [1], the only recurrences $u(r, s)$ for which u_8 has no primitive prime divisor are $u(\pm 2, -7)$ and $u(\pm 1, -2)$. By inspection, we see that $v_4(\pm 2, -7) = 2$, while $v_4(\pm 1, -2) = 1$. We further observe that if $r = \pm 1$ and $s = -2$, then

$$u_8 = u_4 = \pm v_2 = \pm 3,$$

while

$$u_2 = v_1 = \pm 1.$$

It now follows from Lemma 3.18 that condition (xiv) holds.

We now assume that $a \geq 5$ and a is odd. By Lemmas 3.15 and 3.17, it follows $a = 5$, 7, or 13 and $u_a = \pm 1$. Then u_a has no primitive prime divisor. By examining Table 1 of [1] and evaluating the term u_a in all recurrences $u(r, s)$ for which u_a has no primitive prime divisor, we see that u_a is Lucasian if and only if one of the conditions (viii), (ix), (x), (xi), (xiii), or (xix) holds.

Finally, we suppose that $a \geq 10$ and a is even. By Lemmas 3.15 and 3.17, we see that $a = 10$, 14, or 26 and either $v_{a/2} = \pm 1$ or $u_{a/2} = \pm 1$. If $v_{a/2} = \pm 1$, then $u_a = u_{a/2}v_{a/2} = \pm u_{a/2}$, and u_a has no primitive prime divisor. By Theorem 2.10, u_a has a primitive prime divisor if $a = 14$ or 26. Thus, $v_{a/2}$ can equal ± 1 only if $a = 10$. From Table 1 or [1], we see that $u_{10}(r, s)$ has no primitive prime divisor only if $r = \pm 2$, $s = -3$, $r = \pm 5, s = -7$, or $r = \pm 5$, $s = -18$. In each case, we observe that $v_5(r, s) \neq \pm 1$. Hence, we must have that $u_{a/2} = \pm 1$. Since $a/2 \geq 5$ and $a/2$ is odd, we obtain all the values of r and s for which $u_{a/2} = \pm 1$ from conditions (viii) - (xi), (xiii), and (xix). By inspection of these recurrences $u(r, s)$ and the corresponding recurrences $v(r, s)$, we obtain the values of $u_a(r, s)$ and ascertain that $|u_a(r, s)| \geq 3$ and $\overline{\omega}(v_{a/2}(r, s)) = a/2$. It follows from Lemma 3.18 that conditions (xv) - (xviii), (xx), and (xxi) hold. The results now follow. □

<u>**Proof of Theorem 2.8:**</u> By Theorem 2.4, if $|v_a| \geq 3$, then $v_a | u_b$ if and only if $2a|b$. Thus, we can asume that $|v_a| = 1$ or 2. First suppose that $v_a = \pm 1$. Then $u_{2a} = \pm u_a$, and u_{2a} is Lucasian if u_a is Lucasian. Using this observation, the proof of Theorem 2.6 determines all instances in which $v_a = \pm 1$ in the course of finding all terms u_{2a} which are Lucasian. It follows from conditions (ii), (vii), and (xiv) of Theorem 2.6 that $v_a = \pm 1$ if and only if conditions (ii), (iii), or (v) of Theorem 2.8 hold.

We now assume that $v_a = \pm 2$. If $a = 1$ and $v_1 = r = \pm 2$, it follows from Lemma 3.5 (iii) that $v_a | u_b$ if and only if $2a|b$. Now suppose that $a = 2$ and $v_2 = r^2 + 2s = \pm 2$. It follows that $v_2 = \pm 2$ if and only if $s = (\pm 2 - r^2)/2$ and r is even. Lemma 3.5 (iii) now implies that $v_2 | u_b$ if and only if condition (iv) of Theorem 2.8 holds.

Next assume that $a = 3$ and $v_3 = r(r^2 + 3s) = \pm 2$. Then $|r| = 1$ or 2. If $r = \pm 1$, then

$$r^2 + 3s = 3s + 1 = \pm 2.$$

Then $3s = 1$, which is impossible, or $3s = -3$, which yields $r = \pm 1$, $s = -1$. However, this case is excluded by Theorem 2.12 (iv), since $v(r, s)$ is nondegenerate.

Finally, assume that $a \geq 4$. By Lemma 3.5, we see that $\omega(2) \leq 3$. Hence $u_{2a} = u_a v_a$ has no primitive prime divisor. From Table 1 of [1], we find that there are exactly 10 Lucas sequences $u(r, s)$ for which some term u_{2n} has no primitive prime divisor for $n \geq 4$. By examining each of these recurrences, we see that $v_a = \pm 2$ for $a \geq 4$ if and only if r and s have the values given in conditions (vi) and (vii) of Theorem 2.8. We note that by Lemma 3.5 (iii), if $|r| = 2$ then $2|u_n$ if and only if $2|n$. The result now follows. □

<u>ACKNOWLEDGMENT</u>

The author wishes to thank the referee for his or her careful reading of the manuscript and suggestions which improved this paper.

REFERENCES

[1] Bilu, Yu, Hanrot, G. and Voutier, P.M. "Existence of Primitive Divisors of Lucas and Lehmer Numbers." *J. Reine Angew. Math.*, Vol. *539* (2001): pp. 75-122.

[2] Carmichael, R.D. "On the Numerical Factors of the Arithmetic Forms $\alpha^n \pm \beta^n$." *Ann. Math.*, (Second Series), Vol. *15* (1913): pp. 30-70.

[3] Hilton, P. and Pedersen, J. "Fibonacci and Lucas Numbers in Teaching and Research." *Journées Mathématiques & Informatique*, Vol. *3* (1991-1992): pp. 36-57.

[4] Hilton, P., Pedersen, J. and Somer, L. "On Lucasian Numbers." *The Fibonacci Quarterly*, Vol. *35.1* (1997): pp. 43-47.

[5] Lucas, E. "Théorie des Fonctions Numériques Simplement Périodiques." *American J. Math*, Vol. *1* (1878): pp. 184-220, 289-321.

[6] McDaniel, W.L. "The *G.C.D.* in Lucas Sequences and Lehmer Number Sequences." *The Fibonacci Quarterly*, Vol. *29.1* (1991): pp. 24-29.

[7] Somer, L. "Divisibility of Terms in Lucas Sequences of the Second Kind by their Subscripts." Applications of Fibonacci Numbers. Volume **6**. Edited by G.E. Bergum, A.N. Philippou, and A.F. Horadam. Dordrecht; Kluwer (1996), pp. 473-486.

[8] Somer, L. Solution to Problem B-825 Proposed by L. Somer. *The Fibonacci Quarterly*, Vol. *35.4* (1997): p. 376.

[9] Ward, M. "Prime Divisors of Second Order Recurring Sequences." *Duke Math J.*, Vol. *21* (1954): pp. 607-614.

AMS Classification Numbers: 11B39

SOME CONSTRUCTIONS AND THEOREMS IN GOLDPOINT GEOMETRY

John C. Turner

1. INTRODUCTION

In [1] Turner introduced a notion which is called *Goldpoint Geometry*. It consists of the study of geometric figures into which golden-mean points have been constructed or introduced. Such points he defined to be 'goldpoints'.

In fact, goldpoint geometry began with a Christmas puzzle. In late 1996, Turner sent the Atanassov family a Christmas card, on which he had drawn a diagram which he called the Fibonacci Goldpoint Star, and had set a puzzle about its crossing points for them to attempt. The puzzle was to determine, out of all the 58 crossing points in the Star, how many of them were goldpoints relative to other pairs of points in the Star. A picture of this Star, with goldpoints indicated by solid dots, may be seen on the front cover of the Proceedings of 'The Eighth International Research Conference on Fibonacci Numbers and Their Applications', pub. Kluwer, 1999.

We shall now define 'goldpoints', and in the subsequent sections we shall present many simple results and examples of studies in goldpoint geometry.

Definition of goldpoints:

In general, P is an interior goldpoint with respect to a line-segment AB if P is an interior golden-mean of the line-segment. There are two candidates for the position of a goldpoint in AB.

There are also two exterior goldpoints relative to AB.

We can achieve a simple definition of all four goldpoints if we assign a sense ($\pm$) to segments in the line of AB, according as they are traversed (or described) in the direction

This paper is in final form and no version of it will be submitted for publication elsewhere.

F.T. Howard (ed.), Applications of Fibonacci Numbers, Volume 9: Proceedings of the Tenth International Research Conference on Fibonacci Numbers and their Applications, 235–245.

$A \longrightarrow B(+)$, or in the direction $B \longrightarrow A(-)$.

Definition:

(i) If AB is a line-segment, and P is a point in the line of AB such that $|AP:PB|$ equals α or $1/\alpha$ (where $\alpha = (1+\sqrt{5})/2$) then P is a *goldpoint* with respect to AB.

(ii) A goldpoint is an *interior goldpoint* if AP/PB is positive, and an *exterior goldpoint* if AP/PB is negative. (AP and PB are to be given senses ($\pm$) as described in the paragraph above.) If $|AP:PB|$ equals α^i or $1/\alpha^i$, we shall call P an ith*-order goldpoint* with respect to AB.

When calculating goldpoint coordinates, or when checking to see whether a given point is a goldpoint, the following lemma is often most useful.

Lemma 1: The interior goldpoints with respect to two points $\underline{A}$ and $\underline{B}$ are $\underline{G} = 1/\alpha\underline{A} + 1/\alpha^2\underline{B}$ and $\underline{H} = 1/\alpha^2\underline{A} + 1/\alpha\underline{B}$.

Proof: This follows from the fact that the weights $1/\alpha$ and $1/\alpha^2$ sum to 1, and form ratios $\alpha:1$ or $1:\alpha$. In calculations, use is often made of the identities $1/\alpha = \alpha - 1$ and $1/\alpha^2 = 2 - \alpha$. □

Whenever we are studying only ratios, we may assume the length of the line-segment AB to be $|AB| = 1$. Then $|AG| = 1/\alpha^2$, $|GH| = 1/\alpha^3$ and $|HB| = 1/\alpha^2$. The situation is shown in Figure 1 below.

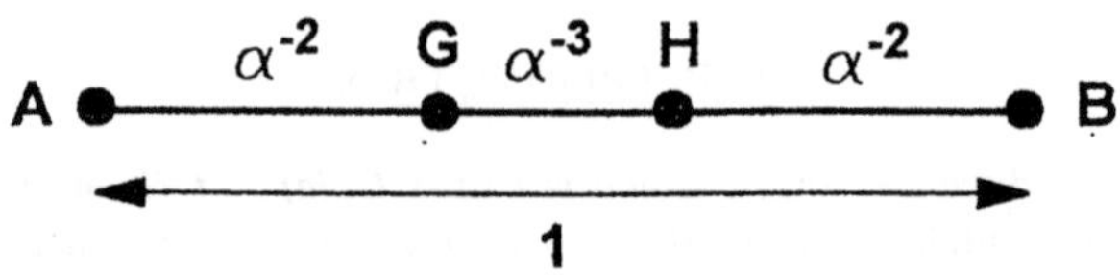

Figure 1. Unit segment AB, and its interior goldpoints

2. CROSS-RATIOS INVOLVING GOLDPOINTS

The following lemma is about cross-ratios in AB, with respect to its goldpoints. It follows directly from formulae given in [4]. A second lemma is given below it, which links goldpoints with harmonic ratios. These lemmas indicate that interesting results in goldpoint geometry may be found as special cases of results in cross-ratio geometry. We do not follow this direction in this paper, but intend to do so later.

Lemma 2: The six possible cross-ratios from $AGHB$ are

$$\pm\alpha^{\pm 1} \text{ and } \alpha^{\pm 2}.$$

Proof: The cross-ratio (AG, HB) is:

$$\frac{AH}{AB} : \frac{GH}{GB} = \frac{1/\alpha}{1} \times \frac{1/\alpha}{1/\alpha^3} = \alpha = +\alpha^{+1}.$$

Then (see [4]) the other possible cross-ratio values are $1/\alpha$, $1/(1-\alpha)$, $(1-\alpha)$, $(\alpha - 1)/\alpha$, $\alpha/(\alpha-1)$, which yield the values given in the lemma.

The next lemma concerns two points placed in AB, which, together with A and B form an harmonic range. The first point is a goldpoint of AB.

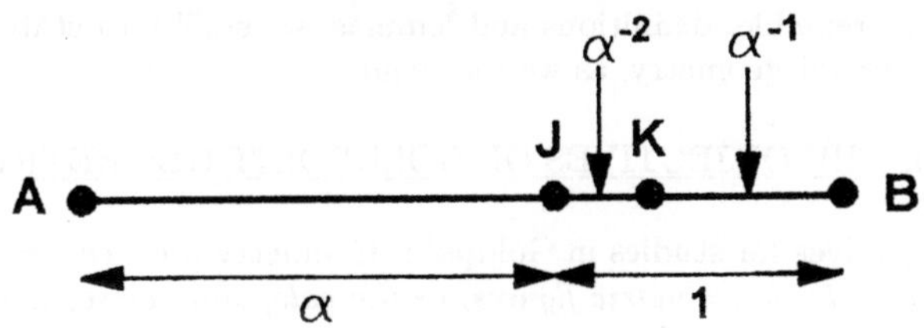

Figure 2. Diagram for Lemma 3

Lemma 3: Let J, K be defined with respect to A, B as follows: J, K are points internal to the segment AB such that:

$$\frac{AJ}{JB} = 1\alpha, \text{ and } \frac{AK}{KB} = 2\alpha.$$

Then
(i) A, J, K, B is a harmonic range [4];
(ii) J is a goldpoint of AB;
(iii) K is a goldpoint of JB;
(iv) J is a 3rd-order goldpoint of AK.

Proof: We may set $|JB| = 1$, and then the diagram of the range is as shown above, since:

$$\begin{aligned}
AJ/JB = \alpha &\Rightarrow AJ = \alpha;\\
\text{and } \; AK/KB &= 2\alpha\\
\Rightarrow AJ + JK &= 2\alpha(JB - JK)\\
\alpha + JK &= 2\alpha - 2\alpha JK\\
\Rightarrow JK &= 1/\alpha^2;\\
\Rightarrow KB &= JB - JK = 1/\alpha;\\
\Rightarrow AK &= \alpha + 1/\alpha^2 = 2,\\
\text{and so } \; AB &= 2 + 1/\alpha = \alpha^2.
\end{aligned}$$

Then for (i), the cross-ratio is:

$$(KA, BJ) = \frac{KB}{KJ} \times \frac{AJ}{AB} = \frac{1/\alpha}{-1/\alpha^2} \times \frac{\alpha}{\alpha^2} = -1,$$

(The other two cross-ratio values are 2 and 1/2.) So, by definition, the range is harmonic [4].

Results (ii), (iii) and (iv) follow directly from the diagram measures. □

In view of Lemma 3, we could call (J, K) a *golden-harmonic pair* of points with respect to the segment AB.

After the above preamble, definitions and lemmas, we shall now state and discuss briefly the objectives of goldpoint geometry, as we see them.

3. THE OBJECTIVES OF GOLDPOINT GEOMETRY

The general objectives for studies in Goldpoint Geometry are: *The discovery and analysis of properties of goldpoints in geometric figures, as the goldpoints arise, naturally or by design, in geometric constructions.*

The occurrences of the goldpoints are either by construction, or else by direct introduction of such points into the figures. Further constructions may then be made by which additional goldpoints arise, thereby extending the scope and interest of the associated figures and analyses. Many examples of these procedures are given in this Section.

By contrast, we point out that so-called Fibonacci mathematics is often concerned with study of pure mathematical or natural objects and processes in which discoveries of golden-means or Fibonacci numbers, and identities involving them, bring surprise, delight and new insights into the objects and processes. The goldpoints occur naturally, not by introduction. Many books have been written which detail studies of this nature, across sundry domains. The book by Huntley [2] is a good example. The web-page maintained by Knott [3] may be consulted for many more examples.

The next three sections present examples of results that we have obtained from our studies in Goldpoint Geometry. The first is no doubt just a variation on a well-studied theme - there surely have been similar constructions known since Hellenic times - but we believe our presentation may be new.

The remaining two sections have novelty, charm and interest in this topic.

4. CONSTRUCTIONS OF LINE-SEGMENTS WITH LENGTHS IN POWERS OF α

The first result concerns the construction of a segment of length α, within the hypotenuse of a $(90°, 60°, 30°)$ triangle. The algorithm is described in some detail, but it is rapidly carried out. Several developments from this construction are later described.

Constructing a segment of length α:

The construction of a segment of length α, as shown in Fig. 3 below, is carried out entirely within the well-known $(90°, 60°, 30°)$ triangle of Mechanics problems fame. It also uses the even more famous $(90°, 45°, 45°)$ triangle, whose hypotenuse wrought havoc to the religious beliefs of the Pythagorean School, some 2500 years ago. We shall use the usual triples of sides to denote these two triangles thus: $T_1 \equiv (1, 2, \sqrt{3})$ and $T_2 \equiv (1, 1, \sqrt{2})$.

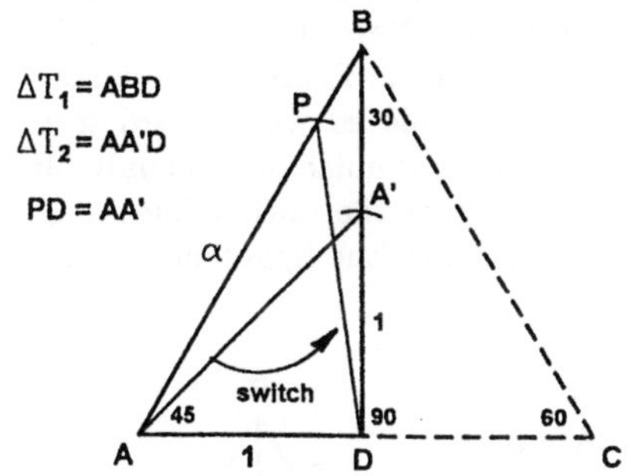

Figure 3. Construction of an α-segment

The algorithm to construct the segment can be described in three lines, and proved in one or two, as we shall see. Its description is:

Embed T_2 in T_1; 'switch' the hypotenuse of T_2 to span from the common right-angle onto the diagonal of T_1. A segment of length α then appears on the latter diagonal!

The entire construction is carried out with straight-edge and compass, as shown in Figure 3. A final swing of the compass can create a goldpoint in a segment of unit length, if such is desired.

The usual method given for constructing a goldpoint in a unit segment is described in [2, page 27] and [3], as it is in all similar texts.

The elegance of our 'new' construction lies in its brevity and wit (if a hypotenuse switch can be so described), and the fact that everything occurs within or between two famous right-angled triangles. The following theorem establishes that we have indeed constructed a segment of length α.

Theorem: $AP = \alpha = (1+\sqrt{5})/2$.

Proof: Let $x = AP$, and consider triangle APD. By the cosine rule, and since $PD = \sqrt{2}, x^2 = PD^2 - AD^2 + 2AP.AD\cos 60° = 1 + x$. Hence $x^2 - x - 1 = 0$. Clearly $x > 1$, hence $x = \alpha$. □

Let us denote the final triangle APD by $T_3(1, \alpha, \sqrt{2})$. A number of interesting consequences follow easily.

Theorem:

(i) The new triangle T_3 lies between T_1 and T_2 in area, with

$$T_2 < T_3 < T_1.$$

(ii) Their areas are in proportions 1: $\sqrt{3}\alpha/2 : \sqrt{3}$.

(iii) $AP/PB = \alpha^3$. Hence P is a 3rd-order goldpoint of AB.

Proof: (i) and (ii) both follow from the fact that all three triangles are on the same base $AD = 1$, and have heights 1, α sin 60° and $\sqrt{3}$ respectively.

(iii) follows from the fact that $PB = 2 - \alpha = \alpha^{-2}$. □

Further constructions within $T_1(1, 2, \sqrt{3})$:

If we use the compasses to mark off just four more points in T_1, we can construct several more segments involving powers of α. Now that the basic idea has been explained with reference to Figure 3, we shall give a second figure which contains all six constructed points; and then we shall list what has been achieved with regard to the six points. Proofs are all simple calculations from triangle and α formulae, and so only hints at them are given.

The constructions will be clear from the dotted lines and small arcs. The points are produced in the following order: A', A'', P (the hypotenuse switch), B', P'.

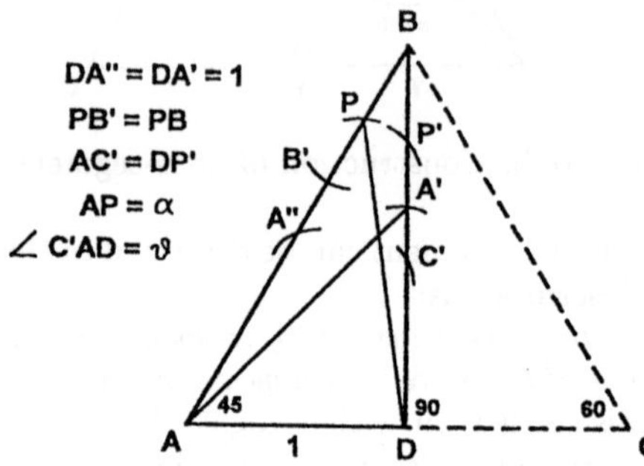

Figure 4. Construction of points and segments in T_1

The following constructions have been achieved:

$$DA' = 1 \quad = DA''$$
$$AP = \alpha \quad = AP', \text{ (proved above)}$$
$$A''P = \alpha^{-1} \quad (AA'' = 1, \text{ since } ADA'' \text{ is equilateral})$$
$$PB = \alpha^{-2} \quad (= 2 - \alpha)$$
$$A''B' = \alpha^{-3} \quad (= \alpha^{-1} - \alpha^{-2})$$

$$DP' = \alpha^{1/2} \quad (= \sqrt{(\alpha^2 - 1)})$$

$$\sin\theta = \alpha^{-1/2} \text{(see below, for segment } C'D)$$

$$P'C' = \alpha^{-3/2} (= \alpha^{1/2} - \alpha^{-1/2}).$$

We can obtain a segment of length $\alpha^{-1/2}$ by setting the compass at $DP' = \sqrt{\alpha}$ and marking a hypotenuse of that length from A to a new point C' on BD, forming another right-triangle ADC'. Then the required segment is $C'D = \sqrt{(\alpha - 1)} = \alpha^{-1/2}$.

By the above we have also achieved, and differently from normal textbook presentations [e.g. [2]], the construction of a goldpoint P in a unit segment. For $A''B$ is a unit segment, and P is an interior goldpoint of it. We observe also that P is an exterior goldpoint with respect to the unit segment $A''A$.

We note that we could easily have constructed a goldpoint of the original unit segment AD, by placing compass on A, radius AP, and marking P'' on AD produced. Then with compass on D and radius DP'', mark point Q between A and D, and that will be the required goldpoint.

To summarize, beginning with the famous mechanics triangle, and making just four or five swings of the compass, we have constructed within the triangle a large number of goldpoints and powers of α, and observed relationships between three fundamental triangles.

5. ON A SQUARE WITH GOLDPOINTS, AND A RELATED STAR

This example of goldpoint geometry begins with a square, and marks two points on each of two adjacent edges. Further constructions are made, which determine two other points; the somewhat surprising conclusion is that these two are goldpoints with respect to a constructed segment if the first pairs of marks are first- or second-order goldpoints within the edges of the square.

Theorem: Let $ABCD$ be a square.

Construction: In the side AB mark two distinct points R, S, symmetrically placed with respect to AB. Similarly, in the side BC mark two distinct points T, U symmetrically placed with respect to BC. Draw DR and CS, and produce them to meet at P. Draw AT and DU, and produce them to meet at Q. Join P to Q.

Let the line PQ cut the lines AB, BC in points V, W respectively.

Propositions:

Case (i): If R, S and T, U are goldpoints of AB, BC respectively, then V, W are goldpoints of segment PQ. (PQ cuts the square internally; and V is a third-order goldpoint of AB.)

Case (ii): If R, S and T, U are second-order goldpoints of AB, BC respectively (e.g. $AS/SB = \alpha^2$), then again V, W are goldpoints of segment PQ. (Now PQ 'cuts' the square externally; and V is an external first-order goldpoint of AB.)

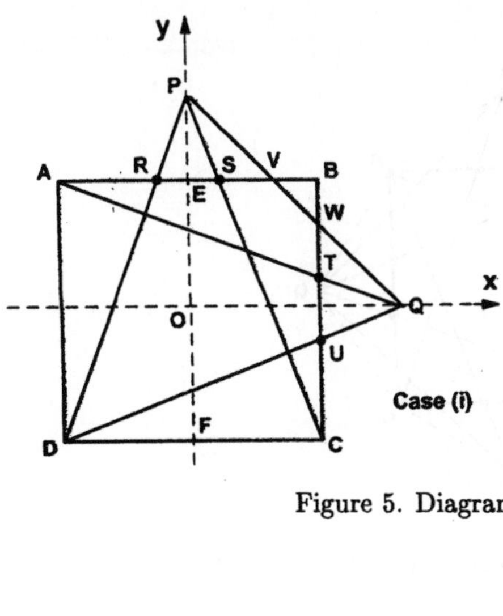

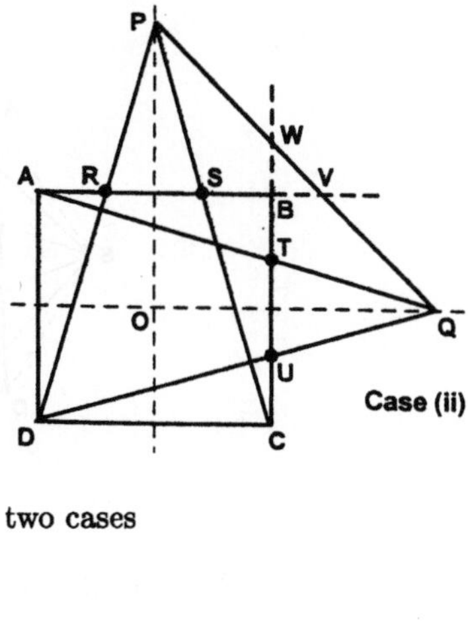

Figure 5. Diagrams for the two cases

Proof of Case (i):

We are concerned with ratios, so we can choose to let the side AB of the square be unity. And we can place Cartesian axes as shown (dotted) with origin at the centre of the square.

Since $|RS| = 1/\alpha^3$, the goldpoint S has coordinates $(1/(2\alpha^3), 1/2)$; and C has coordinates $(1/2, -1/2)$. So the line CS has gradient $(1/2 - (-1/2))/(1/(2\alpha^3) - 1/2) = -\alpha^2$.

Therefore $PF = \alpha^2 FC = (1/2)\alpha^2$.

So $PE = PF - EF = (1/2)\alpha^2 - 1 = 1/(2\alpha)$.

Now, by symmetry, PQ has gradient $-45°$, hence

$$PV = \sqrt{2}PE = 1/(\sqrt{2}\alpha), \text{ and } PW = \sqrt{2}(1/2) = 1/\sqrt{2} = VQ.$$

Hence $PV/VQ = 1/\alpha$, so V is a goldpoint of PQ.

Similarly, W is the other goldpoint of PQ.

The length of PQ is $PV + PW = (1/\alpha + 1)/\sqrt{2} = \alpha/\sqrt{2}$. It is evident that V is internal to $AB(EV = PE = 1/(2\alpha) < 1/2)$. And since $AV/VB = (1/2+1/(2\alpha))/(1/2-1/(2\alpha)) = \alpha^3$, we find that V is a third-order goldpoint of AB. □

Proof of Case (ii):

Similar Cartesian analysis establishes the claims of Case (ii).

Corollary to Case (i): Produce AB and DU to meet at K. Let the intersection of DU and CP be J. Then: $BK = \alpha$ (from $AK/AD = BK/BU$); and the circle drawn on SK as diameter passes through J, since $\angle SJK = 90°$. [N.B. we shall see in Section 6 that this circle is the B-ring, or goldpoint ring, of AB. Note that K is an exterior goldpoint of AB.]

The following figure shows what is achieved by carrying out the constructions used for Case (i) for *all* sides of the square $ABCD$.

It is evident that $PQRS$ is a square; and all its goldpoints are constructed, being the points where its sides cut the sides of $ABCD$.

The interior of the diagram displays an attractive star-like figure, with eight vertices. It is not fully regular, since $ABCD$ and $PQRS$ are not equivalent squares. Some measurements from the diagram, given that $|AB| = 1$, are: $PQ = \sqrt{2}.PO = \sqrt{2}.(\alpha/2) = \alpha/\sqrt{2}$: hence the square $PQRS$ has area $\alpha^2/2$, and diagonal α.

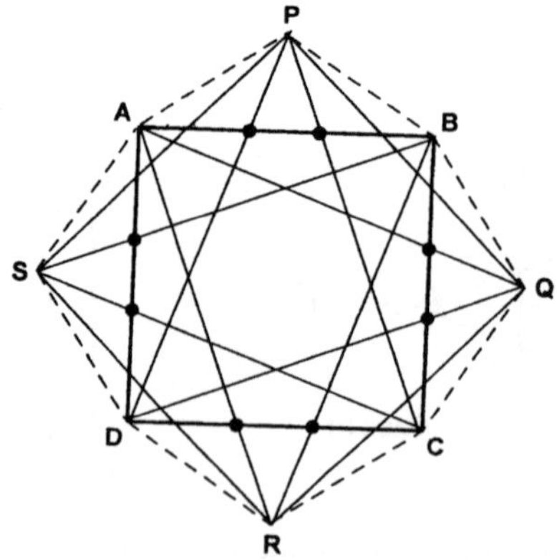

Figure 6. Square and Star constructed from a *SGP*

Further developments: We could apply the case (i) constructions directly to the new square $PQRS$ and its goldpoints, and obtain a third square. This would be rotated through 45° anti-clockwise from $PQRS$, and hence would be in similar position, and concentric with, the original square $ABCD$. Its area would be $\alpha^4/4$.

Evidently we could repeat this process again and again, indefinitely. The result would be a sequence of squares, all with centre O and alternately having sides parallel to $ABCD$ or $PQRS$. Their areas would form the sequence: $1, \alpha^2/2, \alpha^4/4, \alpha^6/8, \ldots$.

A moment's thought, too, reveals a method for constructing a sequence of squares with goldpoints which starts with $ABCD$ and whose members also rotate about O but which shrink in size from one to the next.

Another result about this Star is the following: Diagonals SB and AQ meet in a point F, say. Then lines DH and CG also meet in F (here we are using G and H to denote the goldpoints of AB.) Moreover, the vertical from centre O to P also passes through F; let this vertical bisect AB in point E; then points E and F are the goldpoints of OP.

Many more things could be said about this Star — indeed, there is an embarrassment of goldpoint riches in it — but we must leave it there.

6. THE GOLDPOINT RINGS OF A LINE SEGMENT

6.1 Definition of goldpoint rings:

In this section we introduce the notion of ***goldpoint rings with respect to a line segment***, which is a generalization of the concept of goldpoints in a line segment. We begin with a definition of goldpoint rings, and then study some of their properties. [N.B. the word 'ring' is used here to denote a special kind of circle.]

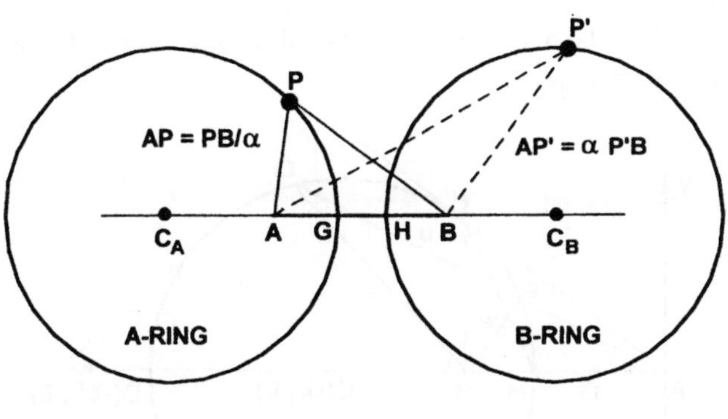

Figure 7. Segment AB with its goldpoint rings

Definition: Let AB be a line segment in a plane, and P be a point in the plane which satisfies one of the following two conditions: (i) $AB/PB = \alpha$; (ii) $AP/PB = 1/\alpha$. Then the two loci of P determined by the conditions are called *the goldpoint rings of* AB.

We prove below that one ring is a circle that contains point A (we shall call this the *A-ring of* AB; and the other is a circle containing point B (which we shall call the *B-ring of* AB). Sometimes we shall refer to them as the α-rings of AB. They are shown in Figure 7.

Note that the goldpoint rings contain the goldpoints (both interior and exterior) of AB. These are, of course, where the loci of P cuts the line of AB produced in both directions.

Note that G and H are the goldpoints of AB. The radius of each goldpoint ring is $|AB|$, as we shall prove shortly.

The locus of P:

We place AB and P in the xy-plane, with origin at A and the x-axis along AB.

Let $\rho = \alpha$ or $1/\alpha$, and $d = AB$.

We obtain P's locus as follows:

$$\begin{aligned} AP^2 &= PB^2\rho^2 \\ x^2 + y^2 &= (x-d)^2\rho^2 + y^2\rho^2 \\ x^2(1-\rho^2) + y^2(1-\rho^2) &= -2dx + d^2\rho^2 \\ x^2 + y^2 + 2dx\rho^2/(1-\rho^2) &= \rho^2/(1-\rho^2)(\rho \neq 1). \end{aligned}$$

Writing $u = \rho^2/(1-\rho^2)$, and completing the square for x we get:

$$(x+du)^2 + y^2 = d^2u(u+1).$$

Thus the locus of P is a circle, having centre $C(-du, 0)$ and radius $d\sqrt{u(u+1)}$. Each value of $\rho(\neq 1)$ determines a circle. We shall call these circles ρ_{AB}-rings. Taking our two particular values of ρ, and letting AB be the interval [0,1], we find that when:

(i) $\rho = \alpha$, the locus is the B-ring, the circle which has centre $C_B = (\alpha, 0)$ and radius 1; [n.b. $u = -\alpha$]

(ii) $\rho = 1/\alpha$, the locus is the A-ring, the circle which has centre $C_A = (-\alpha^{-1}, 0)$ and radius 1. [n.b. $u = \alpha^{-1}$]

These results confirm the diagrams of the complementary α-rings in Fig. 7.

Two theorems about B-rings:

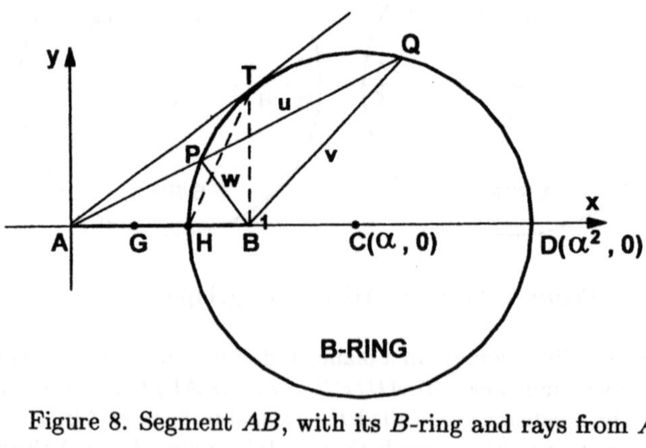

Figure 8. Segment AB, with its B-ring and rays from A

We have discovered many interesting theorems about α-rings, but two must suffice for this paper. They demonstrate properties of chords in the B-ring formed when rays from A intercept the B-ring.

Constructions for the theorem:

$AB = [0,1]$, and G, H are its interior goldpoints.

The B-ring has centre $C(\alpha, 0)$, radius 1; and it cuts the x-axis at the goldpoints $H(1/\alpha, 0)$ and $D(\alpha^2, 0)$ of AB.

APQ is any ray from A which cuts the B-ring in P, Q.

AT is a tangent to the B-ring. $PQ = u,\ QB = v,\ BP = w$.

Theorem:

$$\begin{array}{lrl} (i) & AT & = \sqrt{\alpha} \\ (ii) & TB & = \alpha^{-1/2} \text{ and } TB \perp AB \\ (iii) & HT & = 2/\alpha \\ (iv) & wv & = 1/\alpha \\ (v) & u/(v-w) & = \alpha, \text{ hence } uvw = v - w. \end{array}$$

Proof:

(i) $AT^2 = AH.AD = \alpha \Rightarrow AT = \sqrt{\alpha}$;

(ii) $AT/TB = \alpha$ therefore $TB = AT/\alpha = \alpha^{-1/2}$; the converse of Pythagoras' theorem on ΔABT shows that $TB \perp AB$;

(iii) $HT^2 = HB^2 + BT^2 = \alpha^{-4} + \alpha^{-1} = 4\alpha^{-2}$;

(iv) and (v) $AP = w\alpha$ and $AP = u = v\alpha$ since P and Q are on the B-ring. Also $AP(AP+u) = AT^2 = \alpha$. Therefore $w\alpha.v\alpha = \alpha \Rightarrow wv = 1/\alpha$ and $u = (v-w)\alpha$. Eliminating α gives $uvw = v - w$. □

Theorem: If ΔPQB is isosceles, then P is a goldpoint of AQ.

Proof: ΔPQB is isosceles if either $u = v$, or $v = w$, or $w = u$. The case $v = w$ is impossible, in view of Thm. 2(v) above.

Case: $u = v$: By Thm. 2(v), $v = \alpha(v - w) \Rightarrow v = w\alpha^2 = u$.

Now $AP = w\alpha$, since P is on the B-ring. Therefore $AP/PQ = w\alpha/u = w\alpha/w\alpha^2 = 1/\alpha$. Hence P is a goldpoint of AQ. (Note that ΔPQB is now $(u, v, w) = \alpha^{1/2}(1, 1, \alpha^{-2})$).

Case $w = u$: $AP/PQ = w\alpha/u = w\alpha/w = \alpha$, hence P is a goldpoint of AQ. (Note that $\Delta PQB = \alpha^{-2}(1, \alpha, 1)$; it is similar to the 'sharp' golden triangle.) □

ACKNOWLEDGMENT

We thank the anonymous referee for his/her careful reading of the manuscript and valuable suggestions for improving it.

REFERENCES

[1] Atanassova, V.K and Turner, J.C. "On Triangles and Squares Marked with Goldpoints — Studies of Golden Tiles." F.T. Howard (Edr.), Applications of Fibonacci Numbers. Volume **8**, Kluwer Academic Press, 1999, 11-26.

[2] Huntley, H.E. The Divine Proportion. Dover Publications Inc., 1970.

[3] Knott, R. *Web page.* 2000 et seq. http://www.mcs.surrey.ac.uk/R.Knott/Fibonacci/

[4] Milne, J.J. 1911: Cross-Ratio Geometry. Cambridge University Press, 1911.

[5] Walser, H. Der Goldene Schnitt [The Golden Section]. vdf. Hochschulverlag, an der ETH, 1996. English edition (translators P. Hilton and J. Pedersen) M.A.A., 2002.

AMS Classification Numbers: 11B37, 11A39, 10A35

SOME APPLICATIONS OF TRIANGLE TRANSFORMATIONS IN FIBONACCI GEOMETRY

John C. Turner

1. INTRODUCTION - TRIANGLE TRANSFORMATIONS

The idea for using triangles to transform line-segments and polygons is surely an ancient one, but it is believed that the approach of this paper, and many of the problems which it tackles, are new.

I have no knowledge of earlier work in the field. I became interested in the possibilities for use of triangle transforms when studying polygons in the Fibonacci honeycomb plane $x + y - z = 0$ (see [4]). I gave the symbol $\prod_0$ to this plane, and $\prod_n$ to the general (parallel) plane $x + y - z = n$. All integer vectors $(x, y, z) \in \boldsymbol{Z}^3$ lie in exactly one of these planes. All Fibonacci vectors (F_{n-1}, F_n, F_{n+1}), and their simple generalizations from sequences $a, b, a + b, \ldots, aF_{n-2} + bF_{n-1}, \ldots$, lie in $\prod_0$: and all such points lie on one and only one Fibonacci vector polygon [4].

Triangle transforms of a line-segment:

Suppose one has a fixed line segment AB, of length $|AB| = d$; and one takes any triangle which has a side of length d, and whose plane contains the line-segment. Then by translations and rotations in the plane, the triangle and the line-segment can be conjoined (jigsaw fashion) so that AB coincides with the side of the triangle which has length d. In fact, there will be two ways of doing this, as indicated in Figure 1. Note that a directive arrow has been added to AB, which in turn induces rotational directions on the conjoined triangles. In the left transform, ABC has positive (anti-clockwise) rotation, whereas in the right transform, ABC' has negative (clockwise) rotation. C and C' are the third vertices of the two transforming triangles.

This paper is in final form and no version of it will be submitted for publication elsewhere.

F.T. Howard (ed.), Applications of Fibonacci Numbers, Volume 9: Proceedings of the Tenth International Research Conference on Fibonacci Numbers and their Applications, 247–261.

Note that the triangle must remain in the plane, during the translations and rotations. As in jigsaw formations, we do not allow the triangle to be turned over in the plane.

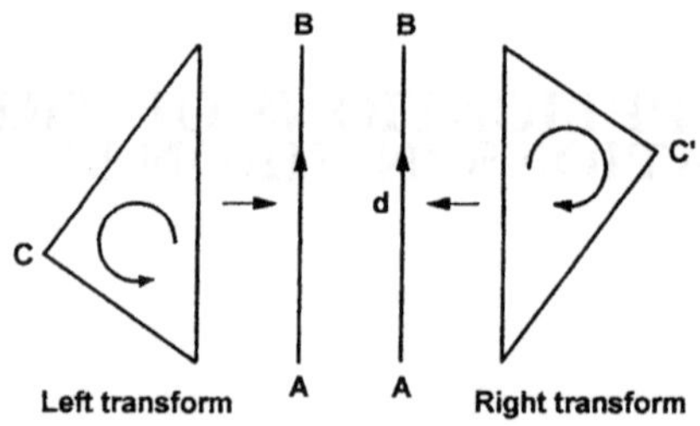

Figure 1. Left and right triangle transforms

The use of the word 'transform' may be confusing to the reader. Indeed, we may think of triangle transforms in two ways; we shall give examples of both uses in this paper. The two ways are:

(i) A triangle conjoined with a line-segment becomes an oriented triangle, fixed along the given line-segment.

(ii) A triangle conjoined with a coplanar line-segment AB, simultaneously in both possible ways, transforms the set of end-points A, B into the set of triangle vertices C, C'.

[Lemma: $AC'BC$ is a parallelogram, so CC' bisects AB.]

Before going on to give examples of triangle transforms, we define constructions of two kinds of triangle which are to be used in the applications. Later we shall use triangle transforms to demonstrate an infinity of proofs of Pythagoras' theorem and then study equilateral triangle transforms in the honeycomb plane. Finally a solution of a related Diophantine equation will be discovered and explained.

2. CONSTRUCTIONS OF SPECIAL TRIANGLES

The golden mean triangle T_α:

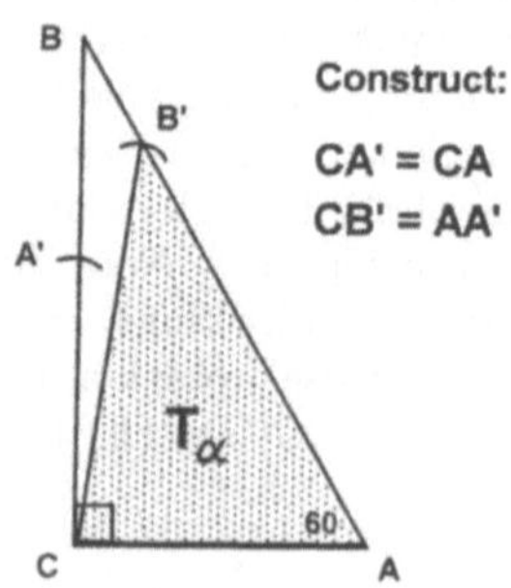

Figure 2. Constructing a T_α triangle

Figure 2 shows how a triangle with sides proportional to $(1, \sqrt{2}, \alpha)$, where α is the golden mean, may be drawn within a $(90°, 60°, 30°)$ triangle. We designate this triangle by T_α; its properties will be proved later, in Section 3. For now we merely record the method of its construction.

Equilateral triangles in $\prod_0$:

When studying triangles in the honeycomb plane [4], we observed that for any two integer vector points A, B in the plane, we could find two other integer vector points C and C' such that triangles ABC and ABC' were both equilateral.

In order to prove this rather surprising observation, we need only give the general formulae for finding the points C and C', and show that they do indeed give two equilateral triangles.

Figure 3 below demonstrates these triangles relative to the general points $A(a_1, a_2, a_1+a_2)$ and $B(b_1, b_2, b_1+b_2)$ where all six coordinates are integers. It is simple to check that the lengths of the sides $|A - B|, \ldots, |C' - B|$ are equal.

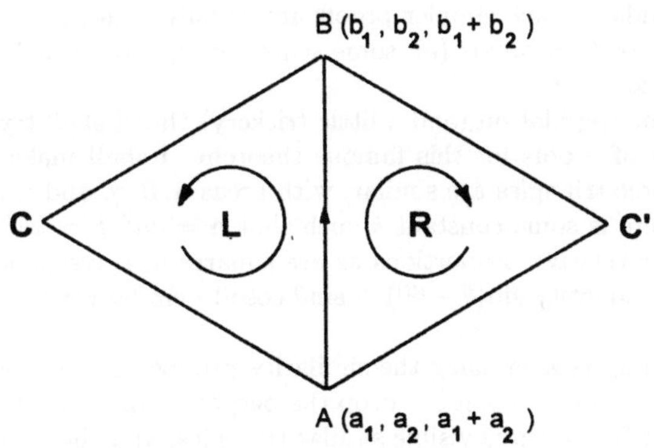

$$C = (a_1 + a_2 - b_2, b_1 + b_2 - a_1, a_2 + b_1)$$
$$C' = (b_1 + b_2 - a_2, a_1 + a_2 - b_1, a_1 + b_2)$$

Figure 3. General equilateral triangles in $\prod_0$

One example towards showing that all six sides of the two triangles are equal will suffice:

$$A - B = (a_1 - b_1, a_2 - b_2, a_1 + a_2 - b_1 - b_2)$$
$$C - A = (a_2 - b_2, b_1 + b_2 - a_1 - a_2, b_1 - a_1)$$

Since the sums of squares of coordinates of each of these vectors is equal, we have $|AB| = |CA|$. All pairs of sides in the triangles may be compared, and found equal in this way.

And so, when working on line-segments with integer end points in $\prod_0$ we can always compute their equilateral triangle transforms in $\mathbf{Z}^3$; we shall refer to these briefly as their ET-transforms. We give examples of their use in Sections 4 and 5.

3. AN INFINITY OF PROOFS OF PYTHAGORAS' THEOREM

A book called *The Pythagorean Proposition* was written by F. Loomis, and first published [1] in 1947. The book listed 367 proofs of the proposition that if a, b, c are the lengths of the sides of a right-angled triangle, with c being the hypotenuse length, then $a^2 + b^2 = c^2$. I have not seen this book, so I do not know how the work that now follows relates to it: if I verify my claim in the Section title, then clearly I am offering something new. Moreover, I introduce the golden mean into my own proof, albeit somewhat obliquely.

It is conjectured that Pythagoras himself (or some member of his Brotherhood) supplied a proof to this Propostion, circa 550 B.C. The proof that is given in Euclid's *Elements of Geometry* (c. 300 B.C.) is reputed to have been discovered by Euclid himself; it appears at the end of Book I (for Book read Chapter) as Proposition 47, with its converse being given as Proposition 48. The theorem is, of course, the most well-known in all of mathematics; and in one or other of its many forms it is a corner-stone of many mathematical subjects. Countless students have wrestled over the past two millennia to grasp Euclid's proof; many (if not most) failed to do so! Nowadays much simpler proofs are usually presented to them.

It is one of the great theorems (on some scales the greatest) of Mathematics. It is well-discussed in [2], for example.

So it is with some trepidation (and a little trickery) that I shall try to make good a claim to supply an infinity of proofs for this famous theorem. I shall make use of the well-known propositon that if three triangles are similar, with areas α, β, γ, and with three corresponding sides, a, b, c, then there is some constant k such that $\alpha = ka^2, \beta = kb^2, \gamma = kc^2$. In brief, the areas are in the same relative proportions as are squares of corresponding sides. I shall also use the trigonometric identity $\sin(\theta - 60) \equiv \sin\theta \cos 60$ - $\sin 60 \cos \theta$.

Notes:

(i) There is nothing new in using the similarity proposition. Indeed, the simplest known proof of Pythagoras' theorem uses it ... drop the perpendicular from the right-angle vertex to the hypotenuse, identify the three visible similar triangles, and the proof follows immediately.

(ii) Both the similarity proposition and the trigonometric identity can be established without appealing to Pythagoras' proposition itself. And so they may be used in a proof of Pythagoras' proposition.

My proof of Pythagoras' Theorem:

In the general part of the proof to be described, three transforming triangles on the sides of ΔABC are of type T_α. The method for constructing them was given in Figure 2, Section 2.

The strategy for effecting the proof is to partition the range of the base angle of ΔABC into three mutually exclusive subranges, and to prove the Pythagoras proposition for each of these three cases.

Theorem: Let ΔABC be a general right-angled triangle having base angle θ and sides a, b, c, as shown in Figure 4(i). Then $a^2 + b^2 = c^2$.

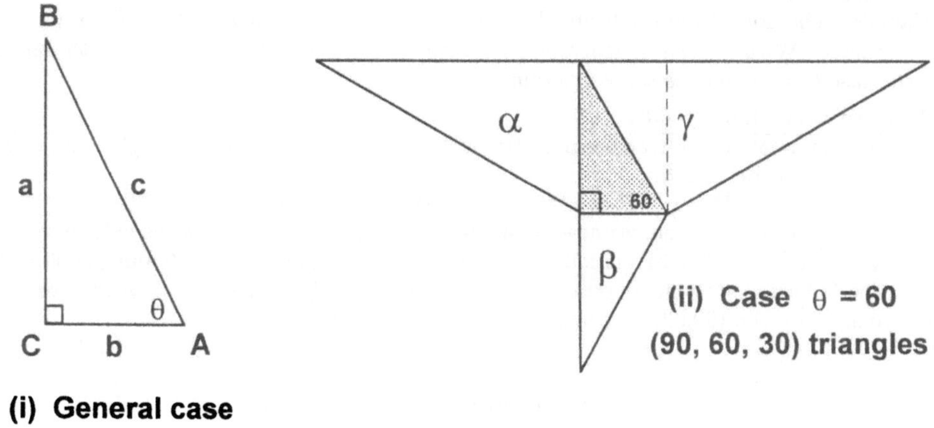

Figure 4. (i) the general triangle; (ii) proof of case $\theta = 60°$

Constructions: Fig. 4(i) shows the general right-triangle, having sides a, b, c; Fig. 4(ii) shows the triangle when $\theta = 60°$, together with transforms of the sides by suitably chosen (90,60,30) triangles.

Proof: Partition the range of θ into the three possible cases

$$(i)\ \theta = 60°;\quad (ii)\ \theta > 60°;\quad (iii)\ \theta < 60°.$$

We shall demonstrate the theorem for each of these cases, and hence effect a proof.

Proof of case (i), $\theta = 60°$:

The figure 4(ii) is sufficient to prove this case. We shall leave it there, as a beautiful visual proof of its case. The reader will note mentally, by observing congruent triangles, how $\alpha + \beta = \gamma$ and a, b, c are corresponding sides in the three transforming (90, 60, 30)-triangles.

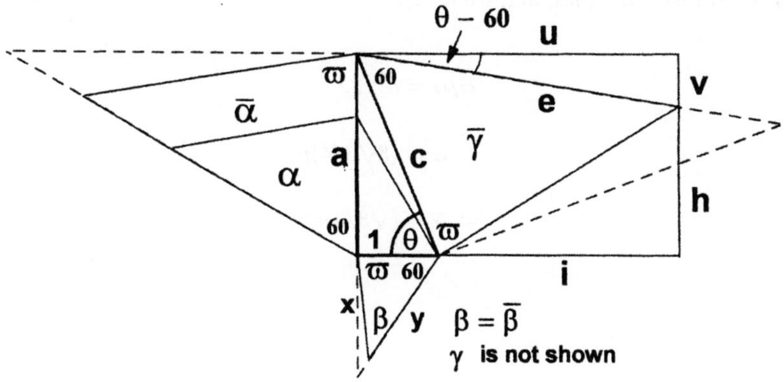

Figure 5. Case (ii), with T_α triangle transforms

Constructions of case (ii), $\theta > 60°$:

Figure 5 shows the general case, and the case $\theta = 60°$, both with sides transformed by T_α triangles. The dotted lines indicate the (90,60,30) triangles from which the T_α triangles were constructed. Without loss of generality, we have taken $|CA| = 1$. Also shown with faint lines, is the case $\theta = 60°$ on the same diagram.

Proof of case (ii), $\theta > 60°$:

Let the areas of the transforming triangles for the $\theta = 60°$ case be α, β, γ and those for the general case be $\overline{\alpha}, \overline{\beta}, \overline{\gamma}$.

We first prove that $\alpha + \beta = \gamma$ with the T_α transforms and with $\theta = 60°$.

Since case (i) is proven, we now know that in the (90,60,30) triangle the sides have lengths $(1, 2, \sqrt{2^2 - 1^2}) = (1, 2, \sqrt{3})$. [It follows that we also know sin 60 $= \sqrt{3}/2$ and cos 60 $= 1/2$.]

The following statements establish that $\alpha + \beta = \gamma$, where these are now the areas of the T_α triangles on the (90,60,30) triangle, as shown dotted in Fig. 5:

$$\begin{aligned} \beta &= (1/2).1.y \sin 60 = (\sqrt{3}/4)y \\ \alpha/\beta &= (\sqrt{3})^2/1^2 \\ \Rightarrow \alpha &= 3\beta = (3\sqrt{3}/4)y \\ \gamma/\beta &= 2^2/1^2 \\ \Rightarrow \gamma &= 4\beta = \sqrt{3}y. \end{aligned} \tag{1}$$

Now, turning to Fig. 5 for the general case ($\theta > 60°$), we obtain formulae for each of the triangle-areas $\overline{\alpha}$ and $\overline{\gamma}$ in terms of side a, and then show that $\overline{\alpha} + \overline{\beta} = \overline{\gamma}$. The proof of Case (ii) is then established by applying the property of similar triangles described above.

It should be noted that this diagram only applies if $\theta + \omega > 90°$. This evidently holds here, because both angles are greater than 60° (it is easy to prove that $\omega > 60$ from the construction diagram for T_α in Fig. 2).

Then, by similar triangles, and from (1),

$$\begin{aligned} \overline{\alpha}/\alpha &= a^2/3, \\ \alpha &= (3\sqrt{3}/4)y, \\ \Rightarrow \overline{\alpha} &= (\sqrt{3}/4)a^2 y. \end{aligned} \tag{2}$$

Now, looking at the triangle transform of the hypotenuse of ΔABC, we see that:

$$\begin{aligned}\overline{\gamma} &= \text{trapezium - two right triangles}\\ &= \frac{1}{2}(a+h)(1+i) - \frac{1}{2}a - \frac{1}{2}ih\\ &= \frac{1}{2}(ai+h) \qquad (3)\\ &= \frac{1}{2}(a(u-1)+a-v)\\ &= \frac{1}{2}(au-v).\end{aligned}$$

And we can find u and v from the upper right triangle as follows:

$$\begin{aligned}u &= e\cos(\theta-60)\\ &= yc[\cos\theta\cos 60 + \sin\theta\sin 60]\\ &= yc\left[\frac{1}{2c}+\frac{a\sqrt{3}}{2c}\right]\\ &= \frac{y}{2}(1+a\sqrt{3}) \qquad (4)\\ \text{and } v &= yc\sin(\theta-60)\\ &= \frac{y}{2}(a-\sqrt{3}).\end{aligned}$$

Putting (3) and (4) together, and using $\overline{\beta} = \beta = (\sqrt{3}/4)y$ from (1), we find:

$$\overline{\alpha} + \overline{\beta} = \overline{\gamma}.$$

The proof of Case (ii) is now completed by applying the property of similar triangles described above.

Proof of case (iii), $\theta < 60°$:

This case is rather more complicated, because differing values of θ can necessitate different kinds of diagram to be used in the proof. To avoid this problem, we note the following:

(a) If $\theta = 30$, then case (i) applies again; and it is already proved.

(b) If $\theta < 30$, then its complement in ΔABC is greater than 60, and case (ii) applies; that is already proved.

(c) If $30 < \theta < 60$, then Figure 5 changes slightly; the horizontal line through B is now *below* the upper line e of ΔABC, and the angle between these two lines is $60 - \theta$. Notice that in this range, $\theta + \omega > 90$, so the transforming triangle on side c still reaches rightwards past the vertical through A.

And exactly the same method of proof as for case (ii) can be used, except for changing $\theta - 60$ to $60 - \theta$ when calculating u and v, and using $a = h - v$ rather than $a = h + v$.

Thus we have proved case (iii) with only one or two minor changes in the diagram and equations being necessary. My proof of Pythagoras' theorem is complete (but admittedly not elegant!).

Measures of T_α:

Now that Pythagoras' theorem is proven, we can calculate the sides and angle of the triangle T_α, and see why the golden mean was mentioned earlier, before the proof began.

Looking at the β-triangle on AC in Fig. 5, and Fig. 2, we see that:

$$\begin{array}{lrl} (i) & x & = \sqrt{1^2 + 1^2} = \sqrt{2} \\ (ii) & y^2 & = x^2 - 1^2 + 2.y.\cos 60 \\ & & = y + 1. \\ & \Rightarrow y & = (1 + \sqrt{5})/2 = \alpha \\ (iii) & \sin \omega & = (\alpha \sin 60)/\sqrt{2} \end{array} \tag{5}$$

Hence the sides of the T_α-triangle are proportional to $(1, \sqrt{2}, \alpha)$, where α is the golden mean. In this round-about and somewhat sneaky way, the golden mean arises in a proof of Pythagoras' proposition.

Examination of the final lines of the proof of case (ii) shows that in fact the quantity $y = \alpha$ played a vanishing (or holding) role in the proof. In the last but one line it cancels out of the equation which gives finally $\overline{\alpha} + \overline{\beta} = \overline{\gamma}$. In a sense, the proof of the proposition is independent of y. This leads to the notion of an infinity of proofs, *which change only in the constructions used to build them upon.* It is in this spirit that I offer the following subsection.

An Infinity of Proofs:

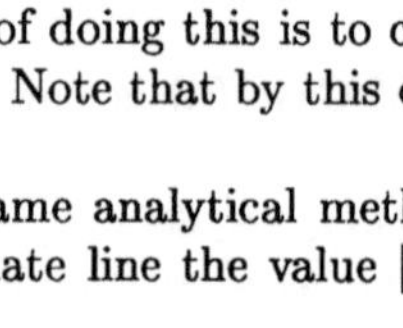

Let the constructions for the proof of the Pythagorean proposition be as in Figure 5 above, **except that** the transforming triangles shall be similar to the ΔAPB shown in the accompanying diagram.

ΔADB is equilateral, on side c; and the point P is chosen to be on line BD produced and with $PB > BD$ by some constructible fraction of c. One way of doing this is to choose a positive integer n and position P such that $|BP| = c(1 + 1/(2^n))$. Note that by this choice we ensure that $\omega > 60°$.

With each of the triangles similar to this one, the same analytical methods of proof as given above can be carried through: and in the penultimate line the value $|BP|$ (unknown) will cancel out.

Hence to each n with $n \in \boldsymbol{N}$, there corresponds a proof of the Pythagorean proposition with its own triangle transformations: hence a potentially infinite number of different proofs has been demonstrated.

Four Visual Proofs of Pythagoras' Theorem:

The above demonstration has already been admitted to be inelegant; it is dreadfully lengthy and somewhat sneaky. It has its interesting moments though; and to the author it constitutes a pleasing closure to a double, as the footnote* below explains.

Before leaving the subject, we show in Figure 6 four visual proofs of Pythagoras' theorem using triangle transformations. Three are for the special cases $\theta = 45°$, (twice), and $\theta = 60°$, and the fourth proves the general case. Check how $\alpha + \beta = \gamma$ in each case.

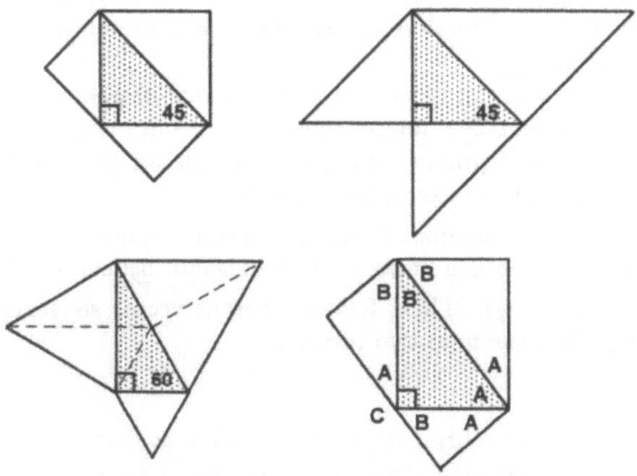

Figure 6. Four visual proofs of Pythagoras' Theorem

4. *ET*-TRANSFORMS OF FIBONACCI VECTOR POLYGONS

Consider the basic Fibonacci vector polygon [4] in $\prod_0$, drawn on the vertices

$$\ldots, \boldsymbol{F}_1(0,1,1), \boldsymbol{F}_2(1,1,2), \boldsymbol{F}_3(1,2,3), \boldsymbol{F}_4(2,3,5), \boldsymbol{F}_5(3,5,8), \ldots$$

Taking the left ET-transforms of the polygon's sides, we get:

$$E_L(\boldsymbol{F}_1\boldsymbol{F}_2) = (0,2,2)$$
$$E_L(\boldsymbol{F}_2\boldsymbol{F}_3) = (0,2,2)$$
$$E_L(\boldsymbol{F}_3\boldsymbol{F}_4) = (0,4,4)$$
$$E_L(\boldsymbol{F}_4\boldsymbol{F}_5) = (0,6,6)$$

*In 1989 a colleague and I published a book on our own methods for fully solving the other major and ancient problem about the right-triangle, that of finding all pythagorean triples [3]. My proof of the geometric proposition in this paper is therefore the second leg of a double on this most famous of equations.

It quickly becomes apparent, when a few more sides are ET-transformed, that all of the sides of the Fibonacci vector polygon are transformed onto the Y-axis; and that the points are themselves in a Fibonacci vector sequence generated by **F(a,b)** with **a** = (0,2,2) and **b**=(0,2,2).

The corresponding right ET-transforms are:

$$\ldots,(1,0,1),(2,1,3),(3,1,4),(5,2,7),\ldots$$

We note that this is another Fibonacci vector sequence, with initial vectors (1,0,1) and (2,1,3); but it is NOT a vector sequence inherent [4] to plane $\prod_0$.

If we take the left ET-transform of this new vector sequence, we find a small surprise. It does not return us to the original vector sequence, but to one which is *twice the original.* Symbolically, $E_L(E_R(\mathcal{F})) = 2\mathcal{F}$.

We shall not pursue these studies of ET-transforms of Fibonacci vector polygons any further, although it is evident that many interesting results await discovery. We conclude this subsection by stating and proving the general result for left ET-transforms of inherent Fibonacci vector sequences [4], as exemplified above.

Theorem: Let **F(a,b)** be an inherent Fibonacci vector sequence in $\prod_0$. Then the left ET-transform of its vector polygon is a linear Fibonacci vector sequence in the Y-axis.

Proof: Let **a**= $(a_1, a_2, a_1 + a_2)$. Then, for an inherent vector sequence in $\prod_0$, we have **b**= $(a_2, a_1 + a_2, a_1 + 2a_2)$. And the next two terms are:

$$\begin{aligned} \boldsymbol{c} &= (a_1 + a_2, a_1 + 2a_2, 2a_1 + 3a_2) \\ \boldsymbol{d} &= (a_1 + 2a_2, 2a_1 + 3a_2, 3a_1 + 5a_2). \end{aligned}$$

Taking left ET-transforms, we obtain:

$$\begin{aligned} E_L(\boldsymbol{ab}) &= (0, 2a_2, 2a_2) &&= 2a_2(0,1,1) \\ E_L(\boldsymbol{bc}) &= (0, 2(a_1 + a_2), 2(a_1 + a_2)) &&= 2(a_1 + a_2)(0,1,1) \\ E_L(\boldsymbol{ca}) &= (0, 2(a_1 + 2a_2 + 2a - 2), 2(a_1 + 2a_2)) &&= 2(a_1 + 2a_2)(0,1,1) \end{aligned}$$

It is evident that this is a sequence on the Y-axis, with the general term being $2(F_n a_1 + F_{n+1}a_2)(0,1,1)$. We should show this to be true by taking **c,d** as the nth and $(n+1)$st terms respectively of **F(a,b)**, and taking the ET-transform of **cd**. It is an elementary exercise to do this.

As a final comment, we remark that it is curious that we have found a way to transform all of the inherent Fibonacci vector polygons onto the same straight line (the Y-axis), when the start of all our investigations (see [3]) in this Part was to 'lift' all of the Fibonacci number sequences (by 'vectorizing') out of the number line and spread them out into the plane $\prod_0$.

5. STUDY OF TRIANGLES IN $\prod_0$ USING *ET*-TRANSFORMS

Recall that an ET-transform of a line-segment is one which uses an equilateral triangle in the transformation.

As we have seen, we can apply the ET-transform to the sides of triangles (or other polygons), and study the geometric figures that can be drawn on the resulting points. Many interesting questions can be asked about these figures, and also about the sets of vertices in them. This section gives some examples of these latter applications.

The *ET*-transform set ε of a triangle:

Suppose we are given an integer triangle ΔABC. Then from each side we can find two *ET*-transform points. We shall call a transform point an *outer* one if the perpendicular from it to its associated side lies wholly outside ΔABC. The others will be called *inner* transform points. We designate the transform points as follows: From sides AB, BC, CA the outers are respectively P, Q, R, and the inners are respectively P', Q', R'. We shall call the following set of integer points

$$\varepsilon = \{A, B, C, P, Q, R, P', Q', R'\} \text{ the } ET\text{-transform set of } \Delta ABC.$$

Now we can pose various types of problem about the set ε, for given types of integer triangle ΔABC. And we shall find that in order to answer them, we can resort to a mixture of geometric and number theoretic methods.

For example, we can ask what kinds of triangle are PQR and P', Q', R', and compare them with ΔABC.

There is one special triangle for which the answer to that is immediate: if ABC is equilateral, then PQR is also equilateral (with double the side-length) whilst P', Q', R' is an equilateral triangle which coincides with ΔCAB.

More generally, we can ask which triples of points in ε are collinear (can more than three points be collinear?); and which triples form triangles which are similar to ΔABC. The following two examples deal with these questions, for special cases of triangles.

The *ET*-transform set of (90,60,30) triangle:

The diagram shows ΔABC, a general $(90°, 60°, 30°)$, together with all its six *ET*-transform points. From the diagram we glean the following information:

$RQ'R'$ and $PQ'P'$ are collinear triples, their lines intersecting at Q', the mid-point of AC. The following eight triangles are congruent to ABC:
$AQ'R, CQ'R, P'BQ, AP'R,$
$AP'C, APR', AQ'R', CQ'R'$.

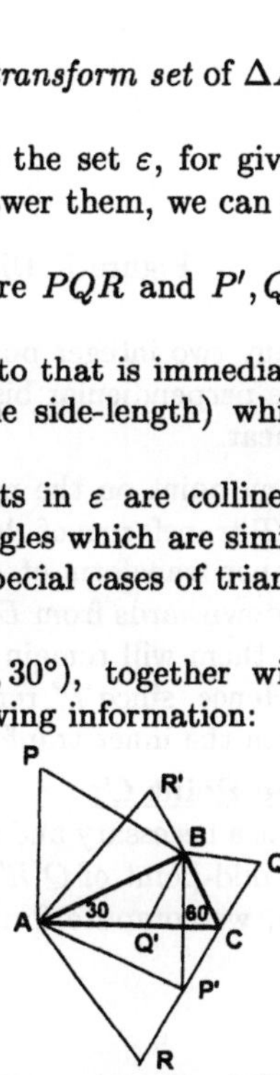

Apart from the equilateral triangles used in the *ET*-transforms, $\Delta R'QQ'$ is also equilateral, with B as its centroid.

The sets of four points A, P, R', B and A, B, C, P' each forms a cyclic quadrilateral.

All of these lines and figures can be specified generally, using the coordinates from $A(a_1, a_2, a_3)$ and $B(b_1, b_2, b_3)$.

Constructing Triangles ABC, for which P', Q', R' are Collinear:

It is possible for the three inner points in ε to be collinear, as the following right-hand diagram shows. Note that the triangle ABC is isosceles. We shall show how to find the correct point C, when any two points A, B are given. And we conjecture that only an isosceles triangle similar to that constructed can have this property.

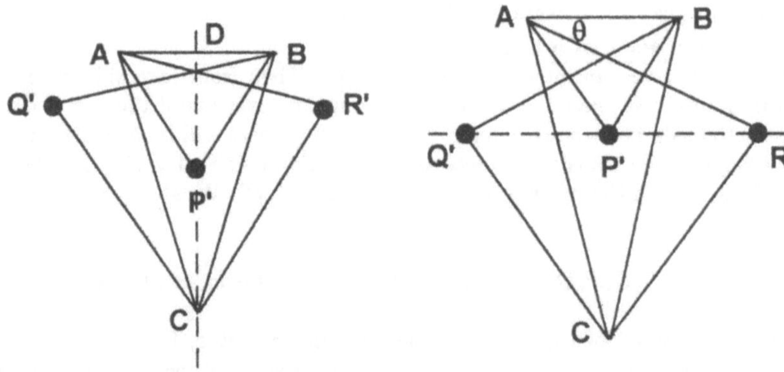

Figure 7. Diagrams for the Theorem below

__Theorem:__ Given any two integer points, say $A(a_1, a_2, a_3)$ and $B(b_1, b_2, b_3)$. Then a point C can be found on the perpendicular bisector of AB such that the three inner transform points of ΔABC are collinear.

__Proof:__ Let C be any point on the perpendicular bisector of AB, as shown in the left-hand diagram. Let the ET-transforms of AC and CB be respectively R' and Q'. And let $\angle BAR' = \angle ABQ' = \theta$. The inner transform of AB is P', and this point must lie on DC, by symmetry.

If C be moved downwards from D, angle θ will increase; points Q' and R' will move apart, but the line joining them will remain parallel to AB. And since θ will increase, this line will move downwards. Hence, since P' remains fixed, there will be one value for θ at which $Q'R'$ will contain P'. Then the inner tranform points of ΔABC will be collinear.

__A Formula for the Point C:__

It is evident that a necessary and sufficient condition for the segment $Q'R'$ to pass through P' is that P' is the mid-point of $Q'R'$.

Using this fact, we compute formulae for the coordinates of C as follows. By the ET-transform formulae,

$$\begin{aligned} P' &= (b_1 + b_2 - a_2, a_1 + a_2 - b_1, b_2 + a_1) \\ Q' &= (c_1 + c_2 - b_2, b_1 + b_2 - c_1, c_2 + b_1) \\ R' &= (a_1 + a_2 - c_2, c_1 + c_2 - a_1, a_2 + c_1) \\ Q' + R' &= (a_1 + a_2 + c_1 - b_2, b_1 + b_2 - a_1 + c_2, a_2 + b_1 + c_1 + c_2) \end{aligned}$$

Applying the condition $Q' + R' = 2P'$, and equating the first two coordinates, we can solve for c_1 and c_2 to get the required coordinates of C:

$$\begin{aligned} c_1 &= 2b_1 + 3b_2 - a_1 - 3a_2 \\ c_2 &= 3a_1 + 2a_2 - 3b_1 - b_2 \\ c_3 &= c_1 + c_2. \end{aligned}$$

There is another point C (say C') which satisfies our initial requirement, which of course is the mirror image (in AB) of the above solution.

Notes:

(i) We conjecture that the above solutions, obtained from isosceles triangles with coordinates calculated by the formulae, are the only possible solutions to the problem of finding a triangle whose inner ET-transform points are collinear.

(ii) Adding the coordinates of A, B, C for the solution triangle, we find that each coordinate sum is divisible by 3, and the resulting point is P'. Thus we have found that P' is the centroid of the solution triangle.

(iii) The sides of the solution triangle are in proportions $1 : \sqrt{7} : \sqrt{7}$. This is found by computing the lengths of the vectors $A - B$ and $B - C$, algebraically from the coordinates of the formulae for A, B and C, thus:

$$\begin{aligned}
BC^2 &= |\boldsymbol{B} - \boldsymbol{C}|^2 \\
&= (2a_1 + 3a_2 - 2b_1 - 3b_2)^2 \\
&\quad + (3a_1 + a_2 - 3b_1 - b_2)^2 + (a_1 - 2a_2 - b_1 + 2b_2)^2 \\
&= a_1^2(4 + 9 + 1) + a_1a_2(12 + 6 - 4) + a_2^2(9 + 1 + 4) \\
&\quad + b_1^2(4 + 9 + 1) + b_1b_2(12 + 6 - 4) + b_2^2(9 + 1 + 4) \\
&\quad - a_1b_1(8 + 18 + 2) - a_1b_2(12 + 6 - 4) \\
&\quad - a_2b_1(12 + 6 - 4) - a_2b_2(18 = 2 + 8) \\
&= 7 \times 2(a_1^2 + a_1a_2 + a_2^2 + b_1^2 + b_1b_2 + b_2^2 \\
&\qquad - 2a_1b_1 - a_1b_2 - a_2b_1 - 2a_2b_2) \\
&= 7|\boldsymbol{A} - \boldsymbol{B}|^2 = 7AB^2, \text{ which was to be demonstrated.}
\end{aligned}$$

An Associated Diophantine Equation:

We can link the triangle calculations from the above subsection with number theory in an immediate and satisfying way, as follows.

The equation just used to find the squares of sides of $(1, \sqrt{7}, \sqrt{7})$ triangles suggests the Diophantine problem: Find all solutions in natural numbers of the equation

$$x^2 + y^2 + z^2 = 7(u^2 + v^2 + w^2) \qquad (*)$$

The parametric solution (possibly not for *all* solutions, but we believe it might be so) is: Take any four natural numbers (p, q, r, s) and compute (x, y, z, u, v, w) by the following formulae (obtained from above, by setting $(a_1, a_2, b_1, b_2) = (p, q, r, s)$):

$$\begin{aligned}
x &= (2p + 3q) - (2r + 3s) \\
y &= (3p + q) - (3r + s) \\
z &= (p - 2q) - (r - 2s) \\
u &= (p - r) \\
v &= (q - s) \\
w &= (p + q) - (r + s).
\end{aligned}$$

We can express this transformation in matrix form thus:

$$(x,y,z,u,v,w) = \begin{bmatrix} 1 & 3 & -1 & -3 \\ -3 & -2 & 3 & 2 \\ -2 & 1 & 2 & -1 \\ 1 & 0 & -1 & 0 \\ 0 & 1 & 0 & -1 \\ 1 & 1 & -1 & -1 \end{bmatrix} \begin{pmatrix} p \\ q \\ r \\ s \end{pmatrix}$$

Examples and notes

(i) Six solutions:

$$\begin{array}{lll} (4,1,5,1,2,1), & (7,7,0,2,3,1), & (7,14,7,4,5,1) \\ (3,1,2,0,1,1), & (10,8,2,4,2,2), & (18,13,5,3,4,7) \end{array}$$

(ii) For any given solution, there are 35 others which are equivalent up to permutations of the x, y, z terms and the u, v, w terms.

For example, (4,1,5,1,2,1)≡(1,4,5,1,1,2).

(iii) We define a primitive solution in the usual way. A solution is ***primitive*** iff $\gcd(x, y, z, u, v, w) = 1$. In examples (i), all solutions are primitive except $(10, 8, 2, 4, 2, 2)$; this is permutation-equivalent to $2(4, 1, 5, 1, 2, 1)$.

(iv) It is curious, with a Diophantine equation of this type, that sometimes two solutions can be added to form another one. For example: $(4,1,5,1,1,2) + (1,2,3,0,1,1) = (5,3,8,1,2,3)$ is a solution to equation $(*)$. But $(4,1,5,1,2,1)+(3,1,2,0,1,1) = (7,2,7,1,3,2)$ is not a solution.

A glance at the sum solution in (iv), viz. (5,3,8,1,2,3), shows that its terms are all Fibonacci numbers. It is natural, since we are working in the honeycomb plane, that we ask if there are classes of solutions of $(*)$ which either are Fibonacci vectors (6-tuples), or which can be characterised in terms of the Fibonacci numbers. One such infinite class of solutions of $(*)$ is the following, $\forall n \in \boldsymbol{Z}$:

$$\{(x, y, z, u, v, w) = (F_{n+3} + F_{n+1}, F_{n+4} - F_{n-1}, F_{n-2}, F_n, F_{n+1}, F_{n+2})\}.$$

This is obtained by putting $(p, q, r, s) = (F_{n+2}, F_{n+3}, F_{n+1}, F_{n+2})$ in the parametric form of the solution.

Expressing the result in another way, we can say that

$$S_n \equiv (F_{n+3} + F_{n+1})^2 + (F_{n+4} - F_{n+1})^2 + F_{n-2}^2$$

is always divisible by 7, and also by

$$T_n \equiv (F_n^2 + F_{n+1}^2 + F_{n+2}^2): \quad \text{and} \quad S_n = 7T_n.$$

This version says something interesting about sums of squares of three consecutive Fibonacci numbers.

Further, 14 divides S_n, since T_n is even (because two of any three consecutive Fibonacci numbers are odd, and the other is even).

One observation is that $S_1 = 2.7^2$ and $S_4 = 2.7^3$; we ask how many S_n are of the form 2.7^i.

6. SUMMARY

This paper has introduced and applied in many ways the idea of *transformation of a line-segment by a triangle.* In particular, it has been used to solve the Pythagorean proposition in a variety of ways; and the use of equilateral triangles to transform polygons has been shown to yield interesting properties of figures (e.g. vector polygons) in the honeycomb plane. In the final section, a study of the point set of ET-transforms of a triangle led to a solution which yielded an infinite set of solutions to a related Diophantine equation. It is clear that fruitful links between geometry, algebra and number theory can be forged by triangle transforms.

REFERENCES

[1] Loomis, E.S. The Pythagorean Proposition. National Council of Teachers of Mathematics, Washington, DC, 1968.

[2] Durham, W. Journey Through Genius: the Great Theorems of Mathematics. Wiley (1990): 89-101.

[3] Schaake, A.G. and Turner, J.C. A New Chapter for Pythagorean Triples. Pub. Department of Mathematics and Statistics, University of Waikato, Hamilton, New Zealand, 1989.

[4] Turner, J.C. "Introduction to a Fibonacci Geometry." Applications of Fibonacci Numbers. Volume **7**, Kluwer (1998): 435-448.

AMS Classification Numbers: 11B37, 11B39, 11A35

CRYPTOGRAPHY AND LUCAS SEQUENCE DISCRETE LOGARITHMS

William A. Webb

1. INTRODUCTION

A number of public key cryptosystems such as El Gamal [2] are based on the difficulty of solving the discrete logarithm problem in certain groups, that is, solving $g^x = h$ for x, where g and h are given group elements. The computational difficulty of the discrete logarithm depends on the representation of the group. The additive version in $\mathbb{Z}_m$ is essentially trivial, involving only the solution of a linear congruence. However, in $\mathbb{F}^*_{p^d}$, the multiplicative group of the finite field with p^d elements, the problem is intractable if p^d is large enough, even though it is isomorphic to the cyclic group $\mathbb{Z}_{p^d-1}$. The computation appears even somewhat more difficult in groups based on elliptic and hyperelliptic curves, with some exceptions such as supersingular curves.

USING LUCAS SEQUENCES

In [4] Smith and Skinner describe two codes they call LUCELG PK (public key) and LUCELG DS (digital signature) based on discrete logarithms for Lucas sequences. Let $\{V_n(P,Q)\}$ denote the Lucas sequence of the second kind: $V_0 = 2,\ V_1 = P, V_n = PV_{n-1} - QV_{n-2}$ for $n \geq 2$. The corresponding Lucas sequence of the first kind $\{U_n(P,Q)\}$ has initial values $U_0 = 0, U_1 = 1$.

These particular codes in [4] require that $Q = 1$. The claim is made that codes based on Lucas sequence discrete logarithms are potentially more secure because there are no known subexponential algorithms in contrast to the finite field discrete logarithm. Bleichenbacher,

This paper is in final form and no version of it will be submitted for publication elsewhere.

F.T. Howard (ed.), Applications of Fibonacci Numbers, Volume 9: Proceedings of the Tenth International Research Conference on Fibonacci Numbers and their Applications, 263–266.

Bosma, and Lenstra [1] point out "Unfortunately, choosing $Q \equiv 1(\text{mod } P)$ also provides the key to an attack on the proposed system." This leaves open the question of whether Lucas sequences with $Q \neq 1$ can be used in cryptography and whether the corresponding discrete logarithm is more difficult.

We will show that public key codes can be constructed even more generally with any linear recurrence sequence of any order. However, the discrete logarithm problem for such sequences can be reduced to a discrete logarithm in a finite field, thus yielding a subexponential algorithm.

The codes LUCELG PK and LUCELG DS use the identity:

$$V_{kn}(P,Q) = V_k(V_n(P,Q),Q^n). \tag{1}$$

In LUCELG PK the public key is $y = V_x(P,1)$ where x is secret. To send the message m, the sender also chooses a random number k and computes $G = V_k(y,1), d_1 = V_k(P,1)$ and $d_2 = Gm$. (All computations are in $\mathbb{F}_p$). Both d_1 and d_2 are sent.

The originator of the code, who knows x, decodes as follows, using (1):

$$V_x(d_1,1) = V_x(V_k(P,1),1) = V_x(V_k(P,1),1^k) = V_{xk}(P,1) = V_k(V_x(P,1),1) = V_k(y,1) = G.$$

Then $m = d_2G^{-1}$.

Note that it is essential that $1 = 1^k$, that is $Q = 1$.

In order to find x and break this code, an eavesdropper must solve $V_x(P,1) = y$ for x. However, since $U_x^2 = (V_x^2 - 4)/D$, where $D = P^2 - 4, U_x$ is also known at least up to $\pm$. In LUCELG DS both $V_x(P,1)$ and $U_x(P,1)$ are explicitly given in the public key, although as noted above, U_x is redundant.

El Gamal codes are based on working in a group. In [4] it is noted that the Lucas functions used are not closed under multiplication, thus making code breaking more difficult. However, there is an underlying group operation present, if we define

$$V_n(P,1) * V_k(P,1) = V_n(V_k(P,1),1) = V_n(V_k(P,1),1^k) = V_{nk}(P,1). \tag{2}$$

Note that we have again used the fact that $Q = 1$.

However, using the well known identities:

$$U_{n+k} = (U_nV_k + U_kV_n)/2 \qquad V_{n+k} = (V_nV_k + DU_nU_k)/2 \tag{3}$$

we can define a group of ordered pairs $\{(U_n, V_n)\}$ in $\mathbb{F}_p$ where

$$(U_n, V_n) * (U_k, V_k) = ((U_nV_k + U_kV_n)/2,\ (V_nV_k + DU_nU_k)/2) = (U_{n+k}, V_{n+k}) \tag{4}$$

for any Lucas sequences $\{U_n\}$ and $\{V_n\}$.

Groups generated by points on an elliptic curve over $\mathbb{F}_p$ can also be described as a set of ordered pairs from F_p with an appropriate multiplication. The standard El Gamal code can be implemented in any of these groups.

THE DISCRETE LOGARITHM FOR LUCAS SEQUENCES

The version of the discrete logarithm problem which appears useful in the cryptographic setting is to find x if $U_x(P,Q) = a$ and $V_x(P,Q) = b$. This form of the discrete logarithm can be reduced to a discrete logarithm in a finite field by the identity:

$$\alpha^n = (PU_n + V_n)/2 - Q\alpha^{-1}U_n. \tag{5}$$

where α is a corresponding characteristic root. Equation (5) is easily derived from the standard power of roots identity. [3] If $\alpha \notin \mathbb{F}_{p^2}$ this is a discrete logarithm problem in $\mathbb{F}_{p^2}$.

This leaves open the questions of whether solving only $V_x(P,Q) = a$ is computationally more difficult, and whether a public key code can be based on this single equation.

HIGHER ORDER SEQUENCES

Consider a kth order linear recurrence

$$z_n = a_1 z_{n-1} + \cdots + a_k z_{n-k}. \tag{6}$$

In particular, let $U_n^{(j)}$ denote the sequence with inital values $U_j^{(j)} = 1$ and $U_n^j = 0$ for $0 \le n < k$ and $n \ne j$. Any sequence satisfying (6) can be written as a linear combination of the $U_n^{(j)}$.

From equation (3.4) in [5] we have the addition identity:

$$U_{m+n}^{(j)} = \sum_{i=0}^{k-1} U_m^{(i)} U_{n+i}^{(j)} \tag{7}$$

Equation (7) can also be used to express $U_{m+n}^{(j)}$ in terms of only products $U_m^{(i)} U_n^{(h)}$.

This in turn can be used to define a group of k-tuples $(U_n^{(0)}, \ldots, U_n^{(k-1)})$ similar to what was done in equation (4).

The corresponding discrete logarithm problem $U_x^{(0)} = b_0, \ldots, U_x^{(k-1)} = b_{k-1}$ seems more complex, but it is still reducable to a finite field problem.

If α is any characteristic root of (6) we can write

$$\begin{bmatrix} U_x^{(0)} \\ \cdot \\ \cdot \\ \cdot \\ U_x^{(k-1)} \end{bmatrix} = M \begin{bmatrix} \alpha^x \\ \cdot \\ \cdot \\ \cdot \\ \cdot \end{bmatrix} = MX$$

where M is a nonsingular $k \times k$ matrix and X is a vector of k linearly inpendent solutions of (7). From

$$X = M^{-1} \begin{bmatrix} U_x^{(0)} \\ \cdot \\ \cdot \\ \cdot \\ U_x^{(k-1)} \end{bmatrix} = M^{-1} \begin{bmatrix} b_0 \\ \cdot \\ \cdot \\ \cdot \\ b_{k-1} \end{bmatrix}$$

we see that α^x can be expressed as a linear combination of the $U_x^{(j)} = b_j$ which is a discrete logarithm problem in a finite field extention of $\mathbb{F}_p$.

Again we have the question of whether a code must have a public key containing k equations or whether a smaller subset might suffice. Identities like (7) seem to indicate that k sequences are needed to define a group operation. Equation (2) seems to give an exception, but recall that in this case the value of $U_n(P, 1)$ was implicitly known. Also, would such forms of the recurrence sequence discrete logarithm be more computationally intractable?

REFERENCES

[1] Bleichenbacher, D. Bosma, W. and Lenstra, A. "Some Remarks on Lucas-Based Cryptosystems." Advances in Cryptology. Crypto '95, Springer-Verlag (1995) pp. 386-396.

[2] ElGamal, T. "A Public-Key Cryptosystem and a Signature Scheme Based on Discrete Logarithms." *IEEE Transactions on Information Theory*, Vol. *31* (1985): pp. 469-472.

[3] Rabinowitz, Stanley. "Generalized Lucas Sequence Identities." International Conference on Fibonacci Numbers and Their Applications.

[4] Smith, P. and Skinner, C. "A Public-Key Cryptosystem and Digital Signature System Based on the Lucas Function Analogue to Discrete Logarithms." Advances in Cryptology - Asiacrypt '94. Springer-Verlag (1995): pp. 357-364.

[5] Zhou, Chizhong. "Constructing Idenitites Involving K^{th}-order F-L Numbers by Using the Characteristic Polynomial." Applications of Fibonacci Numbers, Volume **8**. Kluwer Academic Publishers, Dordrecht, The Netherlands, 1999, 369-379.

AMS Classification Numbers: 11B39, 11Y16

DIVISIBILITY OF AN F-L TYPE CONVOLUTION

Michael Wiemann and Curtis Cooper

1. MOTIVATION

Sometimes when working on one problem, another problem and solution are found. The divisibility result in this paper is a consequence of attempts to prove some conjectures of Melham [9] related to the sum

$$L_1 L_3 \cdots L_{2m+1} \sum_{k=1}^{n} F_{2k}^{2m+1},$$

where m is a nonnegative integer and n is a positive integer. Here, we use the usual notation for Fibonacci and Lucas numbers, i.e.

$$F_0 = 0, \quad F_1 = 1, \quad \text{and} \quad F_n = F_{n-1} + F_{n-2}, \quad \text{for} \quad n \geq 2$$

and

$$L_0 = 2, \quad L_1 = 1, \quad \text{and} \quad L_n = L_{n-1} + L_{n-2}, \quad \text{for} \quad n \geq 2.$$

When $m = 2$, Melham found that

$$L_1 L_3 L_5 \sum_{k=1}^{n} F_{2k}^5 = 4F_{2n+1}^5 - 15F_{2n+1}^3 + 25F_{2n+1} - 14.$$

This paper is in final form and no version of it will be submitted for publication elsewhere.

F.T. Howard (ed.), *Applications of Fibonacci Numbers, Volume 9: Proceedings of the Tenth International Research Conference on Fibonacci Numbers and their Applications*, 267–287.

To prove this result we will use the identity

$$F_m^5 = \frac{1}{25}\left(F_{5m} - 5(-1)^m F_{3m} + 10F_m\right)$$

(proved using Binet's formula), a result by Melham [9] that if m is an odd integer

$$L_m \sum_{k=1}^{n} F_{2mk} = F_{m(2n+1)} - F_m$$

(proved using Binet's formula and summing the resulting geometric series), and the well-known identities [6]

$$F_{5n} = 25F_n^5 + 25(-1)^n F_n^3 + 5F_n \quad \text{and} \quad F_{3n} = 5F_n^3 + 3(-1)^n F_n.$$

Substituting these in turn into our sum we obtain

$$L_1L_3L_5 \sum_{k=1}^{n} F_{2k}^5 = L_1L_3L_5 \sum_{k=1}^{n} \frac{1}{25}\left(F_{10k} - 5F_{6k} + 10F_{2k}\right)$$

$$= \frac{1}{25} L_1L_3L_5 \left(\sum_{k=1}^{n} F_{10k} - 5\sum_{k=1}^{n} F_{6k} + 10\sum_{k=1}^{n} F_{2k}\right)$$

$$= \frac{1}{25}\left(L_1L_3(F_{10n+5} - F_5) - 5L_1L_5(F_{6n+3} - F_3) + 10L_3L_5(F_{2n+1} - F_1)\right)$$

$$= \frac{1}{25}\left(L_1L_3F_{10n+5} - L_1L_3F_5 - 5L_1L_5F_{6n+3} + 5L_1F_3L_5\right.$$

$$\left. + 10L_3L_5F_{2n+1} - 10F_1L_3L_5\right)$$

$$= \frac{1}{25}\left(L_1L_3(25F_{2n+1}^5 - 25F_{2n+1}^3 + 5F_{2n+1}) - L_1L_3F_5\right.$$

$$\left. - 5L_1L_5(5F_{2n+1}^3 - 3F_{2n+1}) + 5L_1F_3L_5 + 10L_3L_5(F_{2n+1}) - 10F_1L_3L_5\right)$$

$$= (L_1L_3)F_{2n+1}^5 - (L_1L_3 + L_1L_5)F_{2n+1}^3$$

$$+ \frac{L_1L_3 + 3L_1L_5 + 2L_3L_5}{5} F_{2n+1} - \frac{L_1L_3F_5 - 5L_1F_3L_5 + 10F_1L_3L_5}{25}$$

$$= 4F_{2n+1}^5 - 15F_{2n+1}^3 + 25F_{2n+1} - 14.$$

In the last step, we note that

$$25 \big| L_1L_3F_5 - 5L_1F_3L_5 + 10F_1L_3L_5. \tag{1}$$

Here, | means divides. This paper will generalize (1).

2. HISTORY AND RESULT

Divisibility of Fibonacci and Lucas numbers has been the topic of much research in the mathematical literature. Some well-known divisibility properties of Fibonacci numbers and Lucas numbers can be found in [3]. For example,

$$\begin{aligned} &F_n|F_m \text{ if and only if } m = kn;\\ &L_n|F_m \text{ if and only if } m = 2kn, \quad n > 1;\\ \text{and } &L_n|L_m \text{ if and only if } m = (2k-1)n, \quad n > 1. \end{aligned}$$

In [8], Matijasevič proved that

$$F_m^2 \big| F_{mr} \text{ if and only if } F_m \big| r.$$

Later, Hoggatt and Bicknell-Johnson [5] extended these results. In [4], Hoggatt and Bergum discovered a number of interesting results. For example, they proved that

$$n = 2 \cdot 3^k \text{ and } k \geq 1 \text{ implies } n|L_n.$$

They also showed that

$$p \text{ is an odd prime and } p|F_n \text{ implies } p^k|F_{np^{k-1}} \text{ for all } k \geq 1.$$

A corollary to this last result is the fact that

$$5^k \big| F_{5^k} \text{ for } k \geq 1.$$

In this paper we will prove the following theorem.

Theorem: Let n be a nonnegative integer. Then

$$5^n \bigg| L_1L_3 \cdots L_{2n+1} \sum_{i=0}^{n} \binom{2n+1}{n-i} (-1)^{n-i} \frac{F_{2i+1}}{L_{2i+1}}. \tag{2}$$

3. LEMMAS

To prove our theorem we will need several lemmas. Some of these lemmas involve the quantity

$$a_{pj} = (-1)^j \sum_{k=j}^{p} (-1)^k 2^{p-k} \binom{p+1}{k+1} \binom{k}{j}, \tag{3}$$

where p and j are positive integers and $1 \le j \le p$. If we list the first few values of a_{pj} we have

1										
4	1									
11	5	1								
26	16	6	1							
57	42	22	7	1						
120	99	64	29	8	1					
247	219	163	93	37	9	1				
502	466	382	256	130	46	10	1			
1013	968	848	638	386	176	56	11	1		
2036	1981	1816	1486	1024	562	232	67	12	1	
4083	4017	3797	3302	2510	1586	794	299	79	13	1 .

This array is part of the sequence A008949 and can be found in [10]. Another notation we will use is $\langle\rangle$. This will denote an Eulerian number [2].

Lemma 1: Let p be a positive integer. Then

$$a_{p1} = \left\langle {p+1 \atop 1} \right\rangle.$$

Lemma 2: Let p and j be positive integers and let $1 \le j \le p$. Then

$$a_{pj} = \sum_{0 \le i \le p-j} \binom{p+1}{i}.$$

Lemma 3: Let n and k be positive integers with $n > k$. Then

$$\sum_{i=0}^{n} \binom{2n+1}{i} (-1)^i (2n-2i+1)^{2k+1} = 0.$$

Lemma 4: Let p and j be positive integers and $1 \leq j \leq p+1$. Then

$$a_{p+1,j} - \binom{p+1}{j} = 2a_{pj}.$$

Here we adopt the convention that $a_{p,p+1} = 0$.

Lemma 5: Let k and p be positive integers with $p \geq 2k$. Then

$$\sum_{j=1}^{p} (-1)^j a_{pj} j^{2k} = 0.$$

4. PROOF OF LEMMA 1

The proof is by induction on p.

Base Step: Since

$$a_{11} = (-1)^1 \sum_{k=1}^{1} (-1)^k 2^{1-k} \binom{2}{k+1} \binom{k}{1}$$

$$= (-1)^1 (-1)^1 2^{1-1} \binom{2}{2} \binom{1}{1} = 1$$

and

$$\left\langle {2 \atop 1} \right\rangle = 1,$$

the result is true for $p = 1$.

Induction Step: Assume the result is true for some positive integer p. Then by properties of binomial coefficients, the induction hypothesis, and a recurrence relation for Eulerian numbers,

we have

$$a_{p+1,1} = -\sum_{k=1}^{p+1}(-1)^k 2^{p+1-k}\binom{p+2}{k+1}\binom{k}{1}$$

$$= -\sum_{k=1}^{p}(-1)^k 2^{p+1-k}\binom{p+1}{k+1}\binom{k}{1} - \sum_{k=1}^{p+1}(-1)^k 2^{p+1-k}\binom{p+1}{k}\binom{k}{1}$$

$$= -2\sum_{k=1}^{p}(-1)^k 2^{p-k}\binom{p+1}{k+1}\binom{k}{1} - \sum_{k=1}^{p+1}(-1)^k 2^{p+1-k}(p+1)\binom{p}{k-1}$$

$$= -2\sum_{k=1}^{p}(-1)^k 2^{p-k}\binom{p+1}{k+1}\binom{k}{1} + (p+1)\sum_{k=0}^{p}(-1)^k 2^{p-k}\binom{p}{k}$$

$$= 2a_{p1} + (p+1)(2-1)^p = 2a_{p1} + (p+1)\cdot 1$$

$$= 2\left\langle {p+1 \atop 1} \right\rangle + (p+1)\left\langle {p+1 \atop 0} \right\rangle = \left\langle {p+2 \atop 1} \right\rangle.$$

Thus, the result is true for $p+1$. By induction, the result is true for all positive integers p.

5. PROOF OF LEMMA 2

We will prove this result in 3 parts. Let

$$c_{pj} = \sum_{0\le i\le p-j}\binom{p+1}{i}.$$

First we will show that for any positive integer p,

$$a_{pp} = c_{pp}.$$

This follows since

$$a_{pp} = (-1)^p\sum_{k=p}^{p}(-1)^k 2^{p-k}\binom{p+1}{k+1}\binom{k}{p}$$

$$= (-1)^p(-1)^p 2^{p-p}\binom{p+1}{p+1}\binom{p}{p} = 1$$

and

$$c_{pp} = \sum_{0 \le i \le p-p} \binom{p+1}{i} = \binom{p+1}{0} = 1.$$

Second we will show that for any positive integer p,

$$a_{p1} = c_{p1}.$$

By Lemma 1

$$a_{p1} = \left\langle {p+1 \atop 1} \right\rangle.$$

By a property of Eulerian numbers

$$c_{p1} = \sum_{0 \le i \le p-1} \binom{p+1}{i} = 2^{p+1} - p - 2 = \left\langle {p+1 \atop 1} \right\rangle.$$

Third we will show that for $p \ge 2$ and $2 \le j \le p$,

$$a_{p+1,j} = a_{pj} + a_{p,j-1}$$

and

$$c_{p+1,j} = c_{pj} + c_{p,j-1}.$$

We see that

$$c_{p+1,j} = \sum_{0 \le i \le p+1-j} \binom{p+2}{i} = \sum_{0 \le i \le p+1-j} \binom{p+1}{i} + \sum_{1 \le i \le p+1-j} \binom{p+1}{i-1}$$

$$= \sum_{0 \le i \le p-(j-1)} \binom{p+1}{i} + \sum_{0 \le i \le p-j} \binom{p+1}{i} = c_{p,j-1} + c_{pj}.$$

We also see (using several binomial coefficient identities and rearranging terms in the sums) that

$$a_{p+1,j} = (-1)^j \sum_{k=j}^{p+1} (-1)^k 2^{p+1-k} \binom{p+2}{k+1}\binom{k}{j}$$

$$= 2^{p+1-j}\binom{p+2}{j+1}\binom{j}{j} + (-1)^j \sum_{k=j+1}^{p} (-1)^k 2^{p+1-k}\binom{p+2}{k+1}\binom{k}{j}$$

$$+ (-1)^j(-1)^{p+1}\binom{p+2}{p+2}\binom{p+1}{j}$$

$$= 2^{p+1-j}\binom{p+1}{j}\binom{j-1}{j-1} + 2^{p+1-j}\binom{p+1}{j+1}\binom{j}{j}$$

$$+ (-1)^j \sum_{k=j+1}^{p} (-1)^k 2^{p+1-k}\left[\binom{p+1}{k}\binom{k-1}{j} + \binom{p+1}{k}\binom{k-1}{j-1} + \binom{p+1}{k+1}\binom{k}{j}\right]$$

$$+ (-1)^j(-1)^{p+1}\binom{p+1}{p+1}\binom{p}{j-1} + (-1)^j(-1)^{p+1}\binom{p+1}{p+1}\binom{p}{j}$$

$$= 2^{p+1-j}\binom{p+1}{j}\binom{j-1}{j-1} + (-1)^j \sum_{k=j+1}^{p} (-1)^k 2^{p+1-k}\binom{p+1}{k}\binom{k-1}{j-1}$$

$$+ (-1)^j(-1)^{p+1}\binom{p+1}{p+1}\binom{p}{j-1} + 2^{p+1-j}\binom{p+1}{j+1}\binom{j}{j}$$

$$+ (-1)^j \sum_{k=j+1}^{p} (-1)^k 2^{p+1-k}\left[\binom{p+1}{k}\binom{k-1}{j} + \binom{p+1}{k+1}\binom{k}{j}\right]$$

$$+ (-1)^j(-1)^{p+1}\binom{p+1}{p+1}\binom{p}{j}$$

$$= 2^{p+1-j}\binom{p+1}{j}\binom{j-1}{j-1} + (-1)^{j-1}\sum_{k=j}^{p-1}(-1)^k 2^{p-k}\binom{p+1}{k+1}\binom{k}{j-1}$$

$$+ (-1)^j(-1)^{p+1}\binom{p+1}{p+1}\binom{p}{j-1} + 2^{p+1-j}\binom{p+1}{j+1}\binom{j}{j}$$

$$+ (-1)^j \sum_{k=j+1}^{p} (-1)^k 2^{p+1-k}\left[\binom{p+1}{k}\binom{k-1}{j} + \binom{p+1}{k+1}\binom{k}{j}\right]$$

$$+ (-1)^j(-1)^{p+1}\binom{p+1}{p+1}\binom{p}{j}$$

$$= (-1)^{j-1} \sum_{k=j-1}^{p} (-1)^k 2^{p-k}\binom{p+1}{k+1}\binom{k}{j-1} + 2^{p+1-j}\binom{p+1}{j+1}\binom{j}{j} - 2^{p-j}\binom{p+1}{j+1}\binom{j}{j}$$

$$+ (-1)^j \sum_{k=j+2}^{p} (-1)^k 2^{p+1-k}\binom{p+1}{k}\binom{k-1}{j} + (-1)^j \sum_{k=j+1}^{p-1} (-1)^k 2^{p+1-k}\binom{p+1}{k+1}\binom{k}{j}$$

$$+ (-1)^j(-1)^p 2\binom{p+1}{p+1}\binom{p}{j} + (-1)^j(-1)^{p+1}\binom{p+1}{p+1}\binom{p}{j}$$

$$= (-1)^{j-1} \sum_{k=j-1}^{p} (-1)^k 2^{p-k}\binom{p+1}{k+1}\binom{k}{j-1} + 2^{p-j}\binom{p+1}{j+1}\binom{j}{j}$$

$$+ (-1)^j \sum_{k=j+1}^{p-1} (-1)^{k+1} 2^{p-k}\binom{p+1}{k+1}\binom{k}{j}$$

$$+ (-1)^j \sum_{k=j+1}^{p-1} (-1)^k 2^{p+1-k}\binom{p+1}{k+1}\binom{k}{j} + (-1)^j(-1)^p\binom{p+1}{p+1}\binom{p}{j}$$

$$= (-1)^{j-1} \sum_{k=j-1}^{p} (-1)^k 2^{p-k}\binom{p+1}{k+1}\binom{k}{j-1} + 2^{p-j}\binom{p+1}{j+1}\binom{j}{j}$$

$$+ (-1)^j \sum_{k=j+1}^{p-1} (-1)^k 2^{p-k}\binom{p+1}{k+1}\binom{k}{j} + (-1)^j(-1)^p\binom{p+1}{p+1}\binom{p}{j}$$

$$= (-1)^{j-1} \sum_{k=j-1}^{p} (-1)^k 2^{p-k}\binom{p+1}{k+1}\binom{k}{j-1} + (-1)^j \sum_{k=j}^{p}(-1)^k 2^{p-k}\binom{p+1}{k+1}\binom{k}{j}$$

$$= a_{p,j-1} + a_{pj}.$$

Thus, by the 3 parts, the two arrays are identical. Therefore, the proof of Lemma 2 is complete.

6. PROOF OF LEMMA 3

Let

$$f(i) = (2n - 2i + 1)^{2k+1}$$

and let $\triangle$ denote the forward-difference operator. Then

$$\begin{aligned}\triangle^{2n+1} f(0) &= \sum_{i=0}^{2n+1} \binom{2n+1}{i} (-1)^i (2n - 2i + 1)^{2k+1} \\ &= 2 \sum_{i=0}^{n} \binom{2n+1}{i} (-1)^i (2n - 2i + 1)^{2k+1}.\end{aligned}$$

But since f is a polynomial in i of degree $2k + 1$ and $n > k$,

$$\triangle^{2n+1} f(0) = 0.$$

Therefore,

$$\sum_{i=0}^{n} \binom{2n+1}{i} (-1)^i (2n - 2i + 1)^{2k+1} = 0.$$

7. PROOF OF LEMMA 4

Let p and j be positive integers and $1 \leq j \leq p + 1$. By Lemma 2

$$a_{pj} = \sum_{0 \leq i \leq p-j} \binom{p+1}{i}.$$

Also, assume $a_{p,p+1} = 0$. Thus,

$$a_{p+1,j} - \binom{p+1}{j} = \sum_{0\le i\le p+1-j} \binom{p+2}{i} - \binom{p+1}{j}$$

$$= \binom{p+2}{0} + \sum_{1\le i\le p+1-j} \binom{p+2}{i} - \binom{p+1}{j}$$

$$= \binom{p+1}{0} + \sum_{1\le i\le p+1-j} \left(\binom{p+1}{i} + \binom{p+1}{i-1}\right) - \binom{p+1}{j}$$

$$= \binom{p+1}{0} + \sum_{1\le i\le p+1-j} \binom{p+1}{i} - \binom{p+1}{p+1-j} + \sum_{1\le i\le p+1-j} \binom{p+1}{i-1}$$

$$= \binom{p+1}{0} + \sum_{1\le i\le p-j} \binom{p+1}{i} + \sum_{0\le i\le p-j} \binom{p+1}{i}$$

$$= 2 \sum_{0\le i\le p-j} \binom{p+1}{i} = 2a_{pj}.$$

8. PROOF OF LEMMA 5

The proof is by induction on p.

Base Step:

We will show that Lemma 5 is true for $p = 2k$. We will do this by solving a sequence of recurrence relations by the perturbation method. Let m be a nonnegative integer. Consider the recurrence relation

$$x_{-1} = 0, \quad \text{and} \quad x_n = n^m - x_{n-1} \quad \text{for} \quad n \ge 0.$$

Let $P_m(n)$ be the solution of this recurrence relation. To describe the solutions to these recurrences we need the following notation. Let $C(n)$ denote a statement which is either true or false, depending on n. Then using APL notation [2] we define

$$[C(n)] = \begin{cases} 1, & \text{if } C(n) \text{ is true} \\ 0, & \text{if } C(n) \text{ is false.} \end{cases}$$

The first 3 recurrence relations and their solutions can be found in Problem 21 of Chapter 2 of [2]. The solutions for $m = 0, 1$ and 2 are

$$P_0(n) = 1 - [n \text{ is odd}]$$

$$P_1(n) = \frac{1}{2}n + \frac{1}{2}[n \text{ is odd}]$$

$$\text{and} \quad P_2(n) = \frac{1}{2}n^2 + \frac{1}{2}n. \tag{4}$$

In using the perturbation method to find the solutions for $m \geq 3$, we obtain the relation

$$P_m(n) = \frac{1}{2}\left((n+1)^m - \sum_{i=1}^{m} \binom{m}{i} P_{m-i}(n)\right). \tag{5}$$

Using this relation, we can compute $P_m(n)$ for $m = 3, 4, \ldots, 12$.

$$P_3(n) = \frac{1}{2}n^3 + \frac{3}{4}n^2 - \frac{1}{4}[n \text{ is odd}]$$

$$P_4(n) = \frac{1}{2}n^4 + n^3 - \frac{1}{2}n$$

$$P_5(n) = \frac{1}{2}n^5 + \frac{5}{4}n^4 - \frac{5}{4}n^2 + \frac{1}{2}[n \text{ is odd}]$$

$$P_6(n) = \frac{1}{2}n^6 + \frac{3}{2}n^5 - \frac{5}{2}n^3 + \frac{3}{2}n$$

$$P_7(n) = \frac{1}{2}n^7 + \frac{7}{4}n^6 - \frac{35}{8}n^4 + \frac{21}{4}n^2 - \frac{17}{8}[n \text{ is odd}]$$

$$P_8(n) = \frac{1}{2}n^8 + 2n^7 - 7n^5 + 14n^3 - \frac{17}{2}n$$

$$P_9(n) = \frac{1}{2}n^9 + \frac{9}{4}n^8 - \frac{21}{2}n^6 + \frac{63}{2}n^4 - \frac{153}{4}n^2 + \frac{31}{2}[n \text{ is odd}]$$

$$P_{10}(n) = \frac{1}{2}n^{10} + \frac{5}{2}n^9 - 15n^7 + 63n^5 - \frac{255}{2}n^3 + \frac{155}{2}n$$

$$P_{11}(n) = \frac{1}{2}n^{11} + \frac{11}{4}n^{10} - \frac{165}{8}n^8 + \frac{231}{2}n^6 - \frac{2805}{8}n^4 + \frac{1705}{4}n^2 - \frac{691}{4}[n \text{ is odd}]$$

$$P_{12}(n) = \frac{1}{2}n^{12} + 3n^{11} - \frac{55}{2}n^9 + 198n^7 - \frac{1683}{2}n^5 + 1705n^3 - \frac{2073}{2}n.$$

Each $P_m(n)$ is a polynomial of degree m plus possibly a term involving $[n \text{ is odd}]$. If we let b_m denote the coefficient in front of the term $[n \text{ is odd}]$ in $P_m(n)$, then we have the table of elements

m	0	1	2	3	4	5	6	7	8	9	10	11	12	$\cdots$
b_m	−1	1/2	0	−1/4	0	1/2	0	−17/8	0	31/2	0	−691/4	0	$\cdots$

By (4) and (5), the values of the b_ms satisfy the conditions $b_0 = -1$ and for $m \geq 1$,

$$b_m = -\frac{1}{2}\sum_{i=0}^{m-1}\binom{m}{i}b_i.$$

Using generating functions, it can be shown that

$$\sum_{k=0}^{\infty} b_k \frac{x^k}{k!} = \frac{-2}{e^x+1}.$$

Since

$$\frac{-2}{e^x+1} + 1 = \frac{e^x-1}{e^x+1}$$

is an odd function it follows that the even subscripted bs are 0, i.e. $b_{2k} = 0$ for $k \geq 1$. Therefore, $P_{2k}(n)$ for $k \geq 1$ is a polynomial of degree $2k$, i.e. it contains no term $[n \text{ is odd}]$.

It should be noted that the Genocchi numbers [1] are defined by

$$\frac{2x}{e^x+1} = \sum_{k=0}^{\infty} G_k \frac{x^k}{k!}.$$

Therefore, for $n \geq 0$

$$b_n = -\frac{1}{n+1}G_{n+1}.$$

Now, using Lemma 2 on the first equality we have

$$\sum_{j=1}^{2k}(-1)^j a_{2k,j} j^{2k} = \sum_{j=1}^{2k}(-1)^j \sum_{i=0}^{2k-j}\binom{2k+1}{i} j^{2k}$$

$$= \sum_{i=0}^{2k-1}\binom{2k+1}{i}\sum_{j=1}^{2k-i}(-1)^j j^{2k}$$

$$= \sum_{i=0}^{2k-1}\binom{2k+1}{i}\sum_{j=0}^{2k-i}(-1)^j j^{2k}$$

$$= \sum_{i=0}^{2k+1}\binom{2k+1}{i}\sum_{j=0}^{2k-i}(-1)^j j^{2k}$$

$$= \sum_{i=0}^{2k+1}\binom{2k+1}{2k+1-i}\sum_{j=0}^{2k-(2k+1-i)}(-1)^j j^{2k}$$

$$= \sum_{i=0}^{2k+1}\binom{2k+1}{i}(-1)^{i+1}\left(\sum_{j=0}^{i-1}(-1)^j j^{2k}(-1)^{i+1}\right)$$

$$= \sum_{i=0}^{2k+1}\binom{2k+1}{i}(-1)^{i+1}P_{2k}(-1+i).$$

But since the last sum is $-\Delta^{2k+1}P_{2k}(-1)$ and P_{2k} is a polynomial of degree $2k$, it follows that the above sum is 0. This completes the proof of the base step.

Induction Step: Next, we will show that if the formula is true for some $p \geq 2k$, then it is true for $p+1$. Suppose that the formula is true for some $p \geq 2k$. We will use the fact that

$$\sum_{j=0}^{p+1}(-1)^{j+1}\binom{p+1}{j} j^{2k} = 0.$$

This can be seen by noting that if $Q(j) = j^{2k}$, then

$$\sum_{j=0}^{p+1}(-1)^{j+1}\binom{p+1}{j} j^{2k} = -\Delta^{p+1}Q(0) = 0$$

since Q is a polynomial in j of degree $2k$ and $p+1 > 2k$. Hence,

$$\begin{aligned}
&\sum_{j=1}^{p+1}(-1)^j a_{p+1,j}j^{2k} \\
&= \sum_{j=1}^{p+1}(-1)^j a_{p+1,j}j^{2k} + \sum_{j=0}^{p+1}(-1)^{j+1}\binom{p+1}{j}j^{2k} \\
&= \sum_{j=1}^{p+1}(-1)^j a_{p+1,j}j^{2k} + \sum_{j=1}^{p+1}(-1)^{j+1}\binom{p+1}{j}j^{2k} \\
&= \sum_{j=1}^{p+1}(-1)^j \left(a_{p+1,j} - \binom{p+1}{j}\right)j^{2k} \\
&= \sum_{j=1}^{p}(-1)^j 2a_{pj}j^{2k} = 2\left(\sum_{j=1}^{p}(-1)^j a_{pj}j^{2k}\right).
\end{aligned}$$

The next to last equality follows from Lemma 4. But the last expression is 0 by our induction hypothesis. Therefore, the result is true for $p+1$. This completes the proof of the induction step.

Thus, by induction, Lemma 5 is proved.

9. PROOF OF THE THEOREM

We begin the proof of (2) by noting that if

$$(x-1)^{2n+1}\Big|(x+1)(x^3+1)\cdots(x^{2n+1}+1)\sum_{i=0}^{n}\binom{2n+1}{n-i}(-1)^{n-i}\frac{x^{2i+1}-1}{x^{2i+1}+1} \tag{6}$$

is true, then (2) is true. Suppose (6) is true and substitute α/β for x in (6), where

$$\alpha = \frac{1+\sqrt{5}}{2} \quad \text{and} \quad \beta = \frac{1-\sqrt{5}}{2}.$$

Using the fact that $\alpha - \beta = \sqrt{5}$ and multiplying (6) by β^{n^2}, (6) becomes

$$5^n\Big|(\alpha+\beta)(\alpha^3+\beta^3)\cdots(\alpha^{2n+1}+\beta^{2n+1})\sum_{i=0}^{n}\binom{2n+1}{n-i}(-1)^{n-i}\frac{\alpha^{2i+1}-\beta^{2i+1}}{\sqrt{5}(\alpha^{2i+1}+\beta^{2i+1})}.$$

But this last result, by the use of Binet's formula [3], i.e.

$$F_n = \frac{\alpha^n - \beta^n}{\sqrt{5}} \quad \text{and} \quad L_n = \alpha^n + \beta^n,$$

is (2) .

Let

$$f(x) = (x+1)(x^3+1)\cdots(x^{2n+1}+1)$$

and

$$g(x) = \sum_{i=0}^{n} \binom{2n+1}{n-i} (-1)^{n-i} \frac{x^{2i+1}-1}{x^{2i+1}+1}.$$

Now, if D denotes the derivative operator, then by applying the product rule j times we obtain the formula

$$D^j f(x)g(x) = \sum_{i=0}^{j} \binom{j}{i} D^i f(x) D^{j-i} g(x). \tag{7}$$

Proving (6) would be equivalent to showing that

$$D^j f(1)g(1) = 0 \text{ for } j = 0, 1, \ldots, 2n. \tag{8}$$

But by (7) we can prove (8) if we can show that

$$g(1) = Dg(1) = D^2 g(1) = \cdots = D^{2n} g(1) = 0. \tag{9}$$

Simplifying $g(x)$ we have

$$\begin{aligned} g(x) &= \sum_{i=0}^{n} \binom{2n+1}{n-i} (-1)^{n-i} \frac{x^{2i+1}-1}{x^{2i+1}+1} \\ &= \sum_{i=0}^{n} \binom{2n+1}{n-i} (-1)^{n-i} \left(1 - \frac{2}{x^{2i+1}+1}\right). \end{aligned} \tag{10}$$

First of all, it is clear that $g(1) = 0$. To compute the pth derivative of $g(x)$ where $1 \le p \le 2n$, we need to find the pth derivative of

$$\frac{1}{x^{2i+1}+1}.$$

Using a result in [7],

$$D^p\left[\frac{1}{x^{2i+1}+1}\right]=\sum_{k=1}^{p}(-1)^k\binom{p+1}{k+1}\frac{1}{(x^{2i+1}+1)^{k+1}}D^p\left[(x^{2i+1}+1)^k\right].$$

We now need the notation for falling factorials [2], i.e.

$$x^{\underline{p}}=x(x-1)\cdots(x-p+1)$$

and the binomial theorem

$$(x^{2i+1}+1)^k=\sum_{j=0}^{k}\binom{k}{j}x^{(2i+1)j}.$$

Thus,

$$D^p\left[\sum_{j=0}^{k}\binom{k}{j}x^{(2i+1)j}\right]=\sum_{j=0}^{k}\binom{k}{j}D^p x^{(2i+1)j}$$

$$=\sum_{j=0}^{k}\binom{k}{j}[(2i+1)j][(2i+1)j-1]\cdots[(2i+1)j-p+1]x^{(2i+1)j-p}$$

$$=\sum_{j=0}^{k}\binom{k}{j}[(2i+1)j]^{\underline{p}}x^{(2i+1)j-p}.$$

It follows that

$$D^p\left[\frac{1}{x^{2i+1}+1}\right]\Bigg|_{x=1}=\sum_{k=1}^{p}(-1)^k\binom{p+1}{k+1}2^{-k-1}\sum_{j=0}^{k}\binom{k}{j}[(2i+1)j]^{\underline{p}}. \tag{11}$$

Next, we will study (11) with $2i+1$ replaced by m, i.e.

$$\sum_{k=1}^{p}(-1)^k\binom{p+1}{k+1}2^{-k-1}\sum_{j=0}^{k}\binom{k}{j}(jm)^{\underline{p}}.$$

Using the fact that $p \geq 1$, so we have no term when $j = 0$, we wish to investigate the sum

$$\sum_{k=1}^{p}(-1)^k \binom{p+1}{k+1} 2^{-k-1} \sum_{j=1}^{k} \binom{k}{j} (jm)^{\underline{p}}. \tag{12}$$

By changing the order of summation, it follows that (12) becomes

$$\sum_{j=1}^{p}(jm)^{\underline{p}} \sum_{k=j}^{p}(-1)^k \binom{p+1}{k+1} \binom{k}{j} 2^{-k-1}$$

$$= \frac{1}{2^{p+1}} \sum_{j=1}^{p}(jm)^{\underline{p}} \sum_{k=j}^{p}(-1)^k 2^{p-k} \binom{p+1}{k+1} \binom{k}{j}.$$

We want to show that the above polynomial in m only contains odd terms, i.e. there are only terms of odd degree in the polynomial. The first few such polynomials are

$$\frac{1}{4}(-m),$$

$$\frac{1}{8}(2m),$$

$$\frac{1}{16}(2m^3 - 8m),$$

$$\frac{1}{32}(-24m^3 + 48m),$$

$$\frac{1}{64}(-16m^5 + 280m^3 - 384m),$$

and $$\frac{1}{128}(480m^5 - 3600m^3 + 3840m),$$

for $p = 1, 2, 3, 4, 5$, and 6, respectively. Now, by (3) we have that the polynomial is

$$D^p \left[\frac{1}{x^m + 1} \right] \Bigg|_{x=1} = \frac{1}{2^{p+1}} \sum_{j=1}^{p}(-1)^j a_{pj} (jm)^{\underline{p}}.$$

Next, we recall the Stirling numbers of the first kind. They are denoted by

$$s(n,k)$$

and count the number of ways to arrange n objects into k cycles [1,2]. A property of Stirling numbers of the first kind is

$$s(n,n-k) = \sum_{0\le i_1<\cdots<i_k\le n-1} i_1\cdots i_k.$$

Thus, we have that

$$x^{\underline{p}} = x(x-1)\cdots(x-p+1) = \sum_{j=0}^{p}(-1)^j s(p,p-j)x^{p-j}.$$

It follows that

$$(jm)^{\underline{p}} = \sum_{k=0}^{p}(-1)^k = s(p,p-k)(jm)^{p-k} = \sum_{k=0}^{p}(-1)^k s(p,p-k)j^{p-k}m^{p-k}. \tag{13}$$

Hence, by using (13) and changing the order of summation, the polynomial in m is

$$\begin{aligned}
&D^p\left[\frac{1}{x^m+1}\right]\Bigg|_{x=1}\\
&= \frac{1}{2^{p+1}}\sum_{j=1}^{p}(-1)^j a_{pj}(jm)^{\underline{p}}\\
&= \frac{1}{2^{p+1}}\sum_{j=1}^{p}(-1)^j a_{pj}\sum_{k=0}^{p}(-1)^k s(p,p-k)j^{p-k}m^{p-k}\\
&= \frac{1}{2^{p+1}}\sum_{k=0}^{p}(-1)^k s(p,p-k)m^{p-k}\sum_{j=1}^{p}(-1)^j a_{pj}j^{p-k}\\
&= \frac{1}{2^{p+1}}\sum_{k=0}^{p}(-1)^{p-k} s(p,k)m^{k}\sum_{j=1}^{p}(-1)^j a_{pj}j^{k}.
\end{aligned}$$

Therefore, for $p \geq 1$ we have by (7) that

$$D^p g(1) = \sum_{i=0}^{n} \binom{2n+1}{n-i} (-1)^{n-i} D^p \left(1 - \frac{2}{g_{2i+1}(x)}\right)\Bigg|_{x=1}$$

$$= \sum_{i=0}^{n} \binom{2n+1}{i} (-1)^{i} D^p \left(1 - \frac{2}{g_{2n-2i+1}(x)}\right)\Bigg|_{x=1}$$

$$= -2\sum_{i=0}^{n} \binom{2n+1}{i} (-1)^{i} D^p \left(\frac{1}{g_{2n-2i+1}(x)}\right)\Bigg|_{x=1}$$

$$= -2\sum_{i=0}^{n} \binom{2n+1}{i} (-1)^{i} \frac{1}{2^{p+1}} \sum_{k=0}^{p} (-1)^{p-k} s(p,k)(2n-2i+1)^k \sum_{j=1}^{p} (-1)^j a_{pj} j^k$$

$$= \frac{-2}{2^{p+1}} \sum_{k=0}^{p} (-1)^{p-k} s(p,k) \sum_{j=1}^{p} (-1)^j a_{pj} j^k \sum_{i=0}^{n} \binom{2n+1}{i} (-1)^i (2n-2i+1)^k.$$

To finish the proof of the Theorem we will prove that the last expression is 0. To do this we will isolate the term when $k = 0$ and the two sums when $0 < 2k+1 \leq p$ and $0 < 2k \leq p$. The term and the two sums are listed below.

$$\frac{-2}{2^{p+1}} (-1)^p s(p,0) \sum_{j=1}^{p} (-1)^j a_{pj} \sum_{i=0}^{n} \binom{2n+1}{i} (-1)^i$$

$$+ \frac{-2}{2^{p+1}} \sum_{0<2k+1\leq p} (-1)^{p-2k-1} s(p, 2k+1) \sum_{j=1}^{p} (-1)^j a_{pj} j^{2k+1}$$

$$\left(\sum_{i=0}^{n} \binom{2n+1}{i} (-1)^i (2n-2i+1)^{2k+1} \right)$$

$$+ \frac{-2}{2^{p+1}} \sum_{0<2k\leq p} (-1)^{p-2k} s(p, 2k) \left(\sum_{j=1}^{p} (-1)^j a_{pj} j^{2k} \right)$$

$$\sum_{i=0}^{n} \binom{2n+1}{i} (-1)^i (2n-2i+1)^{2k}.$$

The term when $k = 0$ is 0 since $s(p, 0) = 0$ for $p \geq 1$. Since $1 \leq p \leq 2n$ and $2k + 1 \leq p$, it follows that $k < n$. Thus by Lemma 3 the first sum is 0. Lemma 5 proves that the second sum is 0.

Summarizing, we have just shown that the term and the two sums are 0. Thus, for $1 \leq p \leq 2n$ we have $D^p g(1) = 0$. Since $g(1) = 0$ we have proved that (6) is true. Therefore, the Theorem is proved.

10. FURTHER QUESTIONS

First of all, we could study the polynomial P_m in Lemma 5. Is there an explicit formula for P_m? Second, in studying (2) we came across the conjecture that

$$(x+1)^n \left| (x+1)(x^3+1)\cdots(x^{2n+1}+1) \sum_{i=0}^{n} \binom{2n+1}{n-i} (-1)^{n-i} \frac{x^{2i+1}-1}{x^{2i+1}+1} \right. .$$

Finally, we could again study Melham's sum

$$L_1 L_3 \cdots L_{2m+1} \sum_{k=1}^{n} F_{2k}^{2m+1},$$

where m is a nonnegative integer and n is a positive integer.

REFERENCES

[1] L. Comtet. Advanced Combinatorics. Dordrecht: D. Reidel, 1974.

[2] R. L. Graham, D. E. Knuth, & O. Patashnik. *Concrete Mathematics.* Reading, Mass.: Addison-Wesley, 1994.

[3] V. E. Hoggatt, Jr. Fibonacci and Lucas Numbers. Boston: Houghton Mifflin, 1969.

[4] V. E. Hoggatt, Jr. & G. E. Bergum. "Divisibility and Congruence Relations." *The Fibonacci Quarterly*, Vol. *12.2* (1974): 189–195.

[5] V. E. Hoggatt, Jr. & Marjorie Bicknell-Johnson. "Divisibility by Fibonacci and Lucas Squares." *The Fibonacci Quarterly, 15.1* (1977): 3–8.

[6] D. Jennings. "Some Polynomial Identities for the Fibonacci and Lucas Numbers." *The Fibonacci Quarterly*, Vol. *31.2* (1993): 134–137.

[7] R. A. Leslie. "How Not to Repeatedly Differentiate a Reciprocal." *American Mathematical Monthly*, Vol. *98* (1991): 732–735.

[8] Y. V. Matijasevič. "Enumerable Sets are Diophantine." *Proc. of the Academy of Sciences of the USSR*, Vol. *11* (1970): No. 2.

[9] R. S. Melham. private communication.

[10] N. J. A. Sloane. *On-Line Encyclopedia of Integer Sequences.* Published electronically at http://www.research.att.com/~njas/sequences.

AMS Classification Numbers: 11B39.

GENERATING FUNCTIONS OF CONVOLUTION MATRICES

Yongzhi (Peter) Yang

1. INTRODUCTION

Hoggatt and Bergum [2] studied the general expression for the entry in the ith row and the jth column of a convolution matrix and obtained row generating functions for the convolution matrix of the sequence $\{1, u_2, u_3, u_4, \ldots\}$. In this paper, we extend the Strong Convolution Decomposition Theorem [3] to a more general case. Based on this extension, we decompose a convolution matrix into a product of a lower trianglular matrix and the upper triangular Pascal-like matrix. This interesting decomposition of a convolution matrix leads a novel approach to the subject proposed in [2]. Using this new method, we obtain a simple explicit formula for entries of a convolution matrix and row generating functions of the convolution matrix of the sequences $\{v_n\}$ and $\{u_n\}$. Moreover, the approach developed here can be easily extended to a rather broad category of integer matrices.

To review, the convolution of two sequences $\{a_n\}$ and $\{b_n\}$, $(n = 1, 2, 3, \ldots)$, is the sequence $\{c_n\}$ where $c_n = \sum_{k=1}^{n} a_k b_{n-k+1}$. The convolution matrix of two sequences $\{a_n\}$ and $\{b_n\}$ is the matrix whose first column is $\{a_n\}$ and whose i^{th} column $(i = 2, 3, \ldots)$ is the convolution sequence of the $(i-1)^{th}$ column with $\{b_n\}$. We say that the convolution matrix of the sequences $\{a_n\}$ and $\{a_n\}$ is the convolution matrix of the sequence $\{a_n\}$. There are many well-known integer matrices which can be written as convolution matrices of some sequences. The rectangular Pascal triangle matrix, for instance, is the convolution matrix of the sequence $\{1, 1, 1, 1, \ldots\}$ and the lower triangular Pascal matrix is the convolution matrix of the sequences $\{1, 1, 1, 1, \ldots\}$ and $\{0, 1, 1, 1, \ldots\}$. Furthermore, the convolution matrices of some particular sequences may have very interesting properties. For example, the convolution matrix of the sequence $\{1, 2, 3, 4, 5, 6, \ldots\}$ is

This paper is in final form and no version of it will be submitted for publication elsewhere.

F.T. Howard (ed.), Applications of Fibonacci Numbers, Volume 9: Proceedings of the Tenth International Research Conference on Fibonacci Numbers and their Applications, 289–295.

$$M = \begin{pmatrix} 1 & 1 & 1 & 1 & 1 & 1 & \dots \\ 2 & 4 & 6 & 8 & 10 & 12 & \dots \\ 3 & 10 & 21 & 36 & 55 & 78 & \dots \\ 4 & 20 & 56 & 120 & 220 & 364 & \dots \\ 5 & 35 & 126 & 330 & 715 & 1365 & \dots \\ 6 & 56 & 252 & 792 & 2002 & 4368 & \dots \\ \vdots & \vdots & \vdots & \vdots & \vdots & \vdots & \ddots \end{pmatrix} \tag{1}$$

It is not difficult to show: (1) The sum of the antidiagonal of the matrix M is the even term of the Fibonacci sequence $\{1, 3, 8, 21, 55, 144, \dots\}$. (2) Each row of the convolution matrix M is clearly the row of the triangle of coefficients of shifted Chebyshev's $S(n, x-2) = U(n, x/2-1)$ polynomials (exponents of x in decreasing order). (3) The determinant of an upper left corner $n \times n$ submatrix of M is $2^{(n-1)n/2}$. We left the proofs of results (1) and (2) as exercises for the interested reader.

Therefore, studying the properties of convolution matrices can be important for understanding the structure of a class of integer matrices.

2. GENERALIZED CONVOLUTION DECOMPOSITION THEOREM

To examine convolution matrices in general, it is convenient to represent a convolution matrix C of two sequences $\{v_n\}$ and $\{u_n\}$ in terms of multiplication of matrices as following:

$$C = [\bar{V}, U\bar{V}, U^2\bar{V}, \dots, U^n\bar{V}, \dots], \tag{2}$$

where

$$U = \begin{pmatrix} u_1 & 0 & 0 & 0 & 0 & 0 & \dots \\ u_2 & u_1 & 0 & 0 & 0 & 0 & \dots \\ u_3 & u_2 & u_1 & 0 & 0 & 0 & \dots \\ u_4 & u_3 & u_2 & u_1 & 0 & 0 & \dots \\ u_5 & u_4 & u_3 & u_2 & u_1 & 0 & \dots \\ u_6 & u_5 & u_4 & u_3 & u_2 & u_1 & \dots \\ \vdots & \vdots & \vdots & \vdots & \vdots & \vdots & \ddots \end{pmatrix} \tag{3}$$

and $\bar{V}$ is a vector $[v_1, v_2, v_3, v_4, \dots]^T$.

In [3] we obtain the Strong Convolution Decomposition Theorem for a convolution matrix of a sequence $\{u_1, u_2, u_3, \dots\}$, where u_1 is a positive integer. We restate the theorem as following:

Strong Convolution Decomposition Theorem: Let $\{u_n\}$ be a sequence whose first term is a positive integer u_1, and let V be the convolution matrix of that sequence. Then $V = SP_U^{u_1}$ for some lower triangular matrix S and the upper triangular Pascal matrix P_U. Moreover, successive columns of S are successive convolutions of the sequence $\{u_n\}$ with $\{0, u_2, u_3, u_4, \dots\}$.

Here we propose to generalize this theorem to any convolution matrix of two sequences $\{v_n\}$ and $\{u_n\}$ using notations developed by Call and Velleman [1].

Let $P_U[x]$ be the infinite dimensional matrix defined by

$$(P_U[x])_{i,j} = \begin{cases} x^{j-i}\binom{j-1}{i-1}, & \text{if } j \geq i \\ 0, & \text{otherwise} \end{cases}, \tag{4}$$

where x is any real number. Then it is proved in [1] that

$$(P_U)^r = P_U[r] \quad \text{and} \quad (P_U^r)^{-1} = P_U[-r] = P_U^{-r}, \tag{5}$$

where $P_U = P_U[1]$, the standard upper triangular Pascal matrix, and r is any real number. If $r = 0$, then we define $P_U^0 = I$, where I is an identity matrix.

Using these facts, we can prove the following generalization of the Strong Convolution Decomposition Theorem in [3]:

Theorem 1(Generalized Convolution Decomposition Theorem): Let $\{v_n\}$ and $\{u_n\}$ be any two sequences of real numbers and let C be the infinite dimensional convolution matrix of $\{v_n\}$ and $\{u_n\}$. Then $C = SP_U^{u_1}$ for some lower triangular matrix S and the upper triangular Pascal matrix P_U. Moreover, successive columns of S are successive convolutions of the sequences $\{v_n\}$ and $\{0, u_2, u_3, u_4, \ldots\}$.

Proof: Using equations (2) and (5),

$$S = C \times P_U^{-u_1}$$

$$= [\bar{V}, U\bar{V}, U^2\bar{V}, \ldots, U^n\bar{V}, \ldots] \begin{pmatrix} \binom{0}{0} & \binom{1}{0}(-u_1)^1 & \binom{2}{0}(-u_1)^2 & \binom{3}{0}(-u_1)^3 & \binom{4}{0}(-u_1)^4 & \cdots \\ 0 & \binom{1}{1} & \binom{2}{1}(-u_1)^1 & \binom{3}{1}(-u_1)^2 & \binom{4}{1}(-u_1)^3 & \cdots \\ 0 & 0 & \binom{2}{2} & \binom{3}{2}(-u_1)^1 & \binom{4}{2}(-u_1)^2 & \cdots \\ 0 & 0 & 0 & \binom{3}{3} & \binom{4}{3}(-u_1)^1 & \cdots \\ 0 & 0 & 0 & 0 & \binom{4}{4} & \cdots \\ \vdots & \vdots & \vdots & \vdots & \vdots & \ddots \end{pmatrix}$$

$$= [\bar{V}, (U - u_1I)\bar{V}, (U - u_1I)^2\bar{V}, \ldots, (U - u_1I)^n\bar{V}, \ldots]$$
$$= [\bar{V}, B\bar{V}, B^2\bar{V}, \ldots, B^n\bar{V}, \ldots] \tag{6}$$

where $B = (U - u_1I)$. Thus, $C = SP_U^{u_1}$ where $S = [\bar{V}, B\bar{V}, B^2\bar{V}, \ldots, B^n\bar{V}, \ldots]$ is the convolution matrix of the sequences $\{v_1, v_2, v_3, v_4, \ldots\}$ and $\{0, u_2, u_3, u_4, \ldots\}$. This completes the proof of the theorem.

When the first term v_1 of the sequence $\{v_1, v_2, v_3, v_4, \ldots\}$ is a positive integer and $\{u_n\} = \{v_n\}$, this theorem agrees with the Strong Convolution Decomposition Theorem in [3]. The following corollary is an immediate conclusion of Theorem 1.

Corollary 1: Let C_m be the upper left corner $m \times m$ submatrix of the convolution matrix C of the sequences $\{v_n\}$ and $\{u_n\}$. Then $|C_m| = v_1^m u_2^{m(m-1)/2}$.

Proof: By Theorem 1, $C = SP_U^{u_1}$. Now, $|P_U^{u_1}| = 1^{u_1} = 1$, so $|C| = |S|$. Since S is a lower triangular with diagonal elements $v_1, v_1u_2, v_1u_2^2, \dots, |S_m| = v_1^m u_2^{1+2+\cdots+(m-1)}$, where S_m is the upper left corner $m \times m$ submatrix of S. Hence, $|C_m| = |S_m| = v_1^m u_2^{m(m-1)/2}$.

Remark: The determinant of any convolution matrix of the sequences $\{v_n\}$ and $\{u_n\}$ is wholly determined by the first term of the sequence $\{v_n\}$ and the second term of the sequence $\{u_n\}$. Corollary 1 is a generalization of the corollary in [3].

3. GENERAL TERM OF A CONVOLUTION MATRIX

In this section, we employ the Generalized Convolution Decomposition Theorem to find the general term of a convolution matrix C of sequence $\{u_n\}$. This approach can be easily extended to the convolution matrix of two sequences $\{v_n\}$ and $\{u_n\}$. Let S_n be the matrix consisting of the first n rows of the matrix S defined previously. It is easy to see that S_n is an $n \times \infty$ convolution matrix of the finite sequences $\{u_1, u_2, u_3, u_4, \dots, u_n\}$ and $\{0, u_2, u_3, u_4, \dots, u_n\}$. If $f(x) = \sum_{i=1}^n u_i x^{i-1}$ is the generating function of the finite sequence $\{u_1, u_2, u_3, u_4, \dots, u_n\}$ and $g(x) = \sum_{i=2}^n u_i x^{i-1}$ is the generating function of the finite sequence $\{0, u_2, u_3, u_4, \dots, u_n\}$ then the generating function for the lth column of S_n is $f(x)g^{l-1}(x) = u_1 g^{l-1}(x) + g^l(x)$. Thus, the entry in the ith row and the jth column of S_n, denoted by $s_{i,j}$, is the coefficient of x^{i-1} in $f(x)g^{j-1}(x)$. By noting that the coefficient of x^k in $(u_2x + u_3x^2 + \cdots + u_nx^{n-1})^m$ is

$$\sum_{\substack{l_2+l_3+\cdots+l_n=m \\ l_i\geq 0;\ i=2,3,\dots,n; \\ 1l_2+2l_3+\cdots+(n-1)l_n=k}} \frac{m!}{l_2!l_3!\dots l_n!} u_2^{l_2} u_3^{l_3} \dots u_n^{l_n}, \tag{7}$$

we have

$$\begin{aligned} s_{n,m} = u_1 & \sum_{\substack{l_2+l_3+\cdots+l_n=m-1 \\ l_i\geq 0;\ i=2,3,\dots,n; \\ 1l_2+2l_3+\cdots+(n-1)l_n=n-1}} \frac{(m-1)!}{l_2!l_3!\dots l_n!} u_2^{l_2} u_3^{l_3} \dots u_n^{l_n}, \\ + & \sum_{\substack{l_2+l_3+\cdots+l_n=m \\ l_i\geq 0;\ i=2,3,\dots,n; \\ 1l_2+2l_3+\cdots+(n-1)l_n=n-1}} \frac{m!}{l_2!l_3!\dots l_n!} u_2^{l_2} u_3^{l_3} \dots u_n^{l_n}. \end{aligned} \tag{8}$$

Let $c_{n,m}$ be the entry in the nth row and the mth column of C. From the Generalized Convolution Decomposition Theorem, we know that

$$\begin{aligned} c_{n,m} &= \sum_{k=1}^{\min\{n,m\}} s_{n,k} \binom{m-1}{k-1} u_1^{m-k} \\ &= \sum_{k=1}^{\min\{n,m\}} \{u_1 \sum_{\substack{l_2+l_3+\cdots+l_n=k-1 \\ l_i \geq 0;\ i=2,3,\ldots,n; \\ 1l_2+2l_3+\cdots+(n-1)l_n=n-1}} \frac{(k-1)!}{l_2! l_3! \ldots l_n!} u_2^{l_2} u_3^{l_3} \ldots u_n^{l_n} \\ &+ \sum_{\substack{l_2+l_3+\cdots+l_n=k \\ l_i \geq 0;\ i=2,3,\ldots,n; \\ 1l_2+2l_3+\cdots+(n-1)l_n=n-1}} \frac{k!}{l_2! l_3! \ldots l_n!}\} \binom{m-1}{k-1} u_1^{m-k}. \end{aligned} \tag{9}$$

4. ROW GENERATING FUNCTIONS FOR A CONVOLUTION MATRIX

Now we are in a position to find the row generating functions of a convolution matrix. Let $F_m(x)$ be the generating function for the mth row of a convolution matrix C of the sequences $\{v_n\}$ and $\{u_n\}$. It is easy to see that

$$F_m(x) = \sum_{k=1}^{\infty} c_{m,k} x^{k-1}. \tag{10}$$

Also, by Theorem 1, $F_m(x)$ equals

$$F_m(x) = [s_{m,1}, s_{m,2}, \ldots, s_{m,m}, 0, 0, \ldots] P_U^{u_1} [1, x, x^2, \ldots, x^m, \ldots]^T. \tag{11}$$

It is not difficult to prove that

$$\begin{aligned} &P_U^{u_1} [1, x, x^2, \ldots, x^m, \ldots]^T \\ &= \left[\frac{1}{(1-u_1x)}, \frac{x}{(1-u_1x)^2}, \frac{x^2}{(1-u_1x)^3}, \ldots, \frac{x^{m-1}}{(1-u_1x)^m}, \ldots\right]^T. \end{aligned} \tag{12}$$

Thus,

$$F_m(x) = [s_{m,1}, s_{m,2}, \ldots, s_{m,m}, 0, 0, \ldots]$$

$$\left[\frac{1}{(1-u_1x)}, \frac{x}{(1-u_1x)^2}, \frac{x^2}{(1-u_1x)^3}, \ldots, \frac{x^{m-1}}{(1-u_1x)^m}, \ldots\right]^T. \tag{13}$$

We summarize this discussion in the following theorem.

Theorem 2: The generating function of the mth row of the convolution matrix C of two sequences $\{v_n\}$ and $\{u_n\}$ is

$$F_m(x) = \sum_{k=1}^{m} \frac{s_{m,k}\, x^{k-1}}{(1-u_1x)^k}, \tag{14}$$

where $s_{m,k}$ is the element in the mth row and the kth column of S and S is the convolution matrix of the sequences $\{v_n\}$ and $\{0, u_2, u_3, u_4, \ldots\}$.

Combining equations (8) and (14) we obtain the generating function of the mth row of the convolution matrix C of the sequence $\{u_n\}$

$$F_m(x) = \sum_{k=1}^{m} \left\{ \sum_{\substack{l_2+l_3+\cdots+l_m=k-1 \\ l_i\geq 0;\ i=2,3,\ldots,m; \\ 1l_2+2l_3+\cdots+(m-1)l_m=m-1}} \frac{(k-1)!}{l_2!l_3!\ldots l_n!} u_2^{l_2} u_3^{l_3} \ldots u_m^{l_m}, \right.$$

$$\left. + \sum_{\substack{l_2+l_3+\cdots+l_m=k \\ l_i\geq 0;\ i=2,3,\ldots,m; \\ 1l_2+2l_3+\cdots+(m-1)l_m=m-1}} \frac{k!}{l_2!l_3!\ldots l_m!} u_2^{l_2} u_3^{l_3} \ldots u_m^{l_m} \right\} \frac{x^{k-1}}{(1-u_1x)^k}. \tag{15}$$

For example, using the formula in equation (15) we obtain the generating function of the nth row of matrix M in equation (1)

$$F_n(x) = \frac{\sum_{k=1}^{\lceil n/2\rceil+1} \binom{n}{2k-1} x^{k-1}}{(1-x)^n}. \tag{16}$$

The reader may wish to fill in the details of the computation.

5. CONCLUSION

The Generalized Convolution Decomposition Theorem allows calculation of row generating functions of convolution matrices. Our matrix decomposition approach differs from the approach in [2]. Indeed we have found that our results yield much insight into the structure of generating functions and can be extended to find the generating functions of a rather broad category of matrices. We hope that the matrix decomposition approach developed here may shed some light on derivations of row or column generating functions of arithmetic-progression matrices [3] and the recursion relation matrices studied by Ollerton and Shannon [4].

REFERENCES

[1] Call, G.S. and Velleman, D.J. "Pascal's Matrices." *The American Mathematical Monthly*, Vol. *100.4* (1993): pp. 372-376.

[2] Hoggatt, V.E., Jr. and Bergum, G.E. "Generalized Convolution Matrices." *The Fibonacci Quarterly*, Vol. *13.3* (1975): pp. 193-198.

[3] Leida, J. and Yang, Y. "Pascal Decompositions of Arithmetic and Convolution Arrays in Matrices." *The Fibonacci Quarterly*, Vol. *40.2* (2002): pp. 136-145.

[4] Ollerton, R.L. and Shannon, A.G. "Some Properties of Generalized Pascal Squares and Triangles." *The Fibonacci Quarterly*, Vol. *36.2* (1998): pp. 98-109.

AMS Classification Numbers: 15A23, 11B25, 11B65

F-L REPRESENTATION OF DIVISION OF POLYNOMIALS OVER A RING

Chizhong Zhou and F. T. Howard

1. INTRODUCTION

The motiviation for this paper is the congruence [3]

$$x^n \equiv F_n x + F_{n-1} \pmod{x^2 - x - 1}.$$

That is, $x^n = (x^2 - x - 1)q(x) + r(x)$, where $q(x)$ is a polynomial with integer coefficients and $r(x) = F_n x + F_{n-1}$. Our goal is to replace x^n by an arbitrary polynomial over a ring, replace $x^2 - x - 1$ by an arbitrary monic polynomial, and to express the coefficients of $q(x)$ and $r(x)$ in terms of F-L numbers, which are analogous to Fibonacci numbers.

We use the notation and terminology in [3] and [4]. In this paper, the symbol R always designates a ring with multiplicative identity 1 and additive identity 0. The distinctions between the ring identities and the integers one and zero will always be clear in context.

Let the sequence $\{w_n\}$ satisfy the recurrence relation

$$w_{n+k} = a_1 w_{n+k-1} + \cdots + a_{k-1} w_{n+1} + a_k w_n \tag{1.1}$$

and the intial conditions

$$w_0 = c_0,\ w_1 = c_1,\ \cdots, w_{k-1} = c_{k-1},$$

where $a_1, \cdots, a_k$ and $c_0, \cdots, c_{k-1}$ are constants in R. We call $\{w_n\}$ a **k'th order F-L sequence (Fibonacci-Lucas sequence)** over R, we call each w_n an **F-L number**, and we call

This paper is in final form and no version of it will be submitted for publication elsewhere.

F.T. Howard (ed.), Applications of Fibonacci Numbers, Volume 9: Proceedings of the Tenth International Research Conference on Fibonacci Numbers and their Applications, 297–308.

$$f(x) = x^k - a_1x^{k-1} - \cdots - a_{k-1}x - a_k \tag{1.2}$$

the **characteristic polynomial** of $\{w_n\}$. The set of F-L sequences over R satisfying (1.1) is denoted by $\Omega(f(x))$.

Let $\{u_n^{(i)}\}_n$ $(0 \le i \le k-1)$ be a sequence in $\Omega(f(x))$ with the intiial conditions $u_n^{(i)} = \delta_{ni}$ for $0 \le n \le k-1$. That is,

$$u_n^{(i)} = \delta_{ni} = \begin{cases} 1 & \text{if } n = i, \\ 0 & \text{if } n \ne i. \end{cases}$$

We call $\{u_n^{(i)}\}$ the ith **basic sequence** in $\Omega(f(x))$, and we call $\{u_n^{(k-1)}\}$ the **principal sequence** in $\Omega(f(x))$.
Example: Let R be the ring of integers. In $\Omega(x^2 - x - 1)$, for $i = 0$ or 1

$$u_{n+2}^{(i)} = u_{n+1}^{(i)} + u_n^{(i)}.$$

Since $u_0^{(0)} = 1$ and $u_1^{(0)} = 0$, we have $u_n^{(0)} = F_{n-1}$. Since $u_0^{(1)} = 0$ and $u_1^{(1)} = 1$, we have $u_n^{(1)} = F_n$. The principal sequence is $\{F_n\}$.

The following lemma was proved in [3].

Lemma 1.1: Let $u_n^{(i)}$ $(0 \le i \le k-1)$ be the ith basic sequence in $\Omega(f(x))$ over the complex field, $f(0) \ne 0$. Then for any integer $n \ge 0$,

$$x^n \equiv \sum_{i=0}^{k-1} u_n^{(i)} x^i \pmod{f(x)}.$$

The lemma suggests that to find the remainder when the polynomial x^n is divided by $f(x)$, we need only calculate some F-L numbers. In section 2, we consider the more general problem of dividing an arbitrary polynomial over a ring by an arbitrary monic polynomial, and expressing the coefficients of the quotient and remainder by F-L numbers. In section 3 we state a divisibility test, which is an application of our main result. In section 4, we find a relationship between two F-L sequences with different characteristic polynomials, and we give an example involving Fibonacci and Tribonacci numbers.

2. MAIN RESULTS AND PROOFS

Our first theorem extends Lemma 1.1 to a ring R and determines the coefficients of the quotient and remainder in terms of F-L numbers.

Theorem 2.1: Let the polynomial $f(x)$ over R be expressed by (1.2), and let $\{u_n\}$ be the principal sequence in $\Omega(f(x))$ over R. Then

$$x^n = f(x)\sum_{i=0}^{n-k} u_{i+k-1}x^{n-k-i} + \sum_{i=0}^{k-1} w_n^{(i)}x^i, \tag{2.1}$$

where

$$w_n^{(i)} = \begin{cases} \delta_{ni} \text{ for } 0 \le n \le k-1 \\ \sum_{j=k-i}^{k} a_j u_{n+k-1-i-j} \text{ for } n \ge k \end{cases} \tag{2.2}$$

$$= u_{n+k-1-i} - \sum_{j=1}^{k-1-i} a_j u_{n+k-1-i-j}. \tag{2.3}$$

Proof: We first prove (2.2). For $0 \le n \le k-1$, (2.2) is clearly true, since $x^n = f(x)\cdot 0 + 1\cdot x^n$. For $n \ge k$, using (1.2) and the fact that $u_{k-1} = 1$, we write

$$r(x) = x^n - f(x)\sum_{i=0}^{n-k} u_{i+k-1}x^{n-k-i} = x^n - \sum_{i=0}^{n-k} u_{i+k-1}x^{n-i} + \sum_{j=1}^{k} a_j \sum_{i=0}^{n-k} u_{i+k-1}x^{n-i-j}$$

$$= -\sum_{i=1}^{n-k} u_{i+k-1}x^{n-i} + \sum_{j=1}^{k} a_j \sum_{i=j}^{n-k+j} u_{i+k-1-j}x^{n-i}.$$

Now using $u_m = 0$ if $0 \le m \le k-2$, we can write

$$r(x) = -\sum_{i=1}^{n-k} u_{i+k-1}x^{n-i} + \sum_{j=1}^{k} a_j \sum_{i=1}^{n-k+j} u_{i+k-1-j}x^{n-i}$$

$$= -\sum_{i=1}^{n-k}\left(u_{i+k-1} - \sum_{j=1}^{k} a_j u_{i+k-1-j}\right)x^{n-i} + \sum_{j=1}^{k} a_j \sum_{i=n-k+1}^{n-k+j} u_{i+k-1-j}x^{n-i}.$$

By (1.1), the first summation on the right side of the last equality is 0. Then

$$r(x) = \sum_{j=1}^{k} a_j \sum_{i=1}^{j} u_{n+i-1-j}x^{k-i} = \sum_{i=1}^{k}\sum_{j=i}^{k} a_j u_{n+i-1-j}x^{k-i} = \sum_{i=0}^{k-1}\sum_{j=k-i}^{k} a_j u_{n+k-i-1-j}x^i.$$

Thus by (1.1) we have, for $n \geq k$

$$w_n^{(i)} = \sum_{j=k-i}^{k} a_j u_{n+k-i-1-j} x^i$$

and (2.2) is proved. To prove (2.3), we let $h_n^{(i)} = u_{n+k-1-i} - \sum_{j=1}^{k-1-i} a_j u_{n+k-1-i-j}$. We again recall that $u_m = 0$ if $0 \leq m \leq k-2$ and $u_{k-1} = 1$. Thus for $n \leq i$, we have $h_n^{(i)} = \delta_{ni} = w_n^{(i)}$. For $n \geq i+1$, it is clear from (1.1) that $h_n^{(i)} = \sum_{j=k-i}^{k} a_j u_{n+k-1-i-j}$, so for $i+1 \leq n \leq k-1$, we have $h_n^{(i)} = 0$, and we see that $h_n^{(i)} = w_n^{(i)}$ for all $n \geq 0$. □

The purpose of the next theorem is to replace x^n in Theorem 2.1 by a general polynomial of degree m.

Theorem 2.2: Let the polynomial $f(x)$ be expressed by (1.2), and let $\{u_n\}$ be the principal sequence in $\Omega(f(x))$. Let $\{w_n^{(i)}\}(0 \leq i \leq k-1)$ be expressed by (2.2) or (2.3). Assume that the polynomial $g(x)$ over the ring R is expressed as

$$g(x) = \sum_{n=0}^{m} b_n x^{m-n}. \tag{2.4}$$

Then we have

$$g(x) = f(x) \sum_{n=0}^{m-k} \left(\sum_{i=k}^{m-n} u_{i-1} b_{m-n-i} \right) x^n + \sum_{i=0}^{k-1} \left(\sum_{n=0}^{m} w_n^{(i)} b_{m-n} \right) x^i. \tag{2.5}$$

Proof: From (2.4) and (2.1) we have

$$g(x) = \sum_{n=0}^{m} x^n b_{m-n} = f(x) \sum_{n=k}^{m} \sum_{i=0}^{n-k} u_{i+k-1} b_{m-n} x^{n-k-i} + \sum_{n=0}^{m} \sum_{i=0}^{k-1} w_n^{(i)} b_{m-n} x^i$$

$$= f(x) \sum_{n=k}^{m} \sum_{i=k}^{n} u_{i-1} b_{m-n} x^{n-i} + \sum_{i=0}^{k-1} \left(\sum_{n=0}^{m} w_n^{(i)} b_{m-n} \right) x^i.$$

The first summation of the last expression can be rewritten as

$$f(x)\sum_{i=k}^{m}\sum_{n=i}^{m}u_{i-1}b_{m-n}x^{n-i}=f(x)\sum_{i=k}^{m}\sum_{n=0}^{m-i}u_{i-1}b_{m-n-i}x^{n}=f(x)\sum_{n=0}^{m-k}\left(\sum_{i=k}^{m-n}u_{i-1}b_{m-n-i}\right)x^{n}$$

and that completes the proof. □

Corollary 2.3: Under the conditions of Theorem 2.2, the quotient $q(x)$ and the remainder $r(x)$ when $g(x)$ is divided by $f(x)$ can be written in the following matrix forms:

$$q(x)=(u_{k-1}\ u_k\ldots u_{m-1})\cdot B\cdot(x^{m-k}\ldots x\ 1)^T,$$

$$r(x)=(1\ x\ldots x^{k-1})\cdot W\cdot(b_m\ b_{m-1}\ldots b_0)^T,$$

where

$$B=\begin{pmatrix} b_0 & b_1 & b_2 & \cdots & b_{m-k}\\ & b_0 & b_1 & \cdots & b_{m-k-1}\\ & & \ddots & \ddots & \vdots\\ & & & b_0 & b_1\\ & & & & b_0\end{pmatrix},$$

$$W=\begin{pmatrix} w_0^{(0)} & w_1^{(0)} & \cdots & w_m^{(0)}\\ w_0^{(1)} & w_1^{(1)} & \cdots & w_m^{(1)}\\ \cdots & \cdots & \cdots & \cdots\\ w_0^{(k-1)} & w_1^{(k-1)} & \cdots & w_m^{(k-1)}\end{pmatrix}. \tag{2.6}$$

Example: Let R be the ring of integer matrices of second order. Let $f(x)=x^3-a_1x^2-a_2x-a_3$ and $g(x)=b_0x^5+b_1x^4+b_2x^3+b_3x^2+b_4x+b_5$, where

$$a_1=\begin{pmatrix}2&3\\1&-1\end{pmatrix},a_2=\begin{pmatrix}1&0\\2&2\end{pmatrix},a_3=\begin{pmatrix}-3&-2\\-2&-1\end{pmatrix},b_0=\begin{pmatrix}1&2\\2&-1\end{pmatrix},b_1=\begin{pmatrix}-10&0\\1&-2\end{pmatrix},$$

$$b_2=\begin{pmatrix}2&-6\\-3&-2\end{pmatrix},b_3=\begin{pmatrix}11&3\\8&1\end{pmatrix},b_4=\begin{pmatrix}-3&5\\-3&3\end{pmatrix},b_5=\begin{pmatrix}0&4\\1&4\end{pmatrix}.$$

Then $k=3$ and $m=5$. By the definition

$$u_0=u_1=\begin{pmatrix}0&0\\0&0\end{pmatrix},\ u_2=\begin{pmatrix}1&0\\0&1\end{pmatrix},\ u_3=a_1,\ u_4=a_1^2+a_2=\begin{pmatrix}8&3\\3&6\end{pmatrix}.$$

By (2.2), for $n \geq 3$ we have $w_n^{(0)} = a_3 u_{n-1}$, $w_n^{(1)} = a_2 u_{n-1} + a_3 u_{n-2}$, $w_n^{(2)} = a_1 u_{n-1} + a_2 u_{n-2} + a_3 u_{n-3}$, and for $0 \leq n \leq 2$, we have $w_n^{(i)} = \delta_{ni} (i = 0, 1, 2)$. Thus

$$w_0^{(0)} = \begin{pmatrix} 1 & 0 \\ 0 & 1 \end{pmatrix}, \; w_1^{(0)} = w_2^{(0)} = \begin{pmatrix} 0 & 0 \\ 0 & 0 \end{pmatrix}, \; w_3^{(0)} = a_3, \; w_4^{(0)} = \begin{pmatrix} -8 & -7 \\ -5 & -5 \end{pmatrix},$$

$$w_5^{(0)} = \begin{pmatrix} -30 & -21 \\ -19 & -12 \end{pmatrix}. \; w_1^{(1)} = \begin{pmatrix} 1 & 0 \\ 0 & 1 \end{pmatrix}, \; w_0^{(1)} = w_2^{(1)} = \begin{pmatrix} 0 & 0 \\ 0 & 0 \end{pmatrix},$$

$$w_3^{(1)} = a_2, \; w_4^{(1)} = \begin{pmatrix} -1 & 1 \\ 4 & 3 \end{pmatrix}, \; w_5^{(1)} = \begin{pmatrix} 0 & -4 \\ 17 & 13 \end{pmatrix}, \; w_2^{(2)} = \begin{pmatrix} 1 & 0 \\ 0 & 1 \end{pmatrix},$$

$$w_0^{(2)} = w_1^{(2)} = \begin{pmatrix} 0 & 0 \\ 0 & 0 \end{pmatrix}, \; w_3^{(2)} = a_1, \; w_4^{(2)} = \begin{pmatrix} 8 & 3 \\ 3 & 6 \end{pmatrix}, \; w_5^{(2)} = \begin{pmatrix} 24 & 25 \\ 9 & 0 \end{pmatrix}.$$

Thus by Corollary 2.3,

$$q(x) = \begin{pmatrix} 1 & 2 \\ 2 & -1 \end{pmatrix} x^2 + \begin{pmatrix} -2 & 1 \\ 0 & 1 \end{pmatrix} x + \begin{pmatrix} -1 & 1 \\ 1 & 0 \end{pmatrix}; \; r(x) = \begin{pmatrix} 3 & 2 \\ -2 & 3 \end{pmatrix} x^2 + \begin{pmatrix} 2 & 1 \\ 1 & 2 \end{pmatrix} x + \begin{pmatrix} 1 & 1 \\ 2 & 2 \end{pmatrix}.$$

In the last example, if we compute $u_n^{(0)}$, we observe that

$$u_4^{(0)} = a_1 a_3 = \begin{pmatrix} -12 & -7 \\ -1 & -1 \end{pmatrix} \neq w_4^{(0)}.$$

Thus in a ring, Lemma 1.1 does not always hold. However, if the ring is commutative, then we can prove that Lemma 1.1 does hold. We need the following:

Lemma 2.4: Let the polynomial $f(x)$ over a commutative ring R be expressed by (1.2). Let a and b be any two elements of R, and let $\{w_n\}$ and $\{b_n\}$ be any two elements of $\Omega(f(x))$. Then $\{aw_n + bh_n\}$ is an element of $\Omega(f(x))$.

Proof: Since $\{w_n\}$ and $\{b_n\}$ are elements of $\Omega(f(x))$, we know

$$w_{n+k} = \sum_{i=1}^{k} a_i w_{n+k-i}; \quad h_{n+k} = \sum_{i=1}^{k} a_i h_{n+k-i}.$$

Since R is commutative, we have

$$aw_{n+k} + bh_{n+k} = \sum_{i=1}^{k} a_i (aw_{n+k-i} + bh_{n+k-i}).$$

This completes the proof. □

Lemma 2.5: Let the polynomial $f(x)$ over a commutative ring R be expressed by (1.2). Let $\{u_n^{(i)}\}(0 \le i \le k-1)$ be the ith basic sequence in $\Omega(f(x))$ over R. Let $w_n^{(i)}$ be expressed by (2.2) or (2.3). Then

$$\{u_n^{(i)}\} = \{w_n^{(i)}\},\ i = 0, 1, \ldots, k-1.$$

Proof: For $1 \le j \le k$, the sequence $\{a_j u_{n+k-i-j}\}_n$ is in $\Omega(f(x))$. Thus by (2.3) and Lemma 2.4, $\{w_n^{(i)}\}$ is an element of $\Omega(f(x))$. Lemma 2.5 now follows from the fact that $\{w_n^{(i)}\}$ has the same initial conditions as $\{u_n^{(i)}\}$. This completes the proof. □

Corollary 2.6: If the ring R is commutative, then (2.1) and (2.5) can be rewritten as

$$x^n = f(x)\sum_{i=0}^{n-k} u_{i+k-1}x^{n-k-i} + \sum_{i=0}^{k-1} u_n^{(i)}x^i,$$

$$g(x) = f(x)\sum_{n=0}^{m-k}\left(\sum_{i=k}^{m-n} u_{i-1}b_{m-n-i}\right)x^n + \sum_{i=0}^{k-1}\left(\sum_{n=0}^{m} u_n^{(i)}b_{m-n}\right)x^i.$$

3. DIVISIBILITY TEST FOR POLYNOMIALS

Theorem 2.2 gives us the following divisibility test. In this section we use the notation $\underline{0}$ for the $k \times 1$ matrix with each entry equal to 0 (the additive identity for R).

Theorem 3.1: Under the conditions of Theorem 2.2, $g(x)$ is divisible by $f(x)$ if and only if

$$\sum_{n=0}^{m} w_n^{(i)}b_{m-n} = 0,\ i = 0, \ldots, k-1.$$

That is,

$$W \cdot (b_m\ b_{m-1} \ldots b_0)^T = \underline{0},$$

where W is defined by (2.6).

Corollary 3.2: Let R be a commutative ring, and let $\{u_n^{(i)}\}$ $(0 \le i \le k-1)$ be the ith basic sequence in $\Omega(f(x))$ over R. Then $g(x)$ is divisible by $f(x)$ if and only if

$$\sum_{n=0}^{m} u_n^{(i)}b_{m-n} = 0, i = 0, \ldots, k-1.$$

Example: Let R be the ring of integers, let $f(x) = x^2 - x - 1$ and let $g(x) = x^4 - 2x^3 - x^2 + 2x + 1$.

Recall that in $\Omega(f(x))$, $u_n^{(0)} = F_{n-1}$ and $u_n^{(1)} = F_n$. The divisibility test says $f(x)$ divides $g(x)$ if and only if $F_{-1} + 2F_0 - F_1 - 2F_2 + F_3 = 0$ and $F_0 + 2F_1 - F_2 - 2F_3 + F_4 = 0$. We easily verify that both equalities are valid.

If the ring R is an integral domain, we can derive a more general test rule. In effect, we can replace W by a more general matrix involving arbitrary $F - L$ sequences in $\Omega(f(x))$. To do this we first prove several lemmas.

Lemma 3.3: Let the polynomial $f(x)$ over a commutative ring R be expressed by (1.2), and let $\{u_n^{(i)}\}$ $(0 \le i \le k-1)$ be the ith basic sequence in $\Omega(f(x))$ over R. Let $\{w_n\}$ be any sequence in $\Omega(f(x))$. Then for $n,\ t \ge 0$

$$w_{n+t} = \sum_{i=0}^{k-1} u_n^{(i)} w_{t+i}.$$

Proof: Let $h_n = \sum_{i=0}^{k-1} u_n^{(i)} w_{t+i}$. Obviously $\{h_n\}$ is an element of $\Omega(f(x))$, and $\{h_n\}$ has the same initial conditions as $\{w_{n+t}\}_n$; i.e., $h_0 = w_t,\ h_1 = w_{t+1},\ \ldots,\ h_{k-1} = w_{t+k-1}$. Therefore $\{h_n\} = \{w_{n+t}\}_n$, and Lemma 3.3 is proved. □

Lemma 3.4: Let R be an integral domain, and let $f(x)$ over R be expressed by (1.2). For $\{w_n\}$ in $\Omega(f(x))$, denote

$$H(w_n) = \begin{pmatrix} w_n & w_{n+1} & \cdots & w_{n+k-1} \\ w_{n+1} & w_{n+2} & \cdots & w_{n+k} \\ \cdots & \cdots & \cdots & \cdots \\ w_{n+k-1} & w_{n+k} & \cdots & w_{n+2k-2} \end{pmatrix} \tag{3.1}$$

Assume that $a_k = -f(0) \neq 0$ and det $H(w_0) \neq 0$. Then for any $n \ge 0$, det $H(w_n) \neq 0$.

Proof: Denote

$$A = \begin{pmatrix} 0 & 1 & 0 & \cdots & 0 \\ 0 & 0 & 1 & \cdots & 0 \\ & & \cdots & & \\ 0 & 0 & 0 & \cdots & 1 \\ a_k & a_{k-1} & \cdots & & a_1 \end{pmatrix}.$$

We observe that det $A = (-1)^{k+1} a_k = (-1)^k f(0) \neq 0$. From (1.1) we have

$$(w_{n+1}\ w_{n+2} \ldots w_{n+k})^T = A \cdot (w_n\ w_{n+1} \ldots w_{n+k-1})^T.$$

This implies $H(w_{n+1}) = A \cdot H(w_n)$. Thus det $H(w_{n+1}) = \det A \cdot \det H(w_n)$. Now R is an integral domain, det $A \neq 0$ and det $H(w_0) \neq 0$; therefore det $H(w_1) = \det A \cdot \det H(w_0) \neq 0$. We can now prove the lemma by induction. □

Lemma 3.5: Let R be an integral domain and let M be a matrix of order k over R. If det $M \neq 0$, then the system of linear homogeneous equations $Mx = \underline{0}$ over R has only the one solution $x = \underline{0}$.

Proof: Suppose there is a solution $x = (x_1, x_2, \ldots, x_k)^T$, and at least one $x_i \neq 0$. Without loss of generality we assume $x_1 \neq 0$. Write $M = (C_1\ C_2 \ldots C_k)$, where C_i denotes the ith column of M. Then $x_1 \cdot \det M = \det(x_1 C_1\ C_2 \ldots C_k)$. In the last determinant we multiply the ith column by $x_i, i = 1, 2, \ldots, k$, and add it to the first column to get $x_1 \cdot \det M = 0$. Since R is an integral domain and since $x_1 \neq 0$, det $M \neq 0$, this is a contradiction, and the proof is complete. □

Theorem 3.6: Let R be an integral domain, let $f(x)$ be defined by (1.2) and let $g(x)$ be defined by (2.4), with the additional condition $f(0) \neq 0$. Assume that $\{w_n\}$ is a given sequence in $\Omega(f(x))$ with det $H(w_0) \neq 0$. Then for any given $t \geq 0$, $g(x)$ is divisible by $f(x)$ if and only if

$$\begin{pmatrix} w_t & w_{t+1} & \cdots & w_{t+m} \\ w_{t+1} & w_{t+2} & \cdots & w_{t+m+1} \\ \cdots & \cdots & \cdots & \cdots \\ w_{t+k-1} & w_{t+k} & \cdots & w_{t+m+k-1} \end{pmatrix} \cdot (b_m b_{m-1} \ldots b_0)^T = \underline{0}. \tag{3.2}$$

Proof: Assume $f(x)$ divides $g(x)$. By Theorem 3.1,

$$H(w_t) \cdot W \cdot (b_m b_{m-1} \ldots b_0)^T = \underline{0}, \tag{3.3}$$

where W is defined by (2.6) and $H(w_t)$ is defined by (3.1). Note that in W, each $w_n^{(i)} = u_n^{(i)}$ since R is commutative. Now (3.2) follows from Lemma 3.3.

Conversely, suppose (3.2) holds. Then (3.3) holds, and by Lemma 3.4 we know det $H(w_t) \neq 0$. Thus by Lemma 3.5, $W \cdot (b_m b_{m-1} \ldots b_0)^T = \underline{0}$. By Corollary 3.2, this implies $f(x)$ divides $g(x)$. □

We note that Theorem 3.6 includes the main result of [1] as a corollary.

4. SUMMATIONS INVOLVING $F - L$ NUMBERS

The following lemma originates from Theorem 2.3 (the Theorem of Constructing Identity, or TCI) in [4], where the conclusion is stated and proved for the complex field. In the same way, we can prove the conclusion holds for the ring R.

Lemma 4.1: (TCI) Let R be a commutative ring and let $f(x)$ over R be expressed by (1.2). Suppose

$$\sum_{i=0}^{s} d_i x^{n_i} \equiv \sum_{j=0}^{t} e_j x^{p_j} (\text{mod } f(x)), \tag{4.1}$$

where $n_i, p_j \geq 0$, and d_i, e_j are independent of $x; i = 0, \ldots, s$ and $j = 0, \ldots, t$. Then for all $\{w_n\}$ in $\Omega(f(x))$ over R, the identity

$$\sum_{i=0}^{s} d_i w_{n_i} = \sum_{j=0}^{t} e_j w p_j \tag{4.2}$$

holds. Conversely, if (4.2) holds for all $\{w_n\}$ in $\Omega(f(x))$ then (4.1) holds.

The following formula is concerned with two $F-L$ sequences with different characteristic polynomials.

Theorem 4.2: Let R be a commutative ring. Let the polynomials $f(x)$ and $g(x)$ over R be expressed by

$$f(x) = x^k - a_1x^{k-1} - \cdots - a_{k-1}x - a_k,$$
$$g(x) = x^m - b_1x^{m-1} - \cdots - b_{m-1}x - b_m,$$

and the remainder of $g(x)$ divided by $f(x)$ by

$$r(x) = \sum_{i=0}^{k-1} r_i x^i.$$

Let $\{u_n\}$ and $\{v_n\}$ be the principal sequences in $\Omega(f(x))$ and $\Omega(g(x))$ over R, respectively. Then

$$\sum_{s=0}^{n}\left(\sum_{i=0}^{k-1} r_i v_{n+m-s+i}\right) u_{s+k-1} = u_{n+m+k} - \sum_{s=0}^{k-1}\left(\sum_{i=s}^{k-1} r_i v_{i-s+m-1}\right) u_{n+k+s}$$

$$-v_{n+m+k}u_{m-1} - \sum_{i=k-1}^{m-2} v_{n+2m+k-i-1} - b_1 v_{n+2m+k-i-2} - \cdots - b_{m-i-1}v_{n+m+k})u_i.$$

Proof: Let $\{v_n^{(i)}\}(0 \le i \le m-1)$ be the ith basic sequence in $\Omega(g(x))$. Then from (2.1) and Lemma 2.5 we have

$$x^n = g(x)\sum_{i=0}^{n-m} v_{i+m-1}x^{n-m-i} + \sum_{i=0}^{m-1} v_n^{(i)}x^i \equiv \sum_{i=0}^{k-1} r_i x^i \sum_{j=0}^{n-m} v_{n-j-1}x^j + \sum_{i=0}^{m-1} v_n^{(i)}x^i$$

$$= \sum_{s=0}^{n-m+k-1}\left(\sum_{i=\max(0,s-n+m)}^{\min(k-1,s)} r_i v_{n-s+i-1}\right) x^s + \sum_{i=0}^{m-1} v_n^{(i)}x^i \pmod{f(x)}.$$

By TCI we have

$$u_n = \sum_{s=0}^{n-m+k-1}\left(\sum_{i=\max(0,s-n+m)}^{\min(k-1,s)} r_i v_{n-s+i-1}\right) u_s + \sum_{i=0}^{m-1} v_n^{(i)}u_i \pmod{f(x)}.$$

Since $u_s = 0$ for $0 \le s < k-1$, the first summation on the right side can be written

$$\sum_{s=k-1}^{n-m+k-1}\left(\sum_{i=\max(0,s-n+m)}^{k-1} r_i v_{n-s+i-1} u_s\right) = \sum_{s=0}^{n-m}\left(\sum_{i=\max(0,s-n+m+k-1)}^{k-1} r_i v_{n-k-s+i} u_{s+k-1}\right)$$

$$= \left(\sum_{s=0}^{n-m-k}\sum_{i=0}^{k-1} + \sum_{s=n-m-k+1}^{n-m}\sum_{i=s-n+m+k-1}^{k-1}\right) r_i v_{n-k-s+i} u_{s+k-1}$$

$$= \sum_{s=0}^{n-m-k}\left(\sum_{i=0}^{k-1} r_i v_{n-k-s+i}\right) u_{s+k-1} + \sum_{s=0}^{k-1}\left(\sum_{i=s}^{k-1} r_i v_{i-s+m-1}\right) u_{n-m+s}.$$

Thus

$$\sum_{s=0}^{n-m-k}\left(\sum_{i=0}^{k-1} r_i v_{n-k-s+i}\right) u_{s+k-1} = u_n - \sum_{s=0}^{k-1}\left(\sum_{i=s}^{k-1} r_i v_{i-s+m-1}\right) u_{n-m+s} - \sum_{i=k-1}^{m-1} v_n^{(i)} u_i.$$

In the last expression, replace n by $n+m+k$. Then use (2.3) and $v_{n+m+k}^{(m-1)} = v_{n+m+k}$ to finish the proof. □

Example: Let $f(x) = x^2 - x - 1$ and $g(x) = x^3 - x^2 - x - 1$. Then $k = 2, m = 3$ and $r(x) = -1$. By Theorem 4.2

$$\sum_{s=0}^{n}(-v_{n+3-s} + 0)F_{s+1} = F_{n+5} - ((-v_2 + 0)F_{n+2} + 0 \cdot F_{n+3}) - v_{n+5}F_2 - (v_{n+6} - v_{n+5})F_1.$$

That is,

$$\sum_{s=0}^{n} v_{n+3-s}F_{s+1} = v_{n+6} - F_{n+5} - F_{n+2}.$$

In this formula, $\{v_n\}$ is the Tribonacci sequence defined by $v_0 = v_1 = 0, v_2 = 1$ and $v_{n+3} = v_{n+2} + v_{n+1} + v_n$.

5. CONCLUDING REMARKS

If the ring R is not commutative, we call the $F - L$ sequence defined by (1.1) the **left** $F - L$ sequence. When we divide $g(x)$ by $f(x)$, we write $g(x) = f(x)q(x) + r(x)$ and call this the **left division**. Similarly, we can define the **right** $F - L$ sequence by

$$w_{n+k} = w_{n+k-1}a_1 + \cdots + w_{n+1}a_{k-1} + w_n a_k$$

and define the **right division** of $g(x)$ by $f(x)$ by means of $g(x) = q(x)f(x) + r(x)$. It is easy to change the conclusions about the left $F - L$ sequence and left division to conclusions about the right $F - L$ sequence and right division. We omit the details.

REFERENCES

[1] Brugia, O. and Filipponi, P. "Polynomial Divisibility in Finite Fields and Recurring Sequences." *The Fibonacci Quarterly*, Vol. *33* (1995): pp. 459-463.

[2] Lidl, R. and Niederreiter, H. "Finite Fields." Encyclopedia of Mathematics and Its Applications. Volume **20**. Edited by Gian-Carlo Rota. Addison Wesley, Reading, Mass, 1983.

[3] Zhou, C.Z. "A Generalization of the 'All or None' Divisibility Property." *The Fibonacci Quarterly*, Vol. *35* (1997): pp. 129-134.

[4] Zhou, C.Z. "Construction Identities Involving k^{th} Order $F - L$ Numbers by Using the Characteristic Polynomial." Applications of Fibonacci Numbers. Volume **8**. Edited by Fredric T. Howard. Kluwer Academic Publishers, Dordrecht, The Netherlands, 1999: pp. 369-379.

AMS Classification Numbers: 11B39, 11C08, 12E05

SUBJECT INDEX

MIX
Papier aus verantwortungsvollen Quellen
Paper from responsible sources
FSC® C105338

If you have any concerns about our products,
you can contact us on
ProductSafety@springernature.com

In case Publisher is established outside the EU,
the EU authorized representative is:
Springer Nature Customer Service Center GmbH
Europaplatz 3, 69115 Heidelberg, Germany

Printed by Libri Plureos GmbH
in Hamburg, Germany